Composites Science and Technology

Series Editor

Mohammad Jawaid, Department of Chemical and Petroleum Engineering, United Arab Emirates University, Al Ain, United Arab Emirates

This book series publishes cutting edge research monographs comprehensively covering topics in the field of composite science and technology. The books in this series are edited or authored by top researchers and professional across the globe. The series aims at publishing state-of-the-art research and development in areas including, but not limited to:

- Conventional Composites from natural and synthetic fibers
- Advanced Composites from natural and synthetic fibers
- Chemistry and biology of Composites and Biocomposites
- Fatigue damage modelling of Composites and Biocomposites
- Failure Analysis of Composites and Biocomposites
- Structural Health Monitoring of Composites and Biocomposites
- Durability of Composites and Biocomposites
- Biodegradability of Composites and Biocomposites
- Thermal properties of Composites and Biocomposites
- Flammability of Composites and Biocomposites
- Tribology of Composites and Biocomposites
- Applications of Composites and Biocomposites

Review Process

The proposal for each volume is reviewed by the main editor and/or the advisory board. The chapters in each volume are individually reviewed single blind by expert reviewers (at least two reviews per chapter) and the main editor.

Ethics Statement for this series can be found in the Springer standard guidelines here - https://www.springer.com/us/authors-editors/journal-author/journal-author-helpdesk/before-you-start/before-you-start/1330#c14214

Swapnila Roy · Shivani Garg · Chin Wei Lai
Editors

Nanocomposites for Sustainable Wastewater Treatment

Performance Evaluation and Future Perspective

Springer

Editors
Swapnila Roy
New Delhi Institute of Management
New Delhi, Delhi, India

Shivani Garg
Institute of Environmental Studies
Kurukshetra University
Kurukshetra, Haryana, India

Chin Wei Lai
Nanotechnology and Catalysis Research
Centre
University of Malaya
Kuala Lumpur, Kuala Lumpur, Malaysia

ISSN 2662-1819 ISSN 2662-1827 (electronic)
Composites Science and Technology
ISBN 978-981-95-1368-0 ISBN 978-981-95-1369-7 (eBook)
https://doi.org/10.1007/978-981-95-1369-7

This Springer imprint is published by the registered company Springer Nature Singapore Pte Ltd.
The registered company address is: 152 Beach Road, #21-01/04 Gateway East, Singapore 189721, Singapore

Competing Interests The authors have no competing interests to declare that are relevant to the content of this manuscript.

Contents

About the Editors

Dr. Swapnila Roy is an accomplished researcher and academician, currently working as an Associate Professor in Research and Innovation at the New Delhi Institute of Management, with over 21 years of academic and research experience. She holds a Ph.D. in Environmental Chemistry from Jadavpur University, Kolkata, and has made significant contributions in the fields of environmental monitoring, water purification, nanomaterials, and green chemistry. Dr. Roy has authored more than 45 peer-reviewed research publications, contributed to several book chapters, and holds an impressive portfolio of 44 national and international patents in cutting-edge areas, including environmental technologies, AI-based systems, and sustainable innovations. With deep expertise in advanced analytical techniques (ICP-OES, AAS, HPLC, FTIR) and a strong track record of leading multidisciplinary research projects, she actively drives innovation in pollutant remediation, circular economy, and smart environmental solutions.

Ms. Shivani Garg is an esteemed Assistant Professor at the Institute of Environmental Studies, Kurukshetra University, Kurukshetra, Haryana, India. As a doctoral candidate, her research focuses on soil science and sustainable agriculture. With a rich background in education and a profound expertise in sustainable development, soil science, environmental issues, climate change, nanotechnology, and energy resource management, she stands as a pivotal figure in the academic realm. Her dedication to environmental compliance, water quality, and environmental impact assessment has been instrumental in shaping her career and academic contributions. With a notable academic record, Er. Garg boasts a commendable 11 h-index, 11 h10-index, and has garnered more than 500 citations for her publications, reflecting the impact and significance of her research endeavors. Her primary research interests revolve around critical areas such as soil sciences, sustainable agriculture, wastewater treatment, energy management, irrigation and water management, biogas studies, waste management, and environmental pollution.

Prof. Ir. Ts. Dr. Chin Wei Lai is currently a Full Professor in Nanotechnology and Catalysis Research Centre (NANOCAT), "National Higher Institution Centre of

Excellence (HiCoE)", University of Malaya. Lai's main research interests are in the areas of chemically modified/functionalized metal oxide photocatalysts and carbon graphene materials, especially for sustainable development to diversify the inclusive science, technology, and innovation in environmental pollution management and energy storage technology. As a scholar and an indication of the global impact of research work (top 2% scientists in a global list), Lai's works have been published in more than 300 refereed international top-tier journals with Scopus h-index of 49, 8 patents (granted 3 patents), 100 book chapters, and 100 international proceedings in materials science, physics, chemistry, and renewable energy researches. Lai's research findings aid in building the fundamental knowledge and groundbreaking industry collaboration experience with many multinational companies/institutions/ NGOs.

Introduction to Nanocomposites in Wastewater Treatment

Swapnila Roy

Abstract Investigating innovative methods for effective water and wastewater treatment has become necessary in order to solve global environmental concerns, such as pollution and water scarcity. Customising nanoparticles and nanocomposites made especially for water purification purposes is one way that nanotechnology offers potential prospects. This study explores the latest developments in nanoparticle-based water treatment technologies, emphasising their inherent qualities, synthesis processes, and range of uses. For the purpose of removing pollutants from water and wastewater matrices, a variety of nanoparticles, including metal nanoparticles and carbon-based nanomaterials, as well as hybrid nanocomposites, including metal/metal oxide-based nanocomposites and polymer-based nanocomposites, were highlighted. Moreover, this chapter clarifies the fundamental principles guiding the methods of pollution removal that are made possible by nanoparticles, including adsorption, catalysis, and membrane filtration for environmental sustainability. Moreover, it delves into the environmental consequences and difficulties linked to the extensive implementation of water treatment methods based on nanoparticles. By integrating earlier research findings, this study provides significant new information on various nanoparticles and composites to mitigate water-related issues. Transferring nanomaterial-based therapy solutions to larger-scale systems poses considerable obstacles, even with encouraging outcomes in the laboratory. To increase the effectiveness of wastewater treatment, research should focus on developing nanomaterials with enhanced selectivity towards target pollutants while minimising interference from other chemicals.

Keywords Nanocomposites · Nanoparticles · Pollutants · Wastewater treatment · Sustainability · Environment friendly

S. Roy (✉)
New Delhi Institute of Management, New Delhi, India
e-mail: swapnilaroy@gmail.com

S. Roy et al. (eds.), *Nanocomposites for Sustainable Wastewater Treatment*, Composites Science and Technology, https://doi.org/10.1007/978-981-95-1369-7_1

1 Background

All life on Earth depends on water, a vital natural resource, to exist. For life in all its manifestations to persist, it must be clean. Maintaining life, maintaining health, and bolstering ecosystems all depend on clean water. Drinking, farming, sanitation, industry, and a host of other human necessities all depend on it. These include dangerous handling of radioactive materials, discharge of industrial waste, poor sewage management, marine contamination, and certain farming practices. Urban sewage discharge, agricultural runoff, and industrial waste all inject a variety of contaminants into natural water sources.

Water quality is deteriorated by these contaminants, rendering it unsuitable for use and consumption (https://www.unwater.org/publications/un-world-water-development-report-2024). Different countries' social values and economic prosperity are negatively impacted by water contamination. According to a recent UN assessment, freshwater scarcity is turning into a worldwide concern for the twenty-first century (UN-Water 2024). Chemicals with non-biodegradable characteristics are commonly discovered in wastewater streams and are considered emerging contaminants since they have the ability to linger in the environment for a long period. These pollutants pose a concern to human health as well as natural ecosystems since they have the ability to spread throughout the food chain. They also endanger the flora, wildlife, and other living things that live in these habitats. Worldwide research is continually being done to determine more efficient and reasonably priced solutions to raise the standard of water. The WHO report 2014 reveals that on water supply and sanitation, 3900 children die every day from infectious illnesses and poor water quality, and 748 million people still lack access to clean drinking water and sanitation (2.5 billion people worldwide). These figures showed that the contamination of water by various contaminants is becoming a worrying global problem. In order to increase the potential of water resources and reduce the difficulties and worries related to water pollution, efficient water treatment techniques have been developed.

Nanocomposite technology is necessary in the water purification process, which includes wastewater desalination and treatment, in order to produce clean water at a reduced cost and with less energy usage by eliminating further environmental consequences (Tang et al. 2015). Recently released data and forecasts from agencies such as the UN and WHO might offer important context to support the applicability of your wastewater treatment efforts. Untreated wastewater is still a significant problem worldwide, especially in metropolitan areas with inadequate infrastructure, where up to 42% of residential wastewater is not properly treated, according to the UN's World Water Development Report 2024. Ecosystems and public health are greatly impacted by this problem, particularly in areas with lower incomes. The necessity for better wastewater monitoring and treatment to safeguard public health and water resources was also highlighted by a 2023 UN-Water study that found that only 56% of water bodies in 120 reporting nations maintained good ambient quality. Wastewater is also becoming more widely recognized as a resource with significant repurposing possibilities. Reusing treated wastewater might help alleviate water shortages by increasing clean water supplies, supplying nutrients for agriculture, and lowering pollution, according to the United Nations University (UNU). In order to fulfil increasing water needs and promote environmental sustainability, water management must transition to a circular economy (UNU 2024, Global Wastewater Status).

One potential remedy for these issues with water contamination is nanocomposites. Nanocomposites may be designed to target and eliminate certain contaminants by using the special qualities of nanomaterials, such as their large surface area, reactivity, and adaptable functionality. They offer economical, environmentally responsible, and effective water treatment solutions.

Nanomaterials are materials having structural elements ranging in size from 1 to 100 nm. When compared to other traditional materials, they have special mechanical, electrical, optical, and magnetic capabilities due to their small size and larger specific surface area. Nanomaterials have been successfully used in several fields recently, including biology, sensing, medicine, and catalysis (Kolya and Kang 2023). They are extensively employed in the treatment of water and wastewater, among other environmental problems.

Because many toxins, such as heavy metals, organic pollutants, inorganic anions, and infections, may be eliminated by nanoparticles (Fang et al. 2017). These days, the best nanomaterials for cleaning water and wastewater include metal oxide nanoparticles, carbon nanotubes (CNTs), zero-valent metal nanoparticles (nZVI), and nanocomposites (Kochkodan and Hilal 2015). One of the best nanomaterials for purifying water is the nZVI. The nZVI contributes to water filtration as an electron donor that facilitates adsorption, co-precipitation processes, high reducing capacity, and the transformation of dangerous metals into safe forms.

The nZVI has been shown to be a useful tool for eliminating a wide range of organic and inorganic contaminants, including radioactive elements, heavy metals, metalloids, organic dyes, polychlorinated compounds, nitrates, phosphates, and perchlorates, and nitroaromatic compounds (Park 2022; Kumar et al. 2015). Other nanoparticles used in water treatment technologies include silver (Ag), iron oxides, zinc oxide (ZnO), titanium dioxide (TiO_2), and carbon nanotubes (CNT) (Zadehahmadi et al. 2023). Due to their extreme toxicity to microorganisms, silver nanoparticles (AgNPs) exhibit strong antibacterial properties against a wide range of germs, including bacteria, fungus, and viruses. AgNPs are the potential antibacterial agents that are widely employed in the disinfection of water. AgNPs have the ability to eradicate bacteria like ampicillin-resistant *E. coli* and methicillin-resistant Staphylococcus aureus. The various ways that nanosilver disinfection works include AgNPs' interaction with DNA, membrane modification, and the enzymatic activity change that leads to its destruction, and AgNP dissolution that can react with thiol sets of enzymes to disable and interrupt the normal functions of the cell. There hasn't been a previously published analysis of nanocomposites' (NCs) use in water and wastewater treatment (Soren et al. 2023).

NCs are created when two or more materials with different physical, chemical, and interfacial characteristics are combined. The special qualities of composite materials, such as their high strength, high durability, high stiffness, gas-barrier properties, corrosion resistance, low density, and heat resistance, provide them with several benefits over other materials.

The special qualities of composites, such as their high strength, high durability, high stiffness, gas-barrier properties, corrosion resistance, low density, and heat resistance, make them superior to traditional materials in many ways. Reinforcing components in a dispersed form and a continuous phase matrix make up composite materials (Ali et al. 2019). The display size of at least one of these multiphase materials from the twenty-first century ranges from 10 to 100 nm. NC materials have become viable solutions to the shortcomings of many manufacturing tools in recent years.

NCs have extensive experience in a variety of disciplines, including wastewater treatment, medication distribution strategies, and life sciences. The nanoparticles in NCs were mixed with various functionalised materials such as polymeric media, activated carbon, multiwall carbon nanotubes, and inexpensive graphene oxide (Ahmed et al. 2021). Applications for NCs include food packaging, coating, biomedical, and anti-corrosion barrier protection. The main emphasis of this chapter is on the fascinating NC kinds and their contemporary uses in water filtration. There was also discussion of the possible uses of nanocomposites in water treatment. With an emphasis on the use of carbon-based nanomaterials, such as metal/metal oxide nanoparticles and carbon-based nanotubes, fullerenes, and their composites, this study carefully looks at the developments in the application of nanomaterials for wastewater treatment (Dutta and Sarkar 2015). Major water pollutants are also covered, and a summary of current advancements in wastewater and water treatment techniques utilising various nanomaterial types is provided.

2 Traditional Methods of Water Purification

Surface water from springs, rivers, and lakes, as well as non-traditional water sources that cannot be used immediately. Seawater, brackish water, and wastewater, for instance, are the main universal water resource potentials. Conventional water purification techniques like coagulation-flocculation, chlorination, and sand filtration have long been proven to be efficient ways to remove impurities from water. Because of their ease of use, dependability, and affordability, these techniques are widely used for large-scale municipal water treatment and are crucial in many developing nations. For instance, sand filtration efficiently eliminates suspended particles, and chlorination provides potent disinfection capabilities that lower microbial contamination. These traditional techniques do, however, have some significant drawbacks. They can find it difficult to eliminate new pollutants that are becoming more prevalent in water sources, such as heavy metals, medications, and other micropollutants. Furthermore, hazardous byproducts from procedures like chlorination might endanger human health if they are exposed for an extended period of time. In light of these difficulties, new technologies—like sophisticated nanocomposite materials—are attracting interest due to their exceptional effectiveness and versatility in dealing with complicated contaminants. In order to satisfy changing water treatment needs, a move towards creative, sustainable solutions is therefore becoming more and more justifiable.

Freshwater resources are becoming more and more contaminated due to rapid population increase, urbanisation, and industrialisation on a global scale. Table 1 lists the traditional water purification methods that are now in use, including air flotation, improved oxidation processes, and coagulation and flocculation (Saleh et al. 2019). These techniques effectively remove the pollutants in a very quiet manner. Nonetheless, these techniques come with a number of drawbacks pertaining to the development of secondary pollutants or increased energy use. As a result, great emphasis should be placed on the development of technologies that are affordable, energy-efficient, and environmentally benign and that can be applied to workable water sanitisation systems. Numerous writers have concentrated on appropriate and financially feasible methods of water purification, such as desalination, reclamation, and water remediation, in order to fulfil the need for clean water standards (Liu and Dong 2020).

3 Nanocomposite

At least one phase of multi-phasic materials, also known as nanocomposites, has dimensions in the nanoscale (10–100 nm) range.

Nowadays, nanocomposite materials have emerged as viable alternatives to the shortcomings of a number of engineering materials. They are regarded as the materials of the twenty-first century.

Materials for nanocomposites can be categorised based on the materials of their dispersion phase and the distributed matrix (Ahmad 2021). Diagram 1 depicts the categorisation of nanocomposites.

Polymer-based nanocomposites (PNCs) are one of the many types of nanocomposites that are now the focus of much study and development. PNCs offer several useful characteristics, including the capacity to form films, dimensional diversity, and active capabilities (Reddy et al. 2015).

3.1 Types of Nanocomposites and How They Are Used to Purify Water

The use of nanoparticles in water management has been linked to a number of real-world issues, including buildup, difficult separation, drainage into contact water, and negative effects on human and environmental health. One practical way to increase the utilisation of nanoparticulate materials is to develop NC materials that are advantageous to the hosts as well as the impregnated nanoparticles (Flowchart).

Table 1 Methods of water purification

Sl. No.	Methodology of water purification	Advantages and limitations
1	Coagulation and flocculation method	**Advantages**: • Efficient at removing suspended particles, turbidity, and some dissolved substances • Helps remove color, taste, and odor issues in water • Often used in large-scale water treatment facilities for bulk purification **Limitations**: • Ineffective for dissolved organic compounds and certain heavy metals • Generates chemical sludge that requires safe disposal • Can be costly and requires skilled operators
2	Distillation	**Advantages**: • Produces highly purified water by removing most contaminants, including heavy metals, salts, and biological impurities • Suitable for small-scale applications where high purity is required **Limitations**: • Highly energy-intensive and costly, making it unsuitable for large-scale or routine water treatment • Slow process, which may not meet the demand for larger communities • Can remove essential minerals, resulting in water that may taste flat
3	Activated carbon adsorption	**Advantages**: • Highly effective at removing organic contaminants, chlorine, and bad odors/tastes • Can adsorb a wide range of pollutants, including some chemicals and volatile organic compounds (VOCs) • Useful in household and portable water purifiers **Limitations**: • Ineffective at removing most heavy metals, pathogens, and nitrates • Limited adsorption capacity; the filters need to be replaced regularly • Can become a breeding ground for bacteria if not maintained properly

(continued)

Table 1 (continued)

Sl. No.	Methodology of water purification	Advantages and limitations
4	Chlorine	**Advantages**: • Effectively kills bacteria, viruses, and other pathogens • Affordable and widely used in municipal water systems • Residual chlorine can help prevent recontamination during distribution **Limitations**: • Ineffective against certain protozoa (e.g., Cryptosporidium) and some chemical pollutants • Can produce harmful byproducts, such as trihalomethanes (THMs), which are potentially carcinogenic • Often leaves an unpleasant taste and odor in the water
5	Filtration	**Advantages**: • Effectively kills bacteria, viruses, and other pathogens • Affordable and widely used in municipal water systems • Residual chlorine can help prevent recontamination during distribution **Limitations**: • Ineffective against certain protozoa (e.g., Cryptosporidium) and some chemical pollutants • Can produce harmful byproducts, such as trihalomethanes (THMs), which are potentially carcinogenic • Often leaves an unpleasant taste and odor in the water

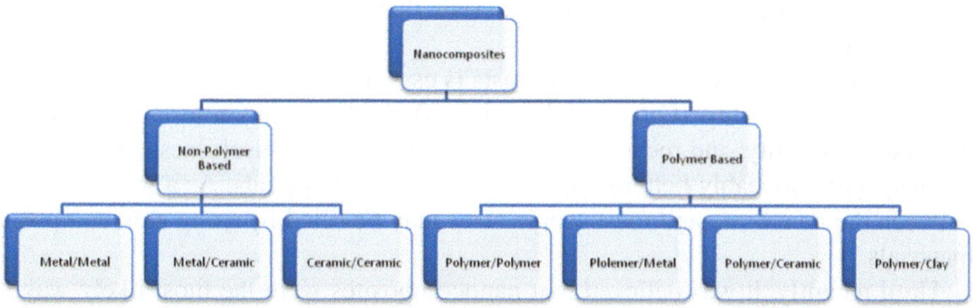

Diagram 1 Classification of nanocomposites

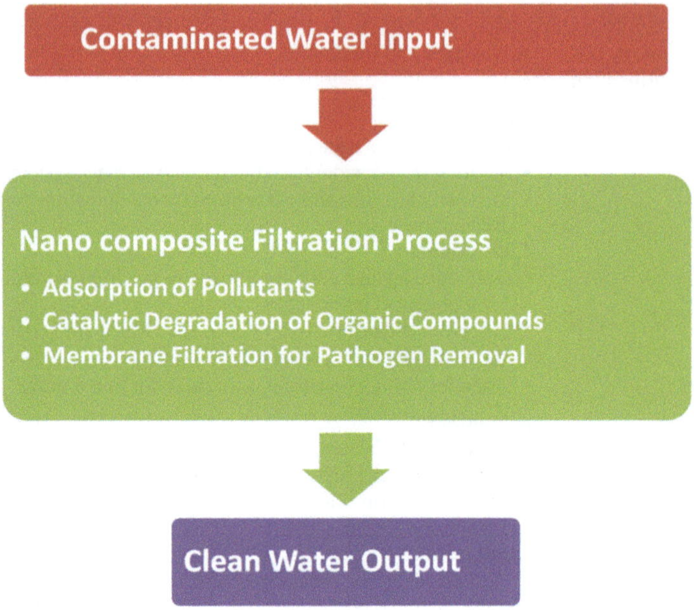

Flowchart: Nanocomposites in water purification

NCs can increase the compatibility of nanotechnology with existing infrastructures and reduce the release of nanoparticles into the environment. Generally speaking, NCs are multiphase solid materials that include porous media, colloids, gels, and copolymers.

The selection of nanocomposites' hosts is essential and even influences the final products' functionality (Gitis and Hankins 2018). When it came to nanoparticle dispersion, stability, and recyclability, the performance and usefulness of nanocomposites were noticeably better than those of free nanomaterials. As a result, the gap between the nanoscale and mesoscopic sizes may be filled by using nanocomposite materials.

Until recently, it was believed that nanocomposites were the most likely way to move water nanotechnology from the lab to large-scale applications (Wu et al. 2019).

3.1.1 Metal

The antibacterial qualities of polyurethane and cellulose acetate impregnated nanosilver-fiber composites have been acknowledged for polymer-supported nanosilver, and these composites exhibit a strong inhibitory effect against both Gram-positive and Gram-negative bacteria.

Strong antibacterial filters are now included in the dispersal of polyurethane foam nanoparticles. Polymeric membranes were demonstrated to have little biofouling and great pathogen eradication effectiveness (Fig. 1) (Sawafta and Shahwan 2023). Additionally, drinking water may be handled using economically viable microfilters made of nanosilver, which is mostly favoured in remote areas. Three distinct techniques were successfully used to manufacture composite beads of silver alginate. To varying degrees, such composite beads equipped with different techniques proved to be excellent cleaners of *E. coli.* Different Organic Dyes have been removed from synthetic waste using a new NC that was synthesised and contained AgNPs and mesoporous alumina (Umar 2018). The improved reactivity, stability, and catalytic capabilities of metal nanocomposites make them essential for water filtration. These nanocomposites may efficiently target pollutants like heavy metals, organic compounds, and pathogens since they frequently contain metals like silver, iron, or titanium mixed with other materials. For instance, iron-based composites are good at adsorbing heavy metals and breaking down organic contaminants, whereas silver-based nanocomposites have potent antibacterial qualities. Furthermore, dangerous organic materials can be broken down by metal oxide nanocomposites like titanium dioxide via photocatalysis, which is triggered by sunshine. Because of these properties, metal nanocomposites are crucial for cutting-edge, environmentally friendly water treatment systems (Muntha et al. 2017). The decreased surface area of bimetallic nanoparticles supported by carbon greatly improves their characteristics. Using ruthenium-palladium bimetallic nanoparticles as reinforcement, NC was effectively developed on the surface carbon matrix, aiding in the treatment of wastewater containing perchlorate as the primary contaminant. Other examples are Au/Pd nanoparticle-reinforced TO_2 NCs, which are produced using a microemulsion method and are highly effective photocatalysts because of their capacity to absorb light (Umar et al. 2019).

3.1.2 Metallic Oxide

Metal oxide nanocomposite (MONC) materials (Fig. 2) are widely used as adsorbents, photocatalysts, and devices to address environmental pollution challenges.

MONCs are used in combination with graphene, silica, carbon nanotubes (CNTs), various oxides, and polymers to eliminate a range of inorganic and organic pollutants. Over the past ten years, there has been a sharp surge in interest in employing metal oxide nanoparticles (ZnO and TiO_2) for photocatalysis as a way of combating worldwide water pollution. Coupling is creating a method to make organic pollutants more destructive when exposed to visible light (Loeb et al. 2019). CNT-reinforced alumina composite was created by elevating CNT over Fe and Ni-doped energetic alumina. Many elements that were able to start high-capacity synthesis had an impact on the composite (Jiang 2021). These elements included amorphous carbon, activated alumina, carbon nanotubes, and different surface functional groups found in the clusters, such as carboxyl, carbonyl, and hydroxyl. The newest and better kind of metal oxide-based composites is called $TiO_{2.}$ Since TiO_2 nanocomposite is safe and can efficiently break down pollutants by photo-oxidative degradation, it has attracted more attention in the water filtration industry. Metallic oxide nanocomposites, such as zinc

Fig. 1 Metal nanocomposite material, depicting nanoparticles embedded in a metal matrix (metal nanoparticles imbedded in a metal matrix provide metal nanocomposite materials their enhanced strength, conductivity, and reactivity. Important elements include homogeneous dispersion, pore size, and nanoparticle size (less than 100 nm). Sol–gel, ball milling, chemical reduction, and electrodeposition are examples of synthesis routes)

Fig. 2 Metallic oxide nanocomposite with nanoscale structures and metallic particles embedded in an oxide matrix (metallic oxide nanocomposites consist of metallic particles (1–100 nm) embedded in an oxide matrix with tunable pore sizes. Key mechanisms include charge transfer and interface effects. Synthesis routes involve sol–gel, co-precipitation, hydrothermal, and chemical vapor deposition methods)

oxide (ZnO) and titanium dioxide (TiO_2), are essential for water filtration because of their photocatalytic qualities. They break down organic contaminants, colours, and medications into innocuous byproducts when exposed to light. Additionally, these nanocomposites have strong adsorption capabilities, which allow them to effectively and sustainably provide safe, clean water by capturing pathogens and heavy metals (Bhat et al. 2018). Together with graphene and carbon nanotubes, iron oxides, also known as Fe_2O_3 and Fe_3O_4, are unique and remarkable magnetic components that create a revolutionary composite.

It has garnered a lot of attention as one of the best smart magnetic metallic oxides because of its remarkable physical and chemical characteristics, as well as other advantages, including its high reversible capacity, abundant supply, affordability, and environmental friendliness. Magnetic nanocomposites may be made by filling a polymer matrix with various Fe and Mn complex nanoparticles. This allows for simple composite separation from aqueous solutions following the sorption process. Several magnetic alginate polymers were created, and batch testing showed that they could extract various heavy metal ions and organic dyes (MB and MO) from aqueous solutions.

Furthermore, because of their magnetic qualities, high surface-to-volume ratio, and quick diffusion rate, nanoclusters (NCs) have demonstrated an extraordinary adsorption ability to fix additional heavy metals and organic dyes. Many researchers are now investigating the use of metal oxide nanoclusters (NCs) in the filtration of water and wastewater (Ahmad et al. 2015). Because heavy metals have a significant impact on both the environment and human health, many scientists are currently focusing on their removal. The generated Zr-MFCs were demonstrated to be effective adsorbents with reasonable adsorption ability and rapid adsorption kinetics for the removal of metal ions and organic dyes from water.

Compared to row MFC-O, the amine-decorated MFCs were much more effective in eliminating metal ions and dyes.

3.1.3 Carbon Nanocomposite

Carbon-based nanomaterials, such as graphene, carbon nanotubes, and fullerenes, are combined with other materials in carbon nanocomposites to improve their characteristics for a variety of uses. They are perfect for energy storage, electronics, environmental protection, and water purification because of their large surface area, conductivity, and stability. Through a simple impregnation-carbonization process, mesoporous carbon containing embedded iron carbide nanoparticles (ICNPs) was successfully synthesised. Mesopores were created utilising an iron pioneer and biomass as the carbon basis using a catalytic graphitisation procedure (Cornwell and Smith 2015). The production of new mesopores was shown to be essential for enhancing the organic dyes' ability to be adsorbed, whereas the cautious elimination of heavy metal ions is carried out by immobile ICNPs. When cutting-edge technologies are combined with tailored nanoparticles in biochar manufacturing schemes, the potential applications of biochar for carbon sequestration, wastewater treatment, and soil fertility upgrading may improve. When cutting-edge technologies are combined with tailored nanoparticles in biochar manufacturing schemes, the potential applications of biochar for carbon sequestration, wastewater treatment, and soil fertility upgrading may improve. In order to create a composite mixture with carbon nanotubes (CNTs) as the main host, two or more constituents must be prepared (Ali et al. 2019). As a result, poly 1,8-diaminonaphthalene/MWCNTs COOH, a hybrid material, was produced that may be employed as an active sorbent for the trace-level separation of Pb^{+2} and Cd^{+2}. Carbon nanocomposites' large surface area, adsorption ability, and improved chemical reactivity make them essential for water filtration. These substances may effectively eliminate impurities, including germs, organic pollutants, and heavy metals. They frequently contain carbon nanotubes, graphene, or activated carbon. They are perfect for sophisticated filtering systems because of their capacity to both adsorb and catalyse processes, providing a viable and efficient alternative for clean water technologies (Hodges et al. 2018).

3.1.4 Polymer as Nanocomposite

When nanoparticles spread throughout a polymer matrix, they form polymer nanocomposites (PNCs), a superior type of instrument that produces novel materials with unique chemical and physical properties.

Given their superior mechanical qualities, chemically bound functional groups, and generally adjustable porosity architectures, polymers are unique supports for nanomaterials (Fig. 3). In the realm of water treatment technology, PNC is one of the several treatment methods that is thought to be an advanced tool for the adsorption of pollutants. PNCs usually combine the main advantages of the nanoparticles and the polymeric matrix.

PNCs can be made by joining nanoparticles into polymer structures or by affixing polymers to nanoparticles (Daer et al. 2015). Attaching polymers to nanoparticles or combining nanoparticles into polymer structures is one potential technique for creating PNCs.

The capacity of some of the nanocomposites to be recycled and regenerated with minimal loss of functionality is essential from an economic standpoint (Gautam and Saini 2020). The PNCs' size allows for easy separation from treated water. Among chitosan's unique properties are its high reactivity, favourable complexation behaviour, and chemical stability.

Chitosan's amino and hydroxyl groups serve as active sites for contaminants in water. CNTs' many and diverse interfaces, which include hydrophobic impact, peppiness interactions, covalent bonding, valence bonding, and electrical connections, allow them to efficiently absorb a wide range of polar organic molecules (Esfahani et al. 2019). Polymer nanocomposites' large surface area, durability, and capacity to contain different nanoparticles such as carbon, clay, and silver make them essential for water purification. By eliminating impurities, including organic molecules, bacteria, and heavy metals, these composites improve filtering efficiency. They are

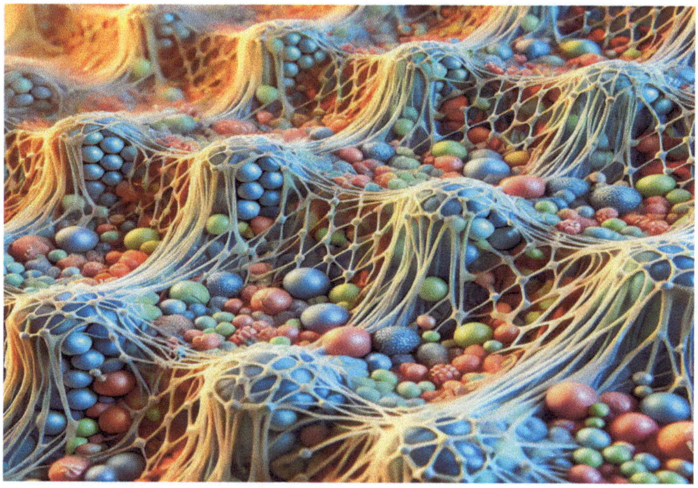

Fig. 3 Polymer-based nanocomposite, showing the flexible polymer matrix with embedded nanoscale particles (polymer-based nanocomposites improve mechanical, thermal, or barrier properties by incorporating nanoparticles (1–100 nm) into a flexible polymer matrix. Interfacial interactions and load transfer are examples of mechanisms. In situ polymerisation, melt compounding, and solution casting are examples of synthesis pathways)

useful in creating cutting-edge, economical, and ecologically friendly water treatment systems because of their adaptable qualities (Samanta et al. 2016). The optimal matrix for polyaniline varnish can be determined using clay.

The synthesis NC's characterisation results showed that the clay sheet was developed layered (Crini et al. 2019). Factors including concentration, pH, and contact time all affected the modified adsorbent's adsorption capacity.

The researchers were informed about a new clay nanocomposite that uses polyaniline to enhance the clay nanocomposite as an adsorbent for lead ion removal from water. Research studies revealed how certain factors, including pH, contact duration, and adsorbent dosage, affected the outcome (Rajendran and Aravindan 2020). The results demonstrated the potent use of NC in the treatment of heavy metal-containing tannery effluent.

3.1.5 Membranes

Membrane technology traps pollutants using porous materials.

One of the many issues with the membranes used to purify water today is the trade relationship between permeability, selectivity, and low fouling resistance. Nanocomposite membranes, or NCMs, are important for the purification and reuse of different water bases, such as drinking water, brackish, marine, and wastewater treatment.

NCM is a novel class of membranes created by combining nanomaterials with complicated elements; it is emerging as a potential solution to issues with membrane

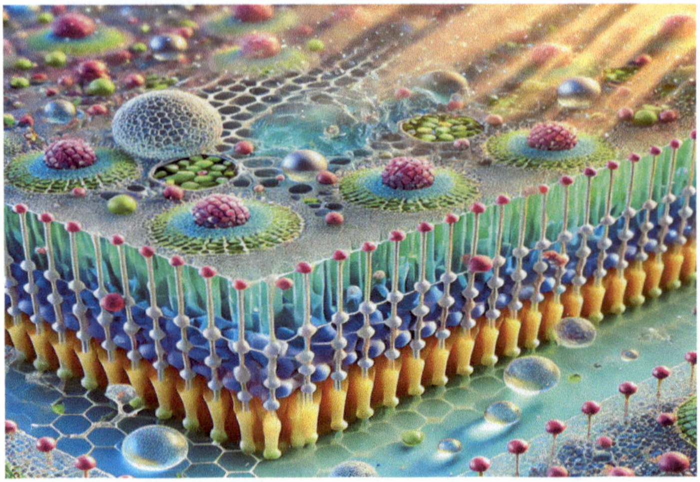

Fig. 4 Nanocomposite membrane, showing its porous structure and embedded nanoparticles (nanocomposite membranes feature porous structures with nanoparticles (1–100 nm) embedded in a matrix, improving separation, filtration, or catalytic functions. Pore size is typically 1–50 nm. Mechanisms include selective adsorption and molecular sieving. Synthesis routes include phase inversion, electrospinning, and in situ polymerisation)

separation (Hogen-Esch et al. 2019) (Fig. 4). The advantages of polymers and nanoparticles are combined in nanocomposite membranes, which improve filtering efficiency and are essential for water purification. These membranes effectively remove impurities such as heavy metals, salts, and organic pollutants because of their increased mechanical strength, chemical stability, and selective permeability. They are perfect for high-performance, environmentally friendly water treatment applications because of their sophisticated qualities (Baruah et al. 2016). Materials having porous architectures and nanoscale inorganic and/or organic solid phases are combined to form NCM (Umar et al. 2015). NCM may be viewed as a new type of filtration instruments with surface-active membranes and hybrid medium membranes. Nonetheless, the majority of researchers have verified that adding nanomaterials to polymers not only changes the assembly and physicochemical properties of membranes, such as hydrophilicity, porosity, and charge density, but also increases their mechanical, chemical, thermal, and bactericidal properties (Larramendy et al. 2015).

3.1.6 Traditional Nanocomposite Membrane

Most researchers have confirmed that the incorporation of nanomaterials into polymers not only modifies the assembly and physicochemical properties of membranes,

Table 2 Traditional nanocomposite membrane

S. No.	Reinforcement	Details
1	Carbon nanotube	Incorporation for enhanced qualities, including strong construction and anti-biofouling
2	Metal oxide (TiO_2, ZnO, SiO_2, Al_2O_3, Fe_3O_4)	Modifies the membrane's physicochemical properties and assembly, including its hydrophilicity, porosity, charge density, and mechanical, chemical, and thermal stability and imparts special qualities to the membranes, such as antifouling and photocatalytic activity (Sekoai et al. 2019)
3	Metals (Ag, Cu, Se)	Functionality of antimicrobials
4	Organic material	Improvements in the resulting membranes' hydrophilicity, sorption capacity, anti-compaction, and antifouling properties
5	Zeolite	Enhancement of hydrophilicity, progression of cross-linking characteristic, and augmentation of membrane rigidity

such as their hydrophilicity, porosity, and charge density, but also their chemical, thermal, and mechanical stability. This results in the membranes exhibiting remarkable properties like bactericidal and photocatalytic properties.

It has been verified by both outlooks that CNM can be made to remove contaminants from water through absorption (Bassin et al. 2021; Epelle et al. 2022). TiO_2 has also been included in a variety of membrane media in the field of CNM research to provide membranes with photocatalytic properties. Since TiO_2 has excellent photocatalytic activity, solidity, and ease of production, it has been widely employed for water treatment (Table 2) (Gupta et al. 2022). It is anticipated that evolving antimicrobial membranes would significantly extend their lifespan and efficiency, which will help offer clean water free of microorganisms.

3.1.7 Thin Film as Nanocomposite

The membrane of thin film nanocomposites (TFNs) has an extremely thin wall sheet positioned over a more permeable auxiliary material. Reverse osmosis or nanofiltration membranes are often used to interfacially synthesise TFN, which is then used to remove organic pollutants such as herbicides and insecticides, desalinate saltwater or brackish water, and remove hardness-causing salts (Kwiatkowski 2023). The thin film membranes are prepared to profit from the characteristics of the nanomaterials by the addition of nanoparticles. Thin film nanocomposites' large surface area and selective permeability make them essential for water filtration. These films improve filtration performance by adding nanoparticles like silica, graphene, or silver, which makes it possible to remove contaminants, including microorganisms, heavy metals, and organic molecules. They are perfect for creating cutting-edge, energy-efficient

Table 3 TFN membranes summarised with a nanocomposite substrate

The kind of reinforcement	Details
Carboxylic MWNTs	Enhanced antioxidative and antifouling qualities
Zeolite	Elimination of salt; combating physical compaction
MWNTs	Boost the substrate's bendable strength and eliminate salt

water treatment systems because of their flexibility and light weight (Sidhu et al. 2019).

The notion of TFN has been introduced by the addition of nanoparticles to external accessories or in between polymerisation pathways through self-assembly. This approach offers potential benefits such as enhanced separation efficiency, decreased fouling, antibacterial activity, and other unique qualities (Table 3).

3.1.8 Nanocomposite Substrate in TFC

The purpose of this membrane is to examine how nanofiller affects the compassionate behaviours of membranes (Zhang et al. 2016). In this category, the postscript substrate used in the IP process to assemble the TFC film was embedded with oxide nanoparticles (Zagklis and Bampos 2022). In this class, oxide nanoparticles were included in the postscript substrate, which was used in the IP process to organise the TFC film. Membranes with NC substrates play a critical role in maintaining high water porosity and can withstand far less physical abuse. The nano-TiO_2-covered nanocomposite membrane exhibited enhanced photoactivity and/or catalysis (Sun et al. 2023; Hirata et al. 2019). AgNP impregnation into the membrane would enable the creation of a thin-film nanocomposite for water treatment that has outstanding antibacterial performance (Yamakata and Junie-Jhon 2019). The creation of multifunctional membranes by incorporating nanoparticles into polymeric or inorganic membranes has been encouraged by a significant number of papers on membrane nanoscience (Gerba and Pepper 2019). Although the addition of metal oxide nanoparticles such as alumina, silica, zeolite, and TiO_2 has helped to increase the water permeability, fouling resistance, or membrane surface hydrophilicity of polymeric ultrafiltration membranes, the additional water-loving metal oxide nanoparticles significantly reduced fouling by enhancing the membrane hydrophilicity.

Alumina, silica, zeolite, and TiO_2 are examples of metal oxide nanoparticles that have been added to polymeric ultrafiltration membranes to increase membrane surface hydrophilicity, water permeability, or fouling resistance (Ying et al. 2017; Mahmoud and Mostafa 2023). TFC improves the membrane's resistance to fouling and lengthens its lifespan by including nanoparticles such as silica or carbon-based compounds. By enabling more effective filtration, they guarantee better water by eliminating impurities like salts, heavy metals, and bacteria (Ji et al. 2023). Research studies revealed that the wastewater discharge from oil refineries and shale gas was difficult to manage due to the presence of oils and salts.

It was challenging to handle this kind of wastewater with traditional membranes due to the membrane's extreme fouling or failure from salts. Another nano composite forward osmosis (NCFO) membrane was created by the researchers to enable direct oil/water separation and desalination (Ibrahim et al. 2016). This NCFO membrane collected a hydrogel separating layer that eliminated salt and prevented oil accumulation on the surface of the polymeric sustaining layer, which GO nanosheets imparted. Strong water heating controlled by the hydrogel separating layer results in improved antifouling competence in a variety of oil and water emulsions.

4 Comparative Study of Water Purification by Nanocomposites

The unique characteristics of nanomaterials, such as high surface area, customisable pore architectures, and reactivity, which improve contamination removal, make nanocomposite-based water purification a revolutionary approach when compared to conventional approaches. Nanocomposites are superior in adsorbing and breaking down a greater variety of contaminants, such as heavy metals, medications, pesticides, and microplastics, in contrast to traditional techniques like sand filtration or activated carbon. This feature makes it possible to remove several pollutants effectively, something that previous technologies frequently can't do at similar scales or efficiencies. Additionally, nanocomposites can be designed to specifically remove particular pollutants, which is useful in environments with different water quality requirements. In addition to requiring fewer dosages and being reusable, many nanocomposites minimise waste and eventually cut operating expenses.

The cost of production, environmental safety, and scalability are still issues, though. However, the enhanced purifying effectiveness and adaptability of nanocomposites promote their investigation as a viable substitute, supporting environmental objectives while resolving shortcomings in current water treatment methods (Table 4).

5 Limitation and Future Scope

The expense involved in using nanomaterials for wastewater treatment is one of the main obstacles. Although nanocomposites have great potential for treating water, their long-term ecological effects and bioaccumulation hazards are crucial factors that need careful research. When nanoparticles from nanocomposites—especially those made of metal and carbon—enter natural water systems, their distinct nanoscale behaviours may result in unanticipated ecological interactions (Ibrahim et al. 2016). According to studies, some nanoparticles have the ability to build up in aquatic creatures, which might result in bioaccumulation and even biomagnification within

Table 4 Comparative study of nanocomposites towards water purification

Nanocomposite type	Composition	Purification mechanism	Advantages	Limitations
Carbon-based nanocomposites	GO, CNTs, activated carbon with metal oxides (e.g., ZnO, TiO_2)	Adsorption, catalytic oxidation	High surface area, excellent adsorption, reusability	Expensive production, difficulty in scaling
Polymer-based nanocomposites	Polymer matrix with embedded nanoparticles (e.g., Ag, TiO_2)	Adsorption, antimicrobial activity	Versatile applications, easy to mold and process	Potential leaching of nanoparticles
Metal–organic frameworks (MOFs)	Metal ions coordinated with organic linkers	Adsorption, catalytic degradation	High porosity, tunable structure for selective adsorption	Stability issues in aqueous environments
Biodegradable nanocomposites	Natural polymers (e.g., chitosan) with metal oxides	Adsorption, coagulation, flocculation	Eco-friendly, non-toxic, and renewable	Limited chemical stability and mechanical strength
Magnetic nanocomposites	Fe_3O_4 nanoparticles combined with other materials	Adsorption, magnetic separation	Easy recovery with magnets, high removal efficiency	High cost, possible toxicity concerns with iron oxides
Zeolite-based nanocomposites	Zeolite framework with metal or metal oxide nanoparticles	Ion exchange, adsorption	Highly selective for heavy metals and ammonium removal	Limited regeneration ability, costly synthesis
Silica-based nanocomposites	Silica nanoparticles functionalised with metals (e.g., Ag)	Adsorption, antimicrobial, catalytic degradation	High chemical stability, good surface functionalisation	High production cost, potential silica leaching
Photocatalytic nanocomposites	TiO_2, ZnO combined with Ag or graphene oxide	Photocatalytic degradation under light exposure	Effective for organic contaminants, self-cleaning ability	Requires UV/visible light for activation
Clay-based nanocomposites	Montmorillonite, kaolinite with metal oxides	Adsorption, ion exchange	Low cost, good cation exchange capacity	Poor mechanical strength and sometimes slow kinetics
Metal nanoparticle composites	Ag, Cu, or Au nanoparticles on carbon or polymer supports	Antibacterial, catalytic, adsorption	Effective against pathogens, high reactivity	Toxicity concerns, high cost of precious metals

food chains. For example, silver nanoparticles, which are frequently employed for their antibacterial qualities, have shown harmful effects on a variety of aquatic species, such as fish and invertebrates, interfering with their ability to reproduce and metabolise.

Enhancing the economic viability of nanomaterial-based treatment methods and promoting their broader use can enhance water quality and accessibility (Jhaveri and Murthy 2016). The environmental endurance of nanocomposites is one of the main issues. Conventional nanocomposite materials cannot break down quickly, which could have long-term ecological consequences.

The safe use and disposal of nanomaterials while minimising any adverse ecological effects requires the establishment of strong monitoring systems and risk assessment procedures (Zadehahmadi et al. 2023). Future studies should concentrate on creating eco-friendly or biodegradable nanocomposites in light of these hazards. After application, biodegradable nanocomposites decompose into innocuous byproducts, potentially offering efficient water treatment without long-term persistence. For example, a possible strategy to lower ecological concerns is to use organic or plant-based polymers that can break down naturally. This strategy might minimise the impact of water treatment technologies while striking a compromise between the requirement for efficient water purification and environmental responsibility. Enhancing nanocomposites' capacity to be recycled and reused is another crucial avenue. Techniques for regenerating and reusing nanocomposites in several water treatment cycles without sacrificing efficiency should be investigated in research. Toxicology research on a range of nanocomposites in diverse environmental environments will also be necessary to comprehend how they behave in practical situations. Important insights may be gained by researching how nanocomposites affect a variety of ecosystems in freshwater and marine settings, as well as under different pH, salinity, and temperature conditions. These initiatives will aid with risk assessment and provide information for regulations that help protect against unforeseen environmental effects. In order to use nanocomposite technology responsibly and make sure that it aids water treatment initiatives without endangering ecological health, it is imperative that these research goals be addressed.

6 Commercialisation Aspects

For wastewater treatment, the commercialisation of sustainable nanomaterials and nanocomposites depends on a number of variables, including environmental impact, cost-effectiveness, scalability, and regulatory compliance. Nanocomposites frequently offer improved adsorption capacity, higher pollution removal efficiency, and reusability when compared to conventional materials. Over time, these features may result in lower operating costs. However, without a large investment, widespread acceptance may be difficult due to the high initial expenses of development and production. Through the sharing of resources and risks, commercialisation initiatives, including license agreements, public–private partnerships, and cooperation with industrial sectors, aid in bridging the gap between research and market. Investors interested in environmentally friendly developments find sustainable nanocomposites intriguing since they also align with governmental incentives targeted at green technologies. However, broad adoption hinges on proving long-term advantages over current systems and guaranteeing adherence to safety and environmental regulations. Nanocomposites can be positioned as competitive, environmentally friendly substitutes in the water treatment sector by resolving these commercialisation issues.

7 Conclusion

The application of nanomaterials has great potential to transform wastewater treatment and solve the world's water shortage issues. The use of nanocomposites in water treatment has numerous important benefits, most notably improved target specificity and efficiency. To specifically target and eliminate pollutants, including heavy metals, organic chemicals, and viruses, these materials may be manufactured at the nanoscale. Nanocomposites' large surface area enables quick pollutant adsorption and degradation, frequently more successfully than with conventional techniques. Furthermore, the ability to customise nanocomposites for certain pollutants makes them incredibly adaptable in tackling a range of environmental issues.

This chapter, which focuses on iron oxide, nanosilver, TiO_2, ZnO, and CuO in addition to other nanoparticles, summarises the advancements made in the use of nanomaterials for wastewater treatment. Furthermore, carbon-based nanomaterials, such as fullerenes, graphene-based nanosheets, nanotubes, and their nanocomposites, were discussed.

Because of their special qualities and great efficacy in removing pollutants, nanomaterials provide creative ways to improve the sustainability and efficiency of wastewater treatment systems. Advanced treatment systems that effectively remove pollutants from wastewater streams can be created by using the extraordinary qualities of nanomaterials, including their increased surface area, improved reactivity, and programmable properties. Additionally, many of the drawbacks of traditional techniques can be overcome by nanomaterial-based therapy approaches. More effective, affordable, and environmentally friendly wastewater treatment methods may be created by addressing these issues and utilising nanotechnology's transformational potential. The necessity for cooperative research efforts is highlighted by the remarkable potential of nanocomposites in water treatment. Collaborative projects can speed up the creation, testing, and safe use of nanocomposite technology by uniting scientists, engineers, business executives, and legislators. This strategy is essential for improving these materials, comprehending their effects on the environment, and cutting production costs so that they may be used more widely. Adopting nanocomposites for wastewater treatment might have a significant worldwide impact, particularly in areas with limited resources. Underprivileged populations might now have easier access to clean water thanks to nanocomposites, which provide an affordable and sustainable way to remediate contaminated water sources. This technology has the potential to boost global environmental sustainability, stimulate economic growth, and enhance public health with wider implementation.

References

Ahmad M (2021) Zeolite nanocomposites for ammonia and heavy metal removal from agricultural wastewater. Chem Eng J 426:131903

Ahmad A, Mohd-Setapar SH, Chuong CS, Khatoon A, Wani WA, Kumar R, Rafatullah M (2015) Recent advances in new generation dye removal technologies: novel search for approaches to reprocess wastewater. RSC Adv 5:30801–30818

Ali I, Ghamdi KA, Wadaani FTA (2019) Advances in iridium nano catalyst preparation, characterization and applications. J Mol Liq 280:274–284

Baruah S, Khan MN, Dutta J (2016) Perspectives and applications of nanotechnology in water treatment. Environ Chem Lett 14:1–14

Bassin JP, Castro FD, Valério RR, Santiago EP, Lemos FR, Bassin ID (2021) The impact of wastewater treatment plants on global climate change. In: Water conservation in the era of global climate change. Elseviers, Amsterdam, The Netherlands, pp 367–410

Bhat AH, Rehman WU, Khan IU, Ahmad S, Ayoub M, Usmani MA (2018) Nanocomposite membrane for environmental remediation. In: Polymer-based nanocomposites for energy and environmental applications. Woodhead Publishing, Cambridge, UK, pp 407–440

Cornwell DJ, Smith DK (2015) Expanding the scope of gels—combining polymers with low-molecular-weight gelators to yield modified self-assembling smart materials with high-Tech applications. Mater Horiz 2:279–293

Crini G, Lichtfouse E, Wilson LD, Morin-Crini N (2019) Conventional and non-conventional adsorbents for wastewater treatment. Environ Chem Lett 17:195–213

Daer S, Kharraz J, Giwa A, Hasan SW (2015) Recent applications of nanomaterials in water desalination: a critical review and future opportunities. Desalination 367:37–48

Dutta A, Sarkar S (2015) Sequencing batch reactor for wastewater treatment: recent advances. Curr Pollut Rep 1:177–190

Epelle EI, Okoye PU, Roddy S, Gunes B, Okolie JA (2022) Advances in the applications of nanomaterials for wastewater treatment. Environments 9:141

Esfahani MR, Aktij SA, Dabaghian Z, Firouzjaei MD, Rahimpour A, Eke J, Escobar IC, Abolhassani M, Greenlee LF, Esfahani AR et al (2019) Nanocomposite membranes for water separation and purification: fabrication, modification, and applications. Sep Purif Technol 213:465–499

Fang X, Li J, Li X, Pan S, Zhang X, Sun X, Wang L (2017) Internal pore decoration with polydopamine nanoparticle on polymeric ultrafiltration membrane for enhanced heavy metal removal. Chem Eng J 314:38–49

Gautam S, Saini G (2020) Use of natural coagulants for industrial wastewater treatment. Glob J Environ Sci Manag 6:553–578

Gerba CP, Pepper IL (2019) Municipal wastewater treatment. In: Environmental and pollution science. Academic Press: Cambridge, MA, USA, pp 393–418

Gitis V, Hankins N (2018) Water treatment chemicals: trends and challenges. J Water Process Eng 25:34–38

Gupta A et al (2022) Graphene oxide-based nanocomposites for organic contaminant removal in wastewater. Water Res 223:118974

Hirata K, Watanabe H, Kubo W (2019) Nanomembranes as a substrate for ultra-thin lightweight devices. Thin Solid Films 676:8–11

Hodges BC, Cates EL, Kim J (2018) Challenges and prospects of advanced oxidation water treatment processes using catalytic nanomaterial. Nat Nanotechnol 13:642–650

Hogen-Esch T, Pirbazari M, Ravindran V, Yurdacan HM, Kim W (2019) High performance membranes for water reclamation using polymeric and nanomaterials. U.S. Patent 10,456,754, 29 October 2019

Ibrahim RK, Hayyan M, Al-saadi MA, Hayyan A, Ibrahim S (2016) Environmental application of nanotechnology; air, soil, and water. Environ Sci Pollut Res 23:13754–13788

Jhaveri JH, Murthy ZVP (2016) A comprehensive review on anti-Fouling nanocomposite membranes for pressure driven membrane separation processes. Desalination 379:137–154

Ji K, Liu C, He H, Mao X, Wei L, Wang H, Zhang M, Shen Y, Sun R, Zhou F (2023) Research progress of water treatment technology based on nanofiber membranes. Polymers 15:741

Jiang L (2021) Magnetic nanocomposites for multi-stage wastewater treatment and recovery of contaminants. Sep Purif Technol 274:118944

Kochkodan V, Hilal N (2015) A comprehensive review on surface modified polymer membranes for biofouling mitigation. Desalination 356:187–207

Kolya H, Kang S (2023) Biopolymer nanocomposites for effective heavy metal adsorption in wastewater treatment. J Environ Chem Eng 11(1):105437

Kumar V, Singh S, Kashyap N, Singla S, Bhadrecha P, Kaur P (2015) Bio remediation of heavy metals by employing resistant microbial isolates from agricultural soil irrigated with industrial wastewater. Orient J Chem 31:357–361

Kwiatkowski C (2023) Advancements in polymeric nanocomposite membranes for wastewater treatment. Membranes 13(4):312

Larramendy M, Soloneski S (eds) Emerging pollutants in the environment: current and further implications; BoD–Books on Demand: Norderstedt, Germany

Liu Y, Dong S (2020) Carbon nanotube-enhanced membranes for toxic substance filtration in wastewater. J Hazard Mater 385:121612

Loeb SK, Alvarez PJ, Brame JA, Cates EL, Choi W, Crittenden J, Dionysiou DD, Li Q, Li-Puma G, Quan X et al (2019) The technology horizon for photocatalytic water treatment: sunrise or sunset? Environ Sci Technol 53:2937–2947

Mahmoud AED, Mostafa E (2023) Nanofiltration membranes for the removal of heavy metals from aqueous solutions: preparations and applications. Membranes 13:789

Muntha ST, Kausar A, Siddiq M (2017) Advances in polymeric nanofiltration membrane: a review. Polym-Plast Technol Eng 56:841–856

Park J (2022) Nanoclay composites for dye and organic contaminant adsorption in textile wastewater treatment. J Water Process Eng 47:102705

Rajendran R, Aravindan M (2020) Silver nanoparticle-based nanocomposites for microbial load reduction in wastewater. Sci Total Environ 728:138595

Reddy AVB, Jaafar J, Majid ZA, Aris A, Umar K, Talib J, Madhavi G (2015) Relative efficiency comparison of carboxymethyl cellulose (cmc) stabilized Fe^0 and Fe^0/Ag nanoparticles for rapid degradation of chlorpyrifos in aqueous solutions. Dig J Nanomater Bios 10:331–340

Saleh A, Parthasarathy P, Irfan M (2019) Advanced functional polymer nanocomposites and their use in water ultra-purification. Trends Environ Anal 24:67–78

Samanta HS, Das R, Bhattachajee C (2016) Influence of nanoparticles for wastewater treatment-a short review. Austin Chem Eng 3:1036–1045

Sawafta R, Shahwan T (2023) Metal oxide nanoparticles in wastewater treatment: heavy metal removal and photocatalytic degradation. Environ Res 220:115114

Sekoai PT, Ouma CNM, Du Preez SP, Modisha P, Engelbrecht N, Bessarabov DG, Ghimire A (2019) Application of nanoparticles in biofuels: an overview. Fuel 237:380–397

Sidhu GK, Singh S, Kumar V, Datta S, Singh D, Singh J (2019) Toxicity, monitoring and biodegradation of organophosphate pesticides: a review. Crit Rev Environ Sci Technol 49:1135–1187

Soren S, Panda P, Chakroborty S (2023) Nanotechnology in water and wastewater treatment. In: Agricultural and environmental nanotechnology: novel technologies and their ecological impact. Springer Nature, New York, NY, USA, pp 127–143

Sun X et al (2023) Photocatalytic degradation using TiO_2 nanocomposites under UV light in wastewater treatment. Appl Catal B 301:120762

Tang C, Wang Z, Petrini'c I, Fane AG, Hélix-Nielsen C (2015) Biomimetic aquaporin membranes coming of age. Desalination 368:89–105

Umar K (2018) Water contamination by organic-pollutants: TiO_2 photocatalysis. In: Modern age environmental problems and their remediation. Springer International Publishing, Cham, Switzerland, pp 95–109

Umar K, Aris A, Parveen T, Jaafar J, Majid ZA, Reddy AVB, Talib J (2015) Synthesis, characterization of Mo and Mn doped Zno and their photocatalytic activity for the decolorization of two different chromophoric dyes. Appl Catal A 505:507–514

Umar K, Ibrahim MNM, Ahmad A, Rafatullah M (2019) Synthesis of Mn-doped TiO_2 by novel route and photocatalytic mineralization/intermediate studies of organic pollutants. Res Chem Intermediat 45:2927–2945

UNU (2024) Global wastewater status

UN-Water (2024) World water development report

Wu Y, Pang H, Liu Y, Wang X, Yu S, Fu D, Wang X (2019) Environmental remediation of heavy metal ions by novel-nanomaterials: a review. Environ Pollut 246:608–620

Yamakata A, Junie Jhon MV (2019) Curious behaviors of photogenerated electrons and holes at the defects on anatase, rutile, and brookite TiO_2 powders: a review. J Photochem Photobiol C Phtotochem Rev 40:234–243

Ying Y, Ying W, Li Q, Meng D, Ren G, Yan R, Peng X (2017) Recent advances of nanomaterial-based membrane for water purification. Appl Mater Today 7:144–158

Zadehahmadi F, Eden NT, Mahdavi H, Konstas K, Mardel JI, Shaibani M, Banerjee PC, Hill MR (2023) Removal of metals from water using MOF-based composite adsorbents. Environ Sci Water Res Technol 9:1305–1330

Zagklis DP, Bampos G (2022) Tertiary wastewater treatment technologies: a review of technical, economic, and life cycle aspects. Processes 10:2304

Zhang Y, Wu B, Xu H, Liu H, Wang M, He Y, Pan B (2016) Nanomaterials-enabled water and wastewater treatment. Nano Impact 3:22–39

Sustainable Nanomaterials: Synthesis and Application of Green Nanocomposites

Shivani Garg

Abstract The growing demand for sustainable materials has spurred significant advancements in nanotechnology, particularly in developing green nanocomposites. These green nanocomposites are used to address environmental issues that are currently in urgent need, offering solutions that combine ecological responsibility with high performance. This chapter provides a comprehensive overview of sustainable nanomaterials, focusing on the synthesis and application of nanocomposites. The fundamental principles of sustainable nanomaterial design, including life cycle assessment and green chemistry approaches, are discussed in this chapter. A diverse array of bio-based and renewable resources is used as matrix materials in these nanocomposites, such as chitosan, cellulose, and alginate, highlighting their characteristics and benefits over conventional petroleum-based materials. Various environmentally friendly synthesis techniques, such as microbial-assisted processes that use the natural capability of microorganisms, plant-mediated ones that utilize plant extract as capping and reducing agents, and enzyme-catalyzed reactions that offer high specificity and mild reaction conditions, are discussed in detail. The use of ionic liquids as green solvents is also discussed, showcasing the potential to replace volatile organic compounds in nanocomposite synthesis. The characterisation methods used to assess the thermal, mechanical, and structural properties of green nanocomposites, as well as techniques for evaluating their biodegradability and environmental impact, are also discussed. This study analyses green nanocomposites that provide essential insights for material scientists, researchers, and industry professionals, emphasising sustainable nanomaterials' important role in advancing environmental stewardship and technological progress.

Keywords Green nanocomposites · Sustainable synthesis · Biopolymers · Eco-friendly nanomaterials · Biodegradability · Environmental applications

---

S. Garg (✉)
Institute of Environmental Studies, Kurukshetra University, Kurukshetra, Haryana, India
e-mail: shivanigarg2015@kuk.ac.in

S. Roy et al. (eds.), *Nanocomposites for Sustainable Wastewater Treatment*, Composites Science and Technology, https://doi.org/10.1007/978-981-95-1369-7_2

1 Introduction

The escalating environmental concerns and increasing demand for sustainable materials have driven significant attention toward the development of green nanomaterials and nanocomposites. These materials represent a crucial intersection between nanotechnology and sustainability, offering solutions that combine environmental responsibility with enhanced performance characteristics (Gupta et al. 2023). The field of sustainable nanomaterials has experienced remarkable growth over the past decade, particularly in developing green nanocomposites that utilize bio-based resources and environmentally benign synthesis methods (Saud et al. 2024). Green nanocomposites, defined as materials that incorporate nanoscale elements derived from sustainable sources or synthesized through environmentally friendly processes, have emerged as promising alternatives to conventional petroleum-based materials (Veeman et al. 2021). These materials typically consist of a bio-based matrix reinforced with nanoparticles or nanostructures, offering improved mechanical, functional, and thermal properties while maintaining biodegradability and reducing environmental impacts (Pandit et al. 2018). Recent advances in synthesis techniques and characterization methods have significantly expanded the potential applications of these materials across various sectors such as environmental remediation, packaging, and biomedical applications (Ahmed et al. 2022). The transition toward sustainable nanomaterials has been driven by several factors, including stringent environmental regulations, growing consumer awareness, and the need to reduce dependency on fossil-based resources. Traditional nanomaterial synthesis often involves hazardous chemicals, high energy consumption, and generates substantial waste, making it environmentally unsustainable (Samuel et al. 2022). Green synthesis techniques incorporate bioresources and eco-friendly reactions and processes, making the synthesis of nanomaterials much more friendly to the environment than conventional chemical synthesis (Kumar et al. 2021). The current research shows a tremendous advancement in creating new synthesis routes for green nanomaterials. Biotechnological processes, including microbial-aided synthesis, plant-mediated conditions, and enzyme-catalyzed transformations, represent new trends in the synthesis of complex molecules (Vadakkan et al. 2024). These approaches not only lead to the minimization of negative effects on the environment but also produce materials that have special characteristics and improved performance. Ionic liquids as green solvents have also attracted more attention, which is a green reaction medium without a significant reduction in process efficiency compared with a traditional organic solvent (Hessel et al. 2022). Sustainable nanomaterials are designed not only during their synthesis but also at the end of the material's life cycle. This has further highlighted the extended role of life cycle assessment (LCA) in addressing environmental features related to emerging nanomaterials from their production to their disposal (Carroccio et al. 2022). This comprehensive assessment is useful to reveal the potential for enhancements and to guarantee that new materials created are environmentally friendly as required (Cassee et al. 2024). However, some issues are still worth discussing to advance the field of sustainable nanomaterials. These are

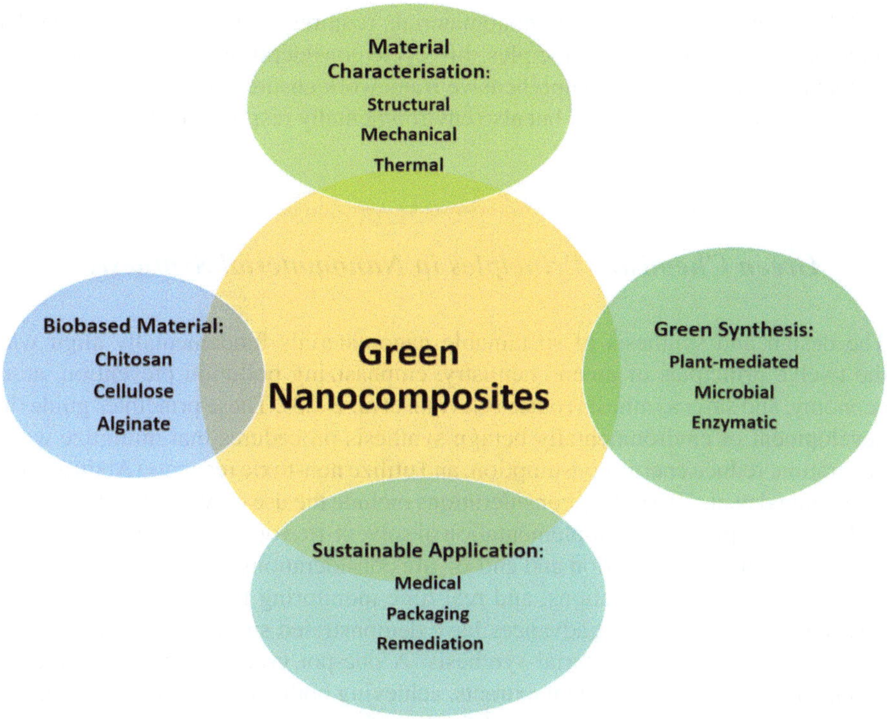

Fig. 1 Overview of sustainable nanomaterials synthesis, components, and applications

the optimization of green synthesis for large-scale industrial production, stability of product quality and properties, and cost issues (Magalhães-Ghiotto et al. 2021). The long-term life cycle assessment, environmental impact, and degradation profiles of these materials cannot be validated for them to be truly green. This chapter provides a comprehensive overview of sustainable nanomaterials, focusing on the synthesis and application of green nanocomposites (Fig. 1). The chapter aims to provide essential insights for researchers, material scientists, and industry professionals working towards advancing environmental stewardship through materials innovation.

2 Fundamental Principles of Sustainable Nanomaterial Design

The design and development of sustainable nanomaterials require a systematic approach that integrates green chemistry principles, life cycle considerations, and environmental impact assessments (Salieri et al. 2021). This comprehensive framework ensures that nanomaterials are not only functionally effective but also environmentally responsible throughout their lifecycle (Stoycheva et al. 2022). The design and development of sustainable nanomaterials requires a systematic approach that integrates green chemistry principles, life cycle considerations, and environmental impact assessments. This comprehensive framework ensures that nanomaterials are not only functionally effective but also environmentally responsible throughout their lifecycle.

2.1 Green Chemistry Principles in Nanomaterial Synthesis

The design and synthesis of sustainable nanomaterials fundamentally align with the twelve principles of green chemistry, emphasizing pollution prevention, atom economy, and safer synthesis routes (Gavilán et al. 2023). These principles guide the development of environmentally benign synthesis procedures that minimize waste generation, reduce energy consumption, and utilize non-toxic reagents (Abdussalam-Mohammed et al. 2020). Key considerations include the use of renewable feedstocks and bio-based precursors, implementation of solvent-free or aqueous-based synthesis routes, design for degradation and end-of-life considerations, maximization of atom economy in synthesis reactions, and real-time monitoring for pollution prevention (Desai et al. 2023). Recent advances have demonstrated successful applications of these principles in nanomaterial synthesis. A one-pot green synthesis method for metal nanomaterials using plant extracts, achieving both reduction and stabilization functions while eliminating the need for harmful reducing agents (Cardoso-Avila et al. 2021).

2.2 Life Cycle Assessment (LCA) Approach

Life cycle assessment serves as a crucial tool in evaluating the environmental sustainability of nanomaterials throughout their lifecycle (Salieri et al. 2018). Current LCA methodologies for nanomaterials focus on four key stages such as raw material extraction and processing, synthesis and manufacturing, use phase, and end-of-life management (Mullen and Morris 2021). A comprehensive LCA framework, especially for green nanocomposites developed, including material and energy inventory analysis, impact assessment methods for nanomaterial releases, end-of-life scenarios for biodegradable materials, and social and economic impact indicators (Gaur 2022).

2.3 Sustainability Metrics and Evaluation Criteria

The evaluation of nanomaterial sustainability requires quantifiable metrics and standardized criteria (García-Quintero and Palencia 2021). A study proposed a comprehensive sustainability assessment framework including environmental metrics (carbon footprint, water consumption, resource depletion, biodegradability rate, and toxicity potential), technical performance metrics (functionality efficiency, durability, recyclability potential, and energy efficiency in use), and economic viability metrics (production costs, scalability potential, and market competitiveness (Dahiya et al. 2020). Recent developments have also focused on developing standardized protocols for evaluating the biodegradation kinetics of green nanocomposites under various environmental conditions (Castro-Aguirre et al. 2018). New technologies have emerged for assessing the long-term stability and ageing behaviours of sustainable nanomaterials, which is crucial for predicting their environmental fate and performance longevity (Gottardo et al. 2021).

2.4 Environmental Impact Considerations

The environmental impact assessment of sustainable nanomaterials has several complex facets that need systematic evaluation throughout their lifecycle (García-Quintero and Palencia 2021). Direct environmental impacts primarily focus on three critical aspects. First, ecotoxicity assessment involves studying the potential harmful effects of nanomaterials on living organisms and ecosystems, including acute and chronic toxicity to various species across different trophic levels (Bour et al. 2015). This pathway analysis in biodegradation studies addresses how these materials decompose within natural environments with specific regard to the decomposition products and their possible environmental persistence (Singh et al. 2021). Atodariya et al. (2024) reported that green nanocomposites obtained from cellulose showed enhanced biodegradation profiles over the typical petroleum-based counterparts, wherein complete degradation occurred within 180 days during composting. Third, the accumulation patterns in environmental matrices, including soil, water, and biological systems, are carefully monitored to understand potential long-term environmental consequences (Solomon et al. 2024).

Another significant advantage of sustainable nanomaterials is related to indirect environmental benefits. Some recent studies have quantified reduced carbon emissions related to green synthesis methods, with up to 45% decreased carbon footprint from conventional processes (Feng et al. 2021). Resource consumption is minimized through efficient synthesis routes and renewable feedstocks, and waste is minimized through atom economy and decreased solvent usage (Khosravi et al. 2024). Brar et al. (2022) stated that implementing green chemistry principles in nanocomposite synthesis resulted in a hazardous waste generation reduction of 60%. Recent research by Osman et al. (2024) has developed new methodologies for assessing the environmental fate of green nanocomposites, especially focusing on their degradation in natural environments and potential impacts on ecosystems. Their work has established protocols for evaluating ecological persistence and transformation under various environmental conditions. Zhang et al. (2019) introduced innovative approaches for real-time monitoring of nanoparticle release and transformation in aquatic environments, providing crucial data for ecological risk assessment.

Advanced life cycle impact assessment techniques have been developed specifically for sustainable nanomaterials using qualitative and quantitative metrics (Pati et al. 2014). These techniques account for various environmental compartments and exposure scenarios, offering comprehensive insight into potential ecological risks and benefits. Improved detection of nanomaterial fate in complex environmental matrices has been possible due to recent advancements in ecological monitoring, leading to more accurate impact assessments (Baalousha et al. 2016).

3 Bio-Based Matrix Materials for Green Nanocomposites

Bio-based matrix materials have emerged as sustainable alternatives to petroleum-based polymers in developing green nanocomposites (Asgher et al. 2020). These materials offer unique advantages, including renewability, biodegradability, and biocompatibility, while maintaining comparable or superior performance characteristics (Liu et al. 2019).

3.1 Natural Polymers

3.1.1 Chitosan-Based Matrices

Chitosan, derived from chitin through deacetylation, has gained considerable attention as a versatile matrix material for green nanocomposites (Mohan et al. 2022). Its unique properties, including antimicrobial activity, biocompatibility, and excellent film-forming capabilities, make it particularly suitable for various applications (Garavand et al. 2022). Studies have recently shown the maximum enhancement of mechanical properties up to 45% via chitosan-based nanocomposites by adding the cellulose nanocrystal; the tensile strength that could be achieved with it was much higher than any other neat chitosan films (Khan et al. 2012; Talebi et al. 2022). Besides, the availability of reactive amino and hydroxyl groups allows surface modifying and crosslinking ability that allows tailoring for several applications (Jouyandeh et al. 2018).

3.1.2 Cellulose and Its Derivatives

The most abundant natural polymer, cellulose, gives a great possibility for developing a green nanocomposite (Shaghaleh et al. 2018). Recent modification and processing technologies of cellulose allowed scientists to obtain different carboxymethyl cellulose, hydroxypropyl methylcellulose, and cellulose acetate (Hosny et al. 2021), among other derivatives. Wang et al. (2018) noted that the HPMC-based nanocomposite has barrier properties better than those of traditional materials due to 70%

decreased oxygen permeability in a package. Cellulose nanofibrils and cellulose nanocrystals have been recently effective reinforcements, delivering both structural and functional characteristics to the resultant nanocomposites (Du et al. 2019).

3.1.3 Alginate and Marine-Derived Polymers

Marine-derived polymers, particularly alginate, have shown promising results in green nanocomposite development. Alginate's unique gel-forming properties and biocompatibility make it ideal for biomedical and food packaging applications (Ahmad et al. 2021). Recent research has focused on developing alginate-based nanocomposites with enhanced mechanical and barrier properties through the incorporation of nano-sized reinforcements (Xie 2024). A study demonstrated that alginate/montmorillonite nanocomposites showed a 200% improvement in water vapour barrier properties while maintaining biodegradability (Zhang and Chen 2023).

3.2 Renewable Resource Utilization

The utilization of renewable resources in nanocomposite development has seen significant advancements in recent years. A comprehensive framework for sustainable resource selection considering factors such as availability and seasonality of raw materials, processing energy requirements, environmental impact of extraction methods, and economic viability of production. Current trends focus on valorizing agricultural and industrial waste streams as sources of bio-based matrices. Liu et al. (2024) successfully developed a nanocomposite using cellulose extracted from agricultural residues, achieving properties comparable to those of virgin material-based composites.

3.3 Comparison with Conventional Matrices

Bio-based matrices offer several advantages over conventional petroleum-based materials, yet some limitations remain. Table 1 explains the advantages and limitations in detail.

4 Environmentally Friendly Synthesis Techniques

Developing environmentally friendly synthesis techniques for nanomaterials represents a significant advancement in sustainable material science. Figure 2 shows the different methods of synthesis of nanocomposites by green synthesis approaches

Table 1 Comparison of bio-based and petroleum-based matrices for nanocomposites

Properties	Bio-based matrices	Petroleum-based matrices	References
Mechanical properties			
Tensile strength	Higher strength in optimized formulations; Enhanced fiber-matrix interfacial bonding	Consistent but lower strength; Well-established performance standards	Wang et al. (2015)
Impact resistance	Excellent energy absorption; Natural fibre reinforcement provides better impact distribution	Moderate to high impact resistance; More brittle behaviour	Kumar et al. (2023)
Flexibility	Higher elasticity in hydrated conditions; Better shape recovery	Standard flexibility; Less responsive to environmental conditions	Zheng et al. (2024)
Environmental performance			
Biodegradability	Naturally biodegradable; Controlled degradation rates possible	Non-biodegradable; Persistent environmental impact	Jha et al. (2024)
Carbon footprint	Significantly lower; Carbon–neutral potential	High carbon footprint; Resource-intensive production	Bhaliya et al. (2024)
Resource sustainability	Renewable resources; Annual crop cycles	Non-renewable petroleum resources	García-Quintero and Palencia (2021)
Processing characteristics			
Processing temperature	Lower processing temperatures; Energy-efficient	Higher processing temperatures required	Cywar et al. (2022)
Processing complexity	May require specific conditions; Moisture sensitivity	Well-established processes; Consistent results	Wang et al. (2021)
Scalability	Improving but still challenging; Batch-to-batch variation	Highly scalable; Consistent quality	Yang et al. (2018)
Functional properties			
Thermal stability	Enhanced stability with proper modifications	Good thermal stability; Often requires additives	He et al. (2021)
Barrier properties	Excellent oxygen and moisture barriers; Natural antimicrobial properties	Good barrier properties; Requires additional treatments	Wang et al. (2018)
Durability	Improving with new formulations; Environment-dependent	Consistent long-term performance	Andrew and Dhakal (2022)

Fig. 2 Illustration of different environmentally friendly synthesis techniques

using natural materials. These methods offer advantages such as mild reaction conditions, reduced environmental impact, and enhanced control over material properties (Ying et al. 2022).

4.1 Microbial-Assisted Synthesis

4.1.1 Bacterial Synthesis Mechanisms

Bacterial synthesis of nanomaterials has emerged as a promising green approach due to its efficiency and environmental compatibility, offering significant advantages in terms of scalability and cost-effectiveness compared to traditional chemical methods (Jeevanandam et al. 2022). Bacteria utilize various mechanisms to synthesize nanoparticles, including extracellular synthesis (bacteria secrete enzymes that reduce metal ions to form nanoparticles in the surrounding medium), intercellular synthesis (nanoparticles are formed within bacterial cells through metabolic processes), and cell surface synthesis (biosorption and reduction of metal ions occur at the bacterial cell surface) (Lahiri et al. 2021). These biological pathways often produce nanoparticles with unique morphologies and surface properties that enhance their functionality in various applications, particularly in biomedical and catalytic processes (Ahmad et al. 2022). The bacterial synthesis pathway also demonstrates remarkable adaptability to different metal precursors and environmental conditions, allowing for fine-tuned control over particle characteristics (Choi and Lee 2020).

Recent studies by Almatroudi (2024) demonstrated that *Pseudomonas aeruginosa* can synthesize silver nanoparticles with controlled size distribution (10–50 nm) under ambient conditions, achieving an impressive 85% yield while operating at room temperature and neutral pH, highlighting the potential of bacterial synthesis for industrial-scale production of nanomaterials.

4.1.2 Fungal-Mediated Processes

Fungal systems offer several significant advantages over bacterial synthesis in the production of nanomaterials, demonstrating remarkable versatility and efficiency in bioaccumulation and biotransformation processes (Adeleke et al. 2024). These systems are characterized by their higher metal tolerance that allows them to withstand and process higher concentrations of metallic ions without compromising their cellular functions (Priyadarshini et al. 2021), enhanced production of reducing enzymes that facilitate more efficient nanoparticle formation (Li et al. 2022a, b), and greater binding capacity coupled with larger surface area due to their extensive hyphal networks and robust cell wall composition (Slavin and Bach 2022). The fungal cellular machinery is particularly adept at controlling particle size and morphology through precise enzymatic reactions and specialized protein interactions, leading to more uniform and stable nanoparticle formations (Hassan et al. 2024). Their exceptional adaptability to various growth conditions and ability to thrive in diverse environmental settings make them ideal for large-scale industrial applications (Rameez et al. 2024). According to a study, *Aspergillus niger* could efficiently produce nanosilver and nanogold particles with unique morphologies while operating in waste metallic solutions, demonstrating both synthesis and remediation capabilities, which underscores the potential of fungal systems in developing sustainable and cost-effective approaches to nanomaterial production while simultaneously addressing environmental concerns through the treatment of metal-contaminated wastewaters (Soni and Prakash, 2014).

4.2 Plant-Mediated Synthesis

Plant extracts have emerged as powerful biological factories for nanomaterials synthesis, demonstrating remarkable versatility and efficiency in the production of diverse nanostructures through their complex biochemical composition and multifunctional properties (Adeyemi et al. 2022). Plant extracts serve multiple critical functions in nanomaterials synthesis. They act as reducing agents due to their rich content of polyphenols and flavonoids, which facilitate the conversion of metal ions into stable nanoparticles through electron transfer mechanisms (Shafey 2020). Proteins and polysaccharides provide stability by forming protective coatings around the newly formed nanoparticles, preventing agglomeration and maintaining their desired size distribution and offering shape-directing properties through various biomolecules that can selectively bind to specific crystal faces, controlling the morphological evolution of the nanostructures (Verma et al. 2020). The synergic interaction between these different plant-derived compounds enables fine-tuned control over particle characteristics, while the abundance and renewable nature of plant resources make this approach particularly attractive for large-scale, environmentally friendly nanomaterial production (Alsaiari et al. 2023). The process can be further optimized by controlling factors such as pH, temperature, and extract concentration, allowing for precise manipulation of nanoparticle size, shape, and surface properties to meet specific application requirements (Harish et al. 2023). Recent advancements in plant-mediated synthesis have also revealed the potential for developing multifunctional nanoparticles that combine the inherent therapeutic properties of plant compounds with the unique physicochemical properties of engineered nanomaterials, opening new avenues for applications in drug delivery, imaging, and theragnostic (Sarvarian et al. 2022).

The sophisticated molecular mechanism underlying plant-mediated nanoparticle synthesis involves a complex interplay of various phytochemical components and biomolecules that work in concert to facilitate controlled nanostructure formation (Puri et al. 2024). The process encompasses several critical stages, including phytochemical reduction, natural capping, and shape control (Puri et al. 2024). Green tea extract could synthesise silver nanoparticles with superior antimicrobial properties while achieving a 90% reduction in chemical reagent use, highlighting the remarkable efficiency and sustainability (Rolim et al. 2019).

4.3 *Enzyme-Catalyzed Reactions*

Enzyme-catalyzed reactions have emerged as a highly sophisticated and precisely controlled approach to nanomaterial synthesis, offering unprecedented control over particle formation and characteristics through the specific catalytic activities of various enzymatic systems (Lakhani et al. 2024). The process employs two main categories of enzymes oxidoreductases (Glucose oxidase, laccase and peroxidases) which facilitate electron transfer reactions and control redox processes during nanoparticle formation (Ratautas and Dagys 2019), and hydrolases (cellulase, chitinase, and proteases) which catalyse the breakdown of specific chemical bonds and enable the controlled release of reactive species (Swetha et al. 2022). Laccase-mediated synthesis could produce uniform gold nanoparticles under mild conditions with precise size control, highlighting the remarkable precision achievable through enzymatic routes (Tocco 2022). The success of enzyme-catalyzed synthesis heavily depends on carefully optimizing reaction conditions, including temperature, pH, enzyme concentration, substrate specificity, and reaction time, all of which must be precisely controlled to achieve desired outcomes (Siedentop et al. 2021). Park and Johnson (2024) successfully developed an optimized protocol using cellulose for the synthesis of cellulose nanocrystals, achieving an impressive 75% yield under mild conditions, demonstrating the industrial viability of enzymatic approaches.

The high specificity of enzymatic reactions enables exceptional control over particle morphology and surface chemistry, while the mild reaction conditions and biocompatibility of enzymes make this approach particularly attractive for biomedical applications (Anwar et al. 2023).

4.4 Green Solvents

Green solvents have revolutionized the landscape of sustainable nanomaterial synthesis by offering environmentally benign alternatives to conventional organic solvents, with ionic liquids emerging as a particularly promising option due to their exceptional properties and versatility (Jhanji 2023). These innovative solvents demonstrate remarkable advantages, including low volatility that minimizes atmospheric emissions, high thermal stability enabling operations across wide temperature ranges, tunable properties that can be optimized for specific applications and excellent recyclability that significantly reduces waste generation and operational costs (Kaur et al. 2022). Recent developments by Singh et al. (2024) demonstrated that choline-based ionic liquids could effectively dissolve cellulose and facilitate nanocomposite formation while being fully recyclable, marking a significant advancement in sustainable materials processing. Several sustainable alternatives to ionic liquids show considerable promise in nanomaterial synthesis. Deep eutectic solvents offer similar advantages to ionic liquids, with potentially lower costs and environmental impact (Kudłak et al. 2015). Supercritical carbon dioxide provides a non-toxic and easily recoverable medium for nanomaterial processing (Xu et al. 2019). Bio-based solvents, derived from renewable resources, align well with green chemistry principles (Silvestri et al. 2021). Water-based systems represent the ultimate in environmental compatibility. The integration of these solvents into nanomaterial synthesis protocols has enabled the development of more sustainable manufacturing processes while often improving product quality and process efficiency through their unique physicochemical properties (Hessel et al. 2022).

5 Characterization and Property Assessment

The characterization and evaluation of sustainable nanomaterial properties consist of complex analytical techniques of value for understanding the composition, behavior, and ecology of the nanomaterial (García-Quintero and Palencia 2021). Advanced methods used in structural characterization include the determination of the crystal structure, high-resolution morphology using TEM, and surface topography analysis using scanning electron microscope (SEM) (Suga et al. 2014). The technique of atomic force microscopy (AFM) is used for surface mapping at the nanoscale (Landoulsi and Dupres 2013). The application of spectroscopic methods-including FTIR and Raman-is used to determine chemical bonding and molecular interaction

(Fakayode et al. 2020). These are the thermal analysis methods: differential scanning calorimetry (DSC), thermogravimetric analysis (TGA), and dynamic mechanical analysis (DMA) which allow for critical knowledge about the stability of the material, phase transitions, decomposition patterns, and the behavior as a function of temperature, which enables the optimization of conditions of processing and application parameters (Leyva-Porras et al. 2019). Mechanical properties characterization is performed using advanced methods, such as nanoindentation, tensile testing, and dynamic mechanical analysis by strength, elasticity, hardness, and fatigue resistance (Golovin 2021). Atomic force microscopy-based force spectroscopy gives insight into nanomechanical behaviors and surface interactions (Zhang et al. 2022). It essentially includes the evaluation of biodegradation through standardized test protocols like ASTM D5338, advanced analytical methods to monitor degradation kinetics, intermediate products, and microbial interactions by employing techniques such as GPC for monitoring molecular weight, and SEM for surface degradation analysis (Yatigala et al. 2024). Environmental impact assessment encompasses diverse methods, including LCA studies, ecotoxicological investigation, and monitoring environmental fate, through sensitive analysis tools like ICP-MS to determine trace metal amounts, LC-MS to find out degradative product concentrations and standardized tests on ecotoxicity about potential risks as well as benefits for the environment besides considering energy consumed, material used and at end-of-life (Li et al. 2022a, b).

6 Applications, Future Perspectives, and Recommendations

The evolution of sustainable nanomaterials presents both promising opportunities and significant challenges across various application domains, with current implementations demonstrating remarkable potential in multiple sectors (Gupta et al. 2023). In environmental remediation, advanced nano-based systems are revolutionizing water treatment and soil decontamination through enhanced adsorption capacity and catalytic activity, while emerging technologies in sustainable packaging utilize nanocomposites to create biodegradable materials with superior barrier properties and mechanical strength, addressing critical waste management issues (Kuhn et al. 2022). Biomedical applications have witnessed groundbreaking developments in drug delivery systems, tissue engineering, and diagnostic tools, leveraging the unique properties of sustainable nanomaterials to enhance therapeutic efficacy while minimizing environmental impact (Wirtu et al. 2025). However, several challenges and limitations persist, including scalability issues, cost considerations, and the need for standardized characterization protocols, while concerns about long-term environmental fate and potential toxicological impacts require careful attention.

Future research directions should focus on developing more efficient synthesis routes, improving control over particle characteristics, and establishing comprehensive safety assessment frameworks, with particular emphasis on understanding the complex interactions between engineered nanomaterials and biological systems

(Albalawi et al. 2021). Industrial scaling considerations demand innovative approaches to process optimization, quality control, and cost reduction while maintaining strict adherence to sustainability principles throughout the manufacturing lifecycle (Leng et al. 2020). Advanced automation and process control systems, coupled with artificial intelligence and machine learning algorithms, could significantly enhance production efficiency and product consistency (Rane et al. 2024). The integration of circular economy principles in nanomaterial production processes presents an exciting opportunity for waste reduction and resource recovery, while the development of hybrid materials combining multiple sustainable components could lead to enhanced functionality and expanded application possibilities (Gilbertson et al. 2024). Recommendations for future development include establishing international standards for sustainability assessment, implementing comprehensive life cycle analysis protocols, fostering collaboration between academia and industry, and developing regulatory frameworks that balance innovation with environmental protection, ensuring the responsible advancement of sustainable nanomaterial technologies.

7 Conclusion

The evolution of sustainable nanomaterial synthesis and applications represents a fundamental advancement in material science, demonstrating remarkable potential for addressing contemporary environmental and technological challenges. Through utilizing bio-based precursors, green synthesis methodologies, and innovative processing techniques, significant progress has been achieved in developing environmentally benign nanomaterials with enhanced performance characteristics. The environmental implications of these developments are profound, offering reduced carbon footprints, minimized toxic waste generation, and improved resource utilization efficiency compared to conventional synthesis methods. The successful implementation of microbial, plant-mediated, and enzyme-catalyzed synthesis routes, coupled with the adoption of green solvents, has established a robust foundation of sustainable nanomaterial production. Future development should emphasize the need for standardized characterization protocols, scaled-up production methodologies, and comprehensive life cycle assessments. Priority should be given to optimizing process parameters for industrial-scale production while maintaining product quality and environmental sustainability. Enhanced collaboration between institutions, industries, and regulatory bodies such as policymakers is crucial for accelerating the transition from laboratory-scale discoveries to commercial applications.

References

Abdussalam-Mohammed W, Ali AQ, Errayes AO (2020) Green chemistry: principles, applications, and disadvantages. Chem Methodol 4(4):408–423

Adeleke BS, Olowe O, Ayilara MS, Fasusi AO, Omotayo OP, Fadiji AE, Babalola OO et al (2024) Biosynthesis of nanoparticles using microorganisms: a focus on endophytic fungi. Heliyon

Adeyemi JO, Oriola AO, Onwudiwe DC, Oyedeji AO (2022) Plant extracts mediated metal-based nanoparticles: synthesis and biological applications. Biomolecules 12(5):627

Ahmad A, Mubarak NM, Jannat FT, Ashfaq T, Santulli C, Rizwan M, Ali S et al (2021) A critical review on the synthesis of natural sodium alginate based composite materials: an innovative biological polymer for biomedical delivery applications. Processes 9(1):137

Ahmad F, Salem-Bekhit MM, Khan F, Alshehri S, Khan A, Ghoneim MM, Elbagory I et al (2022) Unique properties of surface-functionalized nanoparticles for bio-application: functionalization mechanisms and importance in application. Nanomaterials 12(8):1333

Ahmed SF, Mofijur M, Rafa N, Chowdhury AT, Chowdhury S, Nahrin M, Ong HC et al (2022) Green approaches in synthesising nanomaterials for environmental nanobioremediation: technological advancements, applications, benefits, and challenges. Environ Res 204:111967

Albalawi F, Hussein MZ, Fakurazi S, Masarudin MJ (2021) Engineered nanomaterials: the challenges and opportunities for nanomedicines. Int J Nanomed:161–184

Almatroudi A (2024) Unlocking the potential of silver nanoparticles: from synthesis to versatile bio-applications. Pharmaceutics 16(9):1232

Alsaiari NS, Alzahrani FM, Amari A, Osman H, Harharah HN, Elboughdiri N, Tahoon MA (2023) Plant and microbial approaches as green methods for the synthesis of nanomaterials: synthesis, applications, and future perspectives. Molecules 28(1):463

Andrew JJ, Dhakal HN (2022) Sustainable biobased composites for advanced applications: recent trends and future opportunities–a critical review. Compos Part C: Open Access 7:100220

Anwar A, Imran M, Iqbal HM (2023) Smart chemistry and applied perceptions of enzyme-coupled nano-engineered assemblies to meet future biocatalytic challenges. Coord Chem Rev 493:215329

Asgher M, Qamar SA, Bilal M, Iqbal HM (2020) Bio-based active food packaging materials: sustainable alternative to conventional petrochemical-based packaging materials. Food Res Int 137:109625

Atodariya J, Agrawal M, Singh A, Patni N (2024) Advancements in green nanocomposites: a comprehensive review on cellulose-based materials in biocomposites and bionanocomposites. In: Rubber based bionanocomposites: characterisation, pp 213–239

Baalousha M, Cornelis G, Kuhlbusch TAJ, Lynch I, Nickel C, Peijnenburg WJGM, Van Den Brink NW (2016) Modeling nanomaterial fate and uptake in the environment: current knowledge and future trends. Environ Sci Nano 3(2):323–345

Bhaliya JD, Shah VR, Mehta S, Kashyap SS, Borah K, Kutcherlapati SR (2024) Life cycle assessments, environmental implications, and toxicological evaluations of functionalized polymer nanocomposites. In: Advances in functionalized polymer nanocomposites. Woodhead Publishing, pp 1031–1066

Bour A, Mouchet F, Silvestre J, Gauthier L, Pinelli E (2015) Environmentally relevant approaches to assess nanoparticles ecotoxicity: a review. J Hazard Mater 283:764–777

Brar KK, Magdouli S, Othmani A, Ghanei J, Narisetty V, Sindhu R, Pandey A et al (2022) Green route for recycling of low-cost waste resources for the biosynthesis of nanoparticles (NPs) and nanomaterials (NMs)-a review. Environ Res 207:112202

Cardoso-Avila PE, Patakfalvi R, Rodríguez-Pedroza C, Aparicio-Fernández X, Loza-Cornejo S, Villa-Cruz V, Martínez-Cano E (2021) One-pot green synthesis of gold and silver nanoparticles using Rosa canina L. extract. RSC Adv 11(24):14624–14631

Carroccio SC, Scarfato P, Bruno E, Aprea P, Dintcheva NT, Filippone G (2022) Impact of nanoparticles on the environmental sustainability of polymer nanocomposites based on bioplastics or recycled plastics–a review of life-cycle assessment studies. J Clean Prod 335:130322

Cassee FR, Bleeker EA, Durand C, Exner T, Falk A, Friedrichs S, Groenewold M et al (2024) Roadmap towards safe and sustainable advanced and innovative materials (Outlook for 2024–2030). Comput Struct Biotechnol J 25:105–126

Castro-Aguirre E, Auras R, Selke S, Rubino M, Marsh T (2018) Impact of nanoclays on the biodegradation of poly (lactic acid) nanocomposites. Polymers 10(2):202

Choi Y, Lee SY (2020) Biosynthesis of inorganic nanomaterials using microbial cells and bacteriophages. Nat Rev Chem 4(12):638–656

Cywar RM, Rorrer NA, Hoyt CB, Beckham GT, Chen EYX (2022) Bio-based polymers with performance-advantaged properties. Nat Rev Mater 7(2):83–103

Dahiya S, Katakojwala R, Ramakrishna S, Mohan SV (2020) Biobased products and life cycle assessment in the context of circular economy and sustainability. Mater Circ Econ 2:1–28

Desai AV, Lizundia E, Laybourn A, Rainer DN, Armstrong AR, Morris RE, Ettlinger R et al (2023) Green synthesis of reticular materials. Adv Funct Mater:2304660

Du H, Liu W, Zhang M, Si C, Zhang X, Li B (2019) Cellulose nanocrystals and cellulose nanofibrils-based hydrogels for biomedical applications. Carbohyd Polym 209:130–144

Fakayode SO, Baker GA, Bwambok DK, Bhawawet N, Elzey B, Siraj N, Warner IM et al (2020) Molecular (Raman, NIR, and FTIR) spectroscopy and multivariate analysis in consumable products analysis1. Appl Spectrosc Rev 55(8):647–723

Feng Z, Adolfsson KH, Xu Y, Fang H, Hakkarainen M, Wu M (2021) Carbon dot/polymer nanocomposites: from green synthesis to energy, environmental and biomedical applications. Sustain Mater Technol 29:e00304

Garavand F, Cacciotti I, Vahedikia N, Rehman A, Tarhan Ö, Akbari-Alavijeh S, Jafari SM et al (2022) A comprehensive review on the nanocomposites loaded with chitosan nanoparticles for food packaging. Crit Rev Food Sci Nutr 62(5):1383–1416

García-Quintero A, Palencia M (2021) A critical analysis of environmental sustainability metrics applied to green synthesis of nanomaterials and the assessment of environmental risks associated with the nanotechnology. Sci Total Environ 793:148524

Gaur R (2022) Environmental impact and life cycle analysis of green nanomaterials. In: Green functionalized nanomaterials for environmental applications. Elsevier, pp 513–539

Gavilán H, Serrano MB, Cabanelas JC (2023) Nanomaterials and their synthesis for a sustainable future. New Mater Circ Econ 149:233–310

Gilbertson LM, Eckelman MJ, Theis TL (2024) Leveraging engineered nanomaterials to support material circularity. Environ Sci: Nano

Golovin YI (2021) Nanoindentation and mechanical properties of materials at submicro-and nanoscale levels: recent results and achievements. Phys Solid State 63:1–41

Gottardo S, Mech A, Drbohlavová J, Małyska A, Bøwadt S, Sintes JR, Rauscher H (2021) Towards safe and sustainable innovation in nanotechnology: state-of-play for smart nanomaterials. NanoImpact 21:100297

Gupta D, Boora A, Thakur A, Gupta TK (2023) Green and sustainable synthesis of nanomaterials: recent advancements and limitations. Environ Res 231:116316

Harish V, Ansari MM, Tewari D, Yadav AB, Sharma N, Bawarig S, Barhoum A et al (2023) Cutting-edge advances in tailoring size, shape, and functionality of nanoparticles and nanostructures: a review. J Taiwan Inst Chem Eng 149:105010

Hassan S, Karaila GK, Singh P, Meenatchi R, Venkateswaran AS, Ahmed T, Selvin J et al (2024) Implications of fungal nanotechnology for sustainable agriculture-applications and future perspectives. Biocatal Agric Biotechnol:103110

He Y, Li H, Luo F, Jin Y, Huang B, Qian Q (2021) Bio-based flexible phase change composite film with high thermal conductivity for thermal energy storage. Compos A Appl Sci Manuf 151:106638

Hessel V, Tran NN, Asrami MR, Tran QD, Long NVD, Escribà-Gelonch M, Sundmacher K et al (2022) Sustainability of green solvents–review and perspective. Green Chem 24(2):410–437

Hosny KM, Alkhalidi HM, Alharbi WS, Md S, Sindi AM, Ali SA, Kurakula M et al (2021) Recent trends in assessment of cellulose derivatives in designing novel and nanoparticulate-based drug delivery systems for improvement of oral health. Polymers 14(1):92

Jeevanandam J, Kiew SF, Boakye-Ansah S, Lau SY, Barhoum A, Danquah MK, Rodrigues J (2022) Green approaches for the synthesis of metal and metal oxide nanoparticles using microbial and plant extracts. Nanoscale 14(7):2534–2571
Jha S, Akula B, Enyioma H, Novak M, Amin V, Liang H (2024) Biodegradable biobased polymers: a review of the state of the art, challenges, and future directions. Polymers 16(16):2262
Jhanji I (2023) The role of ionic liquids in green solvent chemistry: properties and applications. Int Res J Eng Appl Sci 11:57–67
Jouyandeh M, Jazani OM, Navarchian AH, Shabanian M, Vahabi H, Saeb MR (2018) Surface engineering of nanoparticles with macromolecules for epoxy curing: development of super-reactive nitrogen-rich nanosilica through surface chemistry manipulation. Appl Surf Sci 447:152–164
Kaur G, Kumar H, Singla M (2022) Diverse applications of ionic liquids: a comprehensive review. J Mol Liq 351:118556
Khan A, Khan RA, Salmieri S, Le Tien C, Riedl B, Bouchard J, Lacroix M et al (2012) Mechanical and barrier properties of nanocrystalline cellulose reinforced chitosan based nanocomposite films. Carbohyd Polym 90(4):1601–1608
Khosravi A, Zarepour A, Iravani S, Varma RS, Zarrabi A (2024) Sustainable synthesis: natural processes shaping the nanocircular economy. Environ Sci Nano 11(3):688–707
Kudłak B, Owczarek K, Namieśnik J (2015) Selected issues related to the toxicity of ionic liquids and deep eutectic solvents—a review. Environ Sci Pollut Res 22:11975–11992
Kuhn R, Bryant IM, Jensch R, Böllmann J (2022) Applications of environmental nanotechnologies in remediation, wastewater treatment, drinking water treatment, and agriculture. Appl Nano 3(1):54–90
Kumar JA, Krithiga T, Manigandan S, Sathish S, Renita AA, Prakash P, Crispin S et al (2021) A focus to green synthesis of metal/metal based oxide nanoparticles: various mechanisms and applications towards ecological approach. J Clean Prod 324:129198
Kumar A, Yadav S, Pramanik J, Sivamaruthi BS, Jayeoye TJ, Prajapati BG, Chaiyasut C (2023) Chitosan-based composites: development and perspective in food preservation and biomedical applications. Polymers 15(15):3150
Lahiri D, Nag M, Sheikh HI, Sarkar T, Edinur HA, Pati S, Ray RR (2021) Microbiologically-synthesized nanoparticles and their role in silencing the biofilm signaling cascade. Front Microbiol 12:636588
Lakhani P, Bhanderi D, Modi CK (2024) Nanocatalysis: recent progress, mechanistic insights, and diverse applications. J Nanopart Res 26(7):148
Landoulsi J, Dupres V (2013) Direct AFM force mapping of surface nanoscale organization and protein adsorption on an aluminum substrate. Phys Chem Chem Phys 15(21):8429–8440
Leng J, Ruan G, Jiang P, Xu K, Liu Q, Zhou X, Liu C (2020) Blockchain-empowered sustainable manufacturing and product lifecycle management in industry 4.0: a survey. Renew Sustain Energy Rev 132:110112
Leyva-Porras C, Cruz-Alcantar P, Espinosa-Solís V, Martínez-Guerra E, Piñón-Balderrama CI, Compean Martínez I, Saavedra-Leos MZ (2019) Application of differential scanning calorimetry (DSC) and modulated differential scanning calorimetry (MDSC) in food and drug industries. Polymers 12(1):5
Li J, Li X, Da Y, Yu J, Long B, Zhang P, Dai SY et al (2022a) Sustainable environmental remediation via biomimetic multifunctional lignocellulosic nano-framework. Nat Commun 13(1):4368
Li Q, Liu F, Li M, Chen C, Gadd GM (2022b) Nanoparticle and nanomineral production by fungi. Fungal Biol Rev 41:31–44
Liu H, Jian R, Chen H, Tian X, Sun C, Zhu J, Wang C et al (2019) Application of biodegradable and biocompatible nanocomposites in electronics: current status and future directions. Nanomaterials 9(7):950
Liu Z, Chen Z, Wang Z, Zhang D, Ni BJ (2024) Recycling and repurposing of waste carbon nanofiber polymers: a critical review. Environ Sci: Nano

Magalhães-Ghiotto GA, de Oliveira AM, Natal JP, Bergamasco R, Gomes RG (2021) Green nanoparticles in water treatment: a review of research trends, applications, environmental aspects, and large-scale production. Environ Nanotechnol Monit Manage 16:100526
Mohan K, Ganesan AR, Ezhilarasi PN, Kondamareddy KK, Rajan DK, Sathishkumar P, Conterno L et al (2022) Green and eco-friendly approaches for the extraction of chitin and chitosan: a review. Carbohyd Polym 287:119349
Mullen E, Morris M (2021) Green nanofabrication opportunities in the semiconductor industry: a life cycle perspective. Nanomaterials 2021(11):1085
Osman AI, Zhang Y, Farghali M, Rashwan AK, Eltaweil AS, Abd El-Monaem EM, Yap PS et al (2024) Synthesis of green nanoparticles for energy, biomedical, environmental, agricultural, and food applications: a review. Environ Chem Lett 22(2):841–887
Pandit P, Nadathur GT, Maiti S, Regubalan B (2018) Functionality and properties of bio-based materials. In: Bio-based materials for food packaging: green and sustainable advanced packaging materials, pp 81–103
Park H, Johnson PA (2024) The development and evaluation of chitosan-coated enzyme magnetic nanoparticles for cellulose hydrolysis. Front Chem Eng 6:1479798
Pati P, McGinnis S, Vikesland PJ (2014) Life cycle assessment of "green" nanoparticle synthesis methods. Environ Eng Sci 31(7):410–420
Priyadarshini E, Priyadarshini SS, Cousins BG, Pradhan N (2021) Metal-Fungus interaction: review on cellular processes underlying heavy metal detoxification and synthesis of metal nanoparticles. Chemosphere 274:129976
Puri A, Mohite P, Maitra S, Subramaniyan V, Kumarasamy V, Uti DE, Atangwho IJ et al (2024) From nature to nanotechnology: the interplay of traditional medicine, green chemistry, and biogenic metallic phytonanoparticles in modern healthcare innovation and sustainability. Biomed Pharmacother 170:116083
Rameez M, Khan N, Ahmad S, Ahmad MM (2024) Bionanocomposites: a new approach for fungal disease management. Biocatal Agric Biotechnol:103115
Rane NL, Paramesha M, Choudhary SP, Rane J (2024) Artificial intelligence, machine learning, and deep learning for advanced business strategies: a review. Partners Univ Int Innov J 2(3):147–171
Ratautas D, Dagys M (2019) Nanocatalysts containing direct electron transfer-capable oxidoreductases: recent advances and applications. Catalysts 10(1):9
Rolim WR, Pelegrino MT, de Araújo Lima B, Ferraz LS, Costa FN, Bernardes JS, Seabra AB et al (2019) Green tea extract mediated biogenic synthesis of silver nanoparticles: characterization, cytotoxicity evaluation and antibacterial activity. Appl Surf Sci 463:66–74
Salieri B, Turner DA, Nowack B, Hischier R (2018) Life cycle assessment of manufactured nanomaterials: where are we? NanoImpact 10:108–120
Salieri B, Barruetabeña L, Rodríguez-Llopis I, Jacobsen NR, Manier N, Trouiller B, Hischier R et al (2021) Integrative approach in a safe by design context combining risk, life cycle and socio-economic assessment for safer and sustainable nanomaterials. NanoImpact 23:100335
Samuel MS, Ravikumar M, John JA, Selvarajan E, Patel H, Chander PS, Chandrasekar N et al (2022) A review on green synthesis of nanoparticles and their diverse biomedical and environmental applications. Catalysts 12(5), 459
Sarvarian P, Samadi P, Gholipour E, Shams Asenjan K, Hojjat-Farsangi M, Motavalli R, Yousefi M et al (2022) Application of emerging plant-derived nanoparticles as a novel approach for nano-drug delivery systems. Immunol Invest 51(4):1039–1059
Saud A, Gupta S, Allal A, Preud'Homme H, Shomar B, Zaidi SJ (2024) Progress in the sustainable development of biobased (nano) materials for application in water treatment technologies. ACS Omega 9(27):29088–29113
Shafey AME (2020) Green synthesis of metal and metal oxide nanoparticles from plant leaf extracts and their applications: a review. Green Process Synth 9(1):304–339
Shaghaleh H, Xu X, Wang S (2018) Current progress in production of biopolymeric materials based on cellulose, cellulose nanofibers, and cellulose derivatives. RSC Adv 8(2):825–842

Siedentop R, Claaßen C, Rother D, Lütz S, Rosenthal K (2021) Getting the most out of enzyme cascades: strategies to optimize in vitro multi-enzymatic reactions. Catalysts 11(10):1183
Silvestri C, Silvestri L, Forcina A, Di Bona G, Falcone D (2021) Green chemistry contribution towards more equitable global sustainability and greater circular economy: a systematic literature review. J Clean Prod 294:126137
Singh AK, Bilal M, Iqbal HM, Raj A (2021) Trends in predictive biodegradation for sustainable mitigation of environmental pollutants: recent progress and future outlook. Sci Total Environ 770:144561
Singh H, Richu, Kumar A (2024) Applications of ionic liquids as antimicrobials and their impact on nanotechnology. In: Nanotechnology based strategies for combating antimicrobial resistance. Springer Nature Singapore, Singapore, pp 393–418
Slavin YN, Bach H (2022) Mechanisms of antifungal properties of metal nanoparticles. Nanomaterials 12(24):4470
Solomon NO, Simpa P, Adenekan OA, Obasi SC (2024) Sustainable nanomaterials' role in green supply chains and environmental sustainability. Eng Sci Technol J 5(5):1678–1694
Soni N, Prakash S (2014) Microbial synthesis of spherical nanosilver and nanogold for mosquito control. Ann Microbiol 64:1099–1111
Stoycheva S, Zabeo A, Pizzol L, Hristozov D (2022) Socio-economic lifecycle-based framework for safe and sustainable design of engineered nanomaterials and nano-enabled products. Sustainability 14(9):5734
Suga M, Asahina S, Sakuda Y, Kazumori H, Nishiyama H, Nokuo T, Terasaki O et al (2014) Recent progress in scanning electron microscopy for the characterization of fine structural details of nano materials. Prog Solid State Chem 42(1–2):1–21
Swetha TA, Mohanrasu K, Bora A, Ananthi V, Arun A (2022) Enzymes incorporated nanotechnology for wastewater treatment. In: Handbook of microbial nanotechnology. Academic Press, pp 415–438
Talebi H, Ghasemi FA, Ashori A (2022) The effect of nanocellulose on mechanical and physical properties of chitosan-based biocomposites. J Elastomers Plast 54(1):22–41
Tocco D (2022) The importance of Laccase immobilisation in biocatalysis: towards biomass pretreatment
Vadakkan K, Rumjit NP, Ngangbam AK, Vijayanand S, Nedumpillil NK (2024) Novel advancements in the sustainable green synthesis approach of silver nanoparticles (AgNPs) for antibacterial therapeutic applications. Coord Chem Rev 499:215528
Veeman D, Shree MV, Sureshkumar P, Jagadeesha T, Natrayan L, Ravichandran M, Paramasivam P (2021) Sustainable development of carbon nanocomposites: synthesis and classification for environmental remediation. J Nanomater 2021(1):5840645
Verma ML, Dhanya BS, Rani V, Thakur M, Jeslin J, Kushwaha R (2020) Carbohydrate and protein based biopolymeric nanoparticles: current status and biotechnological applications. Int J Biol Macromol 154:390–412
Wang B, Torres-Rendon JG, Yu J, Zhang Y, Walther A (2015) Aligned bioinspired cellulose nanocrystal-based nanocomposites with synergetic mechanical properties and improved hygromechanical performance. ACS Appl Mater Interfaces 7(8):4595–4607
Wang J, Gardner DJ, Stark NM, Bousfield DW, Tajvidi M, Cai Z (2018) Moisture and oxygen barrier properties of cellulose nanomaterial-based films. ACS Sustain Chem Eng 6(1):49–70
Wang F, Wang W, Zhang C, Tang J, Zeng X, Wan X (2021) Scalable manufactured bio-based polymer nanocomposite with instantaneous near-infrared light-actuated targeted shape memory and remote-controlled accurate self-healing. Compos B Eng 219:108927
Wirtu SF, Jule LT, Nagaraj N (2025) Nanomaterials for biomedical applications: drug delivery, biosensors, and tissue engineering. In: Exploring nanomaterial synthesis, characterization, and applications. IGI Global, pp 125–150
Xie F (2024) Alginate-based nanocomposites for food preservation: recent progress showcasing heightened material properties and functionalities. Adv Nanocompos

Xu Y, Musumeci V, Aymonier C (2019) Chemistry in supercritical fluids for the synthesis of metal nanomaterials. React Chem Eng 4(12):2030–2054

Yang W, Fortunati E, Luzi F, Kenny JM, Torre L, Puglia D (2018) Lignocellulosic based bionanocomposites for different industrial applications. Curr Org Chem 22(12):1205–1221

Yatigala NS, Bajwa DS, Bajwa SG, Chanda S (2024) Biodegradability of green composites: mechanisms and evaluation methods. In: Green micro-and nanocomposites. Jenny Stanford Publishing, pp 11–69

Ying S, Guan Z, Ofoegbu PC, Clubb P, Rico C, He F, Hong J (2022) Green synthesis of nanoparticles: current developments and limitations. Environ Technol Innov 26:102336

Zhang M, Chen H (2023) Development and characterization of starch-sodium alginate-montmorillonite biodegradable antibacterial films. Int J Biol Macromol 233:123462

Zhang M, Yang J, Cai Z, Feng Y, Wang Y, Zhang D, Pan X (2019) Detection of engineered nanoparticles in aquatic environments: current status and challenges in enrichment, separation, and analysis. Environ Sci Nano 6(3):709–735

Zhang Y, Zhu X, Chu C, Xiao X, Chen B (2022) Applications of atomic force microscopy-based imaging and force spectroscopy in assessing environmental interfacial processes. Crit Rev Environ Sci Technol 52(14):2421–2452

Zheng G, Kang X, Ye H, Fan W, Sonne C, Lam SS, Ge S et al (2024) Recent advances in functional utilisation of environmentally friendly and recyclable high-performance green biocomposites: a review. Chin Chem Lett 35(4):108817

Mechanism of Pollutant Removal by Nanocomposites

Dipti Grover, Shakuntla Duman, Neha Kalonia, Sandeep Kumar, and Harpreet Singh

Abstract Nanocomposites have emerged to be very effective materials for the removal of pollutants because of their remarkable physicochemical characteristics. These materials integrate the enhanced properties of nanomaterials, such as the high surface area, variable pore size, and higher reactivity with the mechanical stiffness of the conventional composites. This chapter provides a comprehensive overview of different types of nanocomposites, including polymer-based, ceramic-based, and metal-based variants, each contributing uniquely to pollutant removal processes. The core mechanisms by which nanocomposites remove pollutants, such as adsorption, photocatalysis, chemical reduction, and membrane filtration, are discussed in detail, highlighting their specific role in treating wastewater. Case studies will showcase the real-world scenario and efficacy of different nanocomposites in removing heavy metals and organic contaminants. In addition to this, the chapter critically addresses the practical challenges associated with stability, durability, economic viability, potential toxicity, and environmental impact of nanocomposites. Overall, this chapter emphasizes the critical role of nanocomposites in advancing eco-friendly and effective wastewater treatment technologies for sustainable environmental applications.

Keywords Treatment-technologies · Pollutant-removal · Nanocomposites · Organic-contaminant · Sustainable applications

D. Grover (✉) · S. Duman · N. Kalonia · S. Kumar
Institute of Environmental Studies, Kurukshetra University Kurukshetra, Thanesar, India
e-mail: mdipti@kuk.ac.in

H. Singh
Department of Chemistry, University of California, Berkeley, USA

S. Roy et al. (eds.), *Nanocomposites for Sustainable Wastewater Treatment*, Composites Science and Technology, https://doi.org/10.1007/978-981-95-1369-7_3

1 Introduction

Nanocomposites have garnered significant attention in environmental science due to their remarkable potential in pollution mitigation, particularly in wastewater treatment. The rapid industrialization and urbanization of recent decades have intensified the demand for effective water treatment solutions. However, there are a number of serious problems with traditional wastewater treatment techniques. Particularly for new pollutants like medications, microplastics, and persistent organic chemicals, these systems frequently fall short of eliminating all contaminants (Kumar et al. 2021). They produce a lot of sludge, which creates disposal issues, are energy-intensive, and demand major infrastructure expenditures (Gupta et al. 2020). Furthermore, their ability to adjust to changing wastewater volumes and pollutant profiles is restricted by their dependence on physical, chemical, or biological processes (Zhao et al. 2019).

Inadequate wastewater treatment has become a major problem on a global scale, particularly in many poor nations. Untreated or inadequately treated wastewater exacerbates pollution and water scarcity, endangering environmental sustainability and human health (World Bank 2022). For example, untreated wastewater is frequently dumped straight into rivers and lakes in low- and middle-income countries, which contributes to the development of waterborne illnesses and the deterioration of ecosystems (UNESCO 2021). Furthermore, methane and nitrous oxide emissions from ineffective treatment systems can exacerbate the effects of climate change worldwide (IPCC 2023). Innovative strategies that strike a balance between effectiveness, scalability, and environmental responsibility are required to address these constraints.

Nanocomposites have emerged as a powerful, eco-friendly alternative, offering enhanced performance in pollutant removal due to their unique nanoscale properties —such as high surface area, adjustable pore size, and enhanced reactivity—with the robustness and mechanical stability typical of conventional composites (Zhang and Wang 2021). This fusion of properties enables nanocomposites to effectively target and eliminate a broad spectrum of pollutants, including heavy metals, organic dyes, and other contaminants, which pose major threats to ecosystems and human health (Yang et al. 2023a, b; Singh et al. 2023; Chen et al. 2022).

The versatility of nanocomposites arises from their diverse structural configurations. Polymer-based, ceramic-based, and metal-based nanocomposites each offer unique advantages; for example, polymeric nanocomposites are highly flexible and easily modified, while ceramic nanocomposites provide excellent thermal stability, and metal-based nanocomposites contribute high mechanical strength and durability (Smith et al. 2021; Li and Zhao 2022). These distinct types allow for tailored applications in pollutant removal, exploiting various mechanisms such as adsorption, photocatalysis, chemical reduction, and membrane filtration (Singh et al. 2022). Adsorption remains a widely employed technique, especially for capturing heavy metals and organic dyes, due to its efficiency and simplicity (Gupta et al. 2023).

Photocatalysis, in particular, has emerged as an environmentally friendly solution. Nanocomposites can serve as photocatalysts that degrade hazardous pollutants under light exposure, turning complex contaminants into simpler, less harmful substances, a process highly valued in sustainable water treatment applications (Zhang et al. 2023). Additionally, membrane filtration technologies that integrate nanomaterials show promising results in removing micro-pollutants and heavy metals, proving to be among the most effective approaches in nanotechnology for environmental applications (Wang and Luo 2024).

However, challenges accompany the deployment of nanocomposites. Concerns about their stability, durability, economic feasibility, and environmental impact—especially in relation to potential toxicity—highlight the need for further research and regulatory assessment to ensure safe, effective applications (Kumar et al. 2023). A unique advantage is that nanocomposites, especially those containing magnetic nanoparticles, can often be separated and recovered after pollutant removal, presenting a practical solution for their reuse in large-scale applications (Liu et al. 2022).

This chapter provides a comprehensive discussion of nanocomposites, including their types, mechanisms for pollutant removal, and applications in pollution degradation, emphasizing their role in sustainable environmental management. Through this, we aim to underscore the critical potential and challenges of nanocomposites in combating global pollution.

2 Types and Properties of Nanocomposites

Nanocomposites are hybrid materials composed of a primary matrix—metal, polymer, or ceramic—integrated with nanoscale reinforcements, resulting in superior mechanical, thermal, and functional properties. These enhanced features make nanocomposites ideal for a range of applications, especially in environmental remediation. Below is an overview of the three primary types: Metal Matrix Nanocomposites (MMNCs), Polymer Matrix Nanocomposites (PMNCs), and Ceramic Matrix Nanocomposites (CMNs) (Fig. 1).

2.1 Metal Matrix Nanocomposites (MMNCs)

MMNCs consist of a ductile metal or alloy matrix reinforced with nanoparticles, leveraging the combined properties of metals and nanomaterials (Malik et al. 2023). These composites are formed by combining at least two distinct phases, both physically and chemically, which work together to provide properties that each phase alone could not achieve. Typically, there are two types of phases present: a fibrous phase and a particle phase, both dispersed within the metallic matrix. MMNCs have a great deal of potential for use in a variety of fields, such as the production of structural

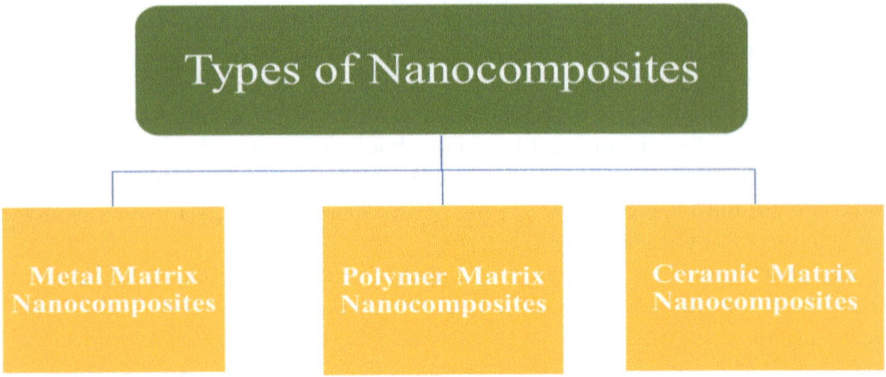

Fig. 1 Types of nanocomposites

materials and the automotive and aerospace sectors (Camargo et al. 2009). MMNCs have some outstanding properties such as higher **strength and stiffness** (Mazaheri et al. 2024), **Wear Resistance** (Sadoun et al. 2020), **Thermal Conductivity and Corrosion Resistance**. Specific reinforcements enable tuning of **electrical and magnetic behaviors** as mentioned in Fig. 2.

According to Khan et al. (2016), Wei et al. (2020), Shameem et al. (2021), the most popular methods for processing metal matrix nanocomposites are spray pyrolysis, liquid metal infiltration, vapour techniques, quick solidification, electrode location, and chemical approaches, such as colloidal and sol–gel procedures. $Fe–Cr/Al_2O_3$, Ni/Al_2O_3, Fe/MgO, Al/CNT, and Mg/CNT are a few examples of popular metal matrix nanocomposites (Fig. 2).

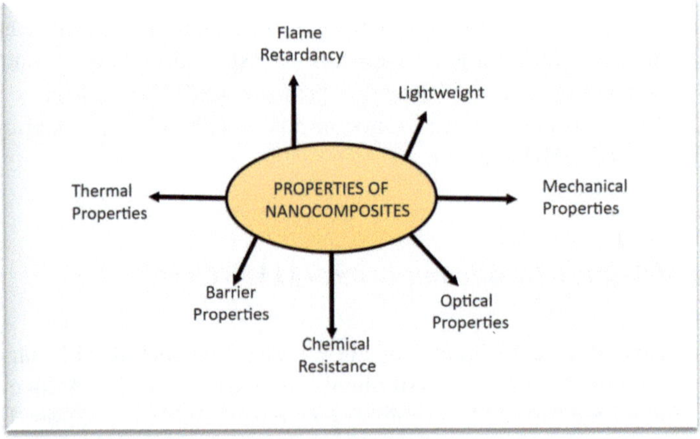

Fig. 2 Significant properties of nanocomposites

2.2 *Polymer Matrix Nanocomposites (PMNCs)*

A polymer matrix and a filler (reinforcement) are the two main components of a conventional polymer composite. In reality, agglomerated nanoparticles are dispersed using polymer as a matrix material, and nano additives are used as reinforcement materials. The most common type of nanocomposite is made up of isolated nanoscale particles that are evenly distributed in a polymer under ideal conditions (Frimpong et al. 2007; Shameem et al. 2021). The additives might be 2D (layered materials like clay), 3D (spherical particles), or 1D (nanotubes and fibres). Outstanding qualities of polymers include their **low cost, ductility, high durability, ease of processing, resistance to corrosion, barrier resistance, flame retardancy, wear resistance, magnetic, electrical, and optical properties** (Mai and Yu 2006).

The mechanical, thermal, and electrical properties of polymers are comparatively inferior to those of ceramics and metals. Additionally, polymers perform poorly in terms of heat resistance, fire resistance, and gas barrier. Because they have a low coordination number, lightweight carbon and hydrogen atoms for their backbone, and are less dense than ceramics and metals, polymers are used as building materials and structural components in lightweight applications like electronics, automotive, defence, and aerospace (Thostenson et al. 2005; Khan et al. 2016).

By incorporating particle nanofiller into polymer matrices to enhance the characteristics of neat polymers, polymer nanocomposites have created new opportunities for a possible class of hybrid materials (Gupta et al. 2009). Among the techniques for creating polymer nanocomposites, the sol–gel procedure appears to be the most promising. Before the nanoparticles were mixed with the polymer gel, they were dispersed at the molecular or near molecular level (Kuo et al. 2005). Better physical characteristics of functional nanocomposites create new opportunities in energy conversion, storage, electronics, and micro-optics.

PMNCs combine a polymer matrix with nanoscale fillers to create materials with significantly enhanced properties (Panigrahy et al. 2024).

2.3 *Ceramic Matrix Nanocomposites (CMNs)*

Ceramic matrix nanocomposites are emerging engineering materials with a variety of uses in the industrial sector as well as in the electrical and mechanical fields. There is a lot of promise in the aluminium oxide (Al_2O_3)/silicon carbide (SiC) system in particular (Khan 2020). All of the research has shown that adding a small volume fraction (~10%) of SiC of a particular size to an Al_2O_3 matrix improves the matrix's characteristics. In certain studies, this toughening mechanism has been explained by the crack-bridging function of nanosized reinforcements. In contrast to ceramic materials that fail quickly, novel nanocomposites with improved failure and high toughness characteristics have been created by integrating high-strength nanofibers into ceramic matrices (Al-Mutairi et al. 2022).

CMNs integrate ceramics with nanoparticles, leveraging the stability of ceramics and the functionality of nanomaterials (Malik et al. 2023). Except that CMNs have a **high surface area** for greater adsorption capacity for pollutants, **enhanced mechanical properties** for improved fracture toughness and wear resistance, and also **thermal and chemical stability** for extreme conditions. Al_2O_3/SiO_2, SiO_2/Ni, Al_2O_3/TiO_2, and Al_2O_3/SiC are a few examples of popular ceramic matrix nanocomposites. Carbon nanotubes have been frequently used in the creation of nanocomposite materials since their discovery. Among the ceramic matrix nanocomposite materials based on carbon nanotubes are Al_2O_3, $MgAl_2O_4$/CNT, and MgO/CNT (Table 1).

Different types of nanocomposites, because of their unique properties, can address environmental challenges in water treatment, air purification, and soil remediation while advancing technologies in energy storage, electronics, and biomedical engineering (Panigrahy et al. 2024; Iravani 2021).

Table 1 Different types of nanocomposites with their key properties and potential applications

Nanocomposite type	Matrix material	Reinforcement material	Key properties	Potential applications
Metal matrix nanocomposites (MMNCs)	Metal or alloy (e.g., Al, Fe, Ni)	Nanoparticles (e.g., Al_2O_3, SiC, WC)	High strength, high modulus, improved wear resistance, enhanced thermal conductivity	Structural components, aerospace, automotive, cutting tools
Polymer matrix nanocomposites (PMNCs)	Polymer (e.g., epoxy, polyester)	Nanoparticles (e.g., clay, carbon nanotubes)	Improved mechanical properties, enhanced thermal stability, better electrical conductivity, increased barrier properties	Electronics, packaging, automotive, aerospace
Ceramic matrix nanocomposites (CMNCs)	Ceramic (e.g., Al_2O_3, SiC)	Nanoparticles (e.g., SiC, CNTs)	High temperature resistance, excellent mechanical strength, improved toughness	Aerospace, automotive, electronics, energy storage

3 Mechanism of Pollutant Removal by Nanocomposites

3.1 Adsorption

Adsorption is an exothermic process where molecules of a gaseous or liquid chemical collect at the surface of an adsorbent (Sing et al. 1998; Kecili and Hussain 2018). The term “adsorbate” refers to the substance that is adsorbed onto the adsorbent. In the process of removing and remediating different organic and inorganic pollutants from the contaminated surface, it is a straightforward procedure due to its low cost, simplicity, high dependability, and ease of operation (Alqadami et al. 2020). Depending on the type of pollutant (adsorbate) and adsorbent utilised, the interaction between the two can be either chemical (strong forces like ionic, metallic or covalent) or physical (weak interactions like van der Waals forces) (Ahmad 2021).

There are two methods that molecules might adsorb to the adsorbent surface: “physical adsorption,” also known as “physisorption,” and “chemical sorption,” also known as “chemisorption.” The way the molecules interact with the surface determines this. Weak forces like van der Waals forces and electrostatic interactions have a role in physical adsorption. In chemical adsorption, the adsorbed molecules and the surface establish strong chemical bonds, such as covalent bonds. Physical adsorption typically results in the production of a thick multilayer on the surface, whereas chemical adsorption is slower and typically results in the formation of a monomolecular layer (monolayer) on the adsorbent surface (Dabrowski 2001; Kecili and Hussain 2018) (Fig. 3).

Nanomaterials are promising adsorbents that can effectively remove contaminants from aqueous solutions (Marcelo et al. 2021). Moges et al. (2022) demonstrated that magnetite graphene oxide (GO–Fe_2O_3) nanocomposite efficiently removes chromium from tannery industrial wastewater, achieving removal up to 90.3%. Graphene oxide is an effective adsorbent due to its multifunctional groups—hydroxyl (−OH), carboxyl (−COOH), and epoxy (C–O–C)—as well as its high surface area

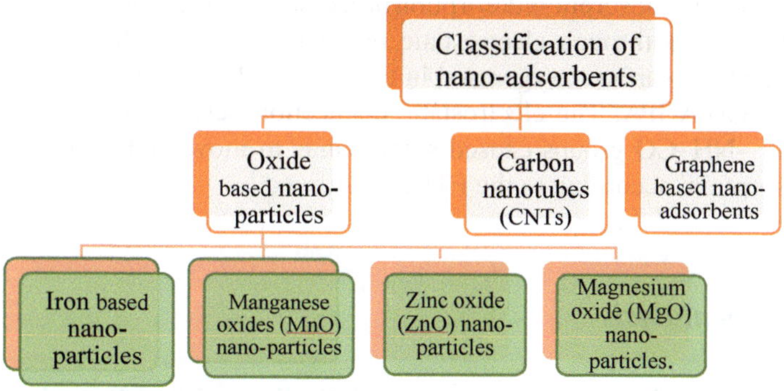

Fig. 3 Classification of nanoadsorbents

and numerous active sites. Abdi and Abedini (2020) also concluded that polymeric nanocomposite based on zeolite imidazolate frameworks (ZIFs) such as PES/ZIF-8/ZIF-67 has excellent efficiency in eliminating Malachite Green dye from colored wastewater up to 99.2% with adsorption capacity of 613.2 mg/g.

John et al. (2024) reported some of the advantages and disadvantages, which are as follows:

(1) Easy to maintain and operate and low-cost for water treatment.
(2) The availability of a variety of effective adsorbents modified to target specific pollutants.
(3) High yield and the potential for excellent performance, combined with flexibility.
(4) Simple design with a fast process.
(5) Different adsorbents have different removal efficiencies (Table 2).

According to Alsuhybani et al. (2020), magnetite nanoparticles (Fe_3O_4) have been regarded as a superior adsorbent when compared to other nonmagnetic adsorbents because of their intriguing properties, which include high surface area, high saturation magnetisation, ease of separation under an external magnetic field, and efficiency and cost effectiveness (Huang and Pan 2016; Zhang et al. 2019). To remove contaminants from wastewater, researchers have created magnetite nanocomposites using a variety of materials, including TAS (Alqadami et al. 2017), trisodium citrate (Alqadami et al. 2016), zeolite (Javanbakht et al. 2016), and BiVO4 layered on charcoal (Kumar et al. 2017). Two carboxyl groups in 1,4-benzene dicarboxylic acid interact with heavy metal ions. It functions as an organic molecule that cross-links with metal ions to create metal–organic frameworks that are used to remove pollutants from wastewater (Ricco et al. 2015; Wang et al. 2009).

Mechanism of adsorption

The primary functional groups on the Fe_3O_4@BDC@AGPA nanocomposite surface in this work are carboxyl (–COOH), amino (–NH2), and amide (–NH–CO) groups. These groups can bind strongly with Pb (II) through electrostatic interactions. Figure 1 depicts the adsorption/desorption processes of Pb(II) ions. The quick adsorption of Pb(II) ions through electrostatic interactions was facilitated by the negative charge of the surface oxygenated in carboxyl and nitrogen in amino groups of Fe_3O_4@BDC@AGPA. The electrostatic connection between Pb(II) and –COOH, $-NH_2$, and –NH–CO groups caused some bands to move and decrease following Pb(II) adsorption, according to the FTIR study.

3.2 *Photocatalysis*

The phenomenon of water splitting photocatalytically on a TiO_2 electrode under ultraviolet (UV) light was discovered by Fujishima and Honda (1972). Photocatalysis is

Table 2 Removal of heavy metals from wastewater by different types of nanoadsorbents (Anjum et al. 2019)

S. No.	Types	Definition
1	Oxide-based nanoparticles	Titanium oxides, zinc oxides, magnesium oxides, manganese oxides, and ferric oxides are typical examples The high BET surface area, low solubility, low environmental impact, and lack of secondary contaminants make these nanoparticles preferred
1.1	Iron-based nanoparticles	Adsorption efficiency on FeO_3 nanoparticles is dependent on temperature, pH, incubation duration, and adsorbent dose By improving the adsorption capacity of Fe_2O_3, surface modifications like those involving 3-aminopropyltrimethoxysilane enable the efficient removal of contaminants Cr^{3+}, Co^{2+}, Ni^{2+}, Cu^{2+}, Cd^{2+}, Pb^{2+}, and As^{3+}
1.2	Manganese oxides (MnO) nanoparticles	Modified forms like hydrous manganese oxide (HMO) adsorb metals like Pb(II), Cd(II), and Zn(II) by ion-exchange. Initial surface adsorption and subsequent intraparticle diffusion are the two phases of adsorption on HMOs
1.3	Zinc oxide (ZnO) nanoparticles	ZnO forms, such as nano-assemblies nanoplates and exhibit greater removal effectiveness for heavy metals. These distinct structures and electropositive nature allow them to efficiently remove metals such as Cu(II), Co^{2+}, Ni^{2+}, Cd^{2+}, Pb^{2+}, Hg^{2+}, and As^{3+}
1.4	Magnesium oxide (MgO) nanoparticles	Adsorption affinity is increased by modified MgO structures such as nanorods, nanobelts, nanowires, nanotubes, nanocubes, and three-dimensional forms Particularly efficient Pb(II) and Cd(II) adsorption has been demonstrated by flowerlike mesoporous MgO

(continued)

Table 2 (continued)

S. No.	Types	Definition
2	Carbon nanotubes (CNTs)	Modified CNTs, such as magnetic and multi-walled CNTs, facilitate separation and enhance dispersion. Numerous changes, including metal impregnation, acid treatment, and functional group grafting, improve the surface properties of CNTs, increasing their adsorption capacity
3	Graphene-based nanoadsorbents	GO is mostly made via the Hummers process, which is superior to other nanomaterials like CNTs because it has more adsorption sites and is easier to synthesise High adsorption capabilities for metals such as Pb^{2+}, Cd^{2+}, and Zn^{2+} are demonstrated by GO, including composites like TiO_2-GO

defined as the amalgamation of photochemistry and catalysis. The word photocatalysis is derived from the Greek language and consists of The prefix photo which means light and Catalysis is the process in which a substance alters the rate at which the reactants undergo a chemical transformation without eventually changing itself, known as catalysis. By reducing the activation energy, the substance that is known to be a catalyst speeds up the process.

Hence, photocatalysis is the simultaneous use of both light and catalysts to speed up a chemical reaction. So, photocatalysis can be defined as "catalysis-driven acceleration of a light-induced reaction" (Saravanan et al. 2017). TiO_2, ZnO, CeO_2, or plasmonic (Ag, Au, Pt, Pd) nanoparticles are the most commonly used inorganic nanoparticles that have strong catalytic properties in UV or visible light (Melinte et al. 2019). Photocatalysts are of two types: Homogenous photocatalytic process and heterogeneous photocatalytic process (Rehman et al. 2009) (Table 3).

Role of nanocomposite in photocatalysis

Photocatalysis is an environmentally friendly process that can be useful in degrading numerous harmful organic pollutants in wastewater (Dey and Gogate 2021). Nanocomposites decrease the bandgap of individual materials and also minimize the process of electron–hole recombination. High surface area is provided by nanocomposites, which facilitate more active sites for catalysis, eventually leading to a faster reaction rate.

Mechanism of Photocatalysis

The catalyst and wavelength or light energy (photon) are the primary factors that influence photocatalytic reactions. Because of their electronic structure, semiconductor materials are mostly used as catalysts that act as sensitizers for light-stimulated redox processes. This structure is defined by an unoccupied conduction band and a filled

Table 3 Comparison between homogeneous photocatalysts and heterogenous photocatalysts

Aspect	Homogenous photocatalysis	Heterogeneous photocatalysis
Phase of the catalyst	The catalyst is in the same phase as the reactants (usually dissolved in the liquid phase)	The catalyst is in a different phase than the reactants (solid catalyst in liquid or gas phase)
Common catalyst	Metal ions (e.g., Fe^{3+}, Cu^{2+}), dissolved ozone, H_2O_2	Solid semiconductors like TiO_2, ZnO, CdS
Catalytic mechanism	Involves dissolved ions or molecules producing reactive species in solution	Involves photon absorption by a solid catalyst, creating electron–hole pairs on the surface
Reactive species	Hydroxyl radicals (•OH), singlet oxygen ($^1O_2$) formed by dissolved catalysts	Reactive species such as hydroxyl radicals (•OH), superoxide ($O_2•^-$) generated on catalyst surface
Reaction example	Fenton reaction (Fe^{2+}/Fe^{3+} with H_2O_2), ozone-based oxidation	UV/TiO_2 degradation of organic pollutants in water and air purification
Environmental application	Wastewater treatment, disinfection using dissolved catalysts	Water and air purification, self-cleaning surfaces, wastewater treatment

valence band (Khan et al. 2015). The following are the fundamental steps in the semiconductor photocatalysis process (Fig. 4):

When light energy in the form of photons strikes the surface of a semiconductor, and the energy of these photons is equal to or exceeds the semiconductor's bandgap energy, the electrons in the valence band become energized. This agitation causes

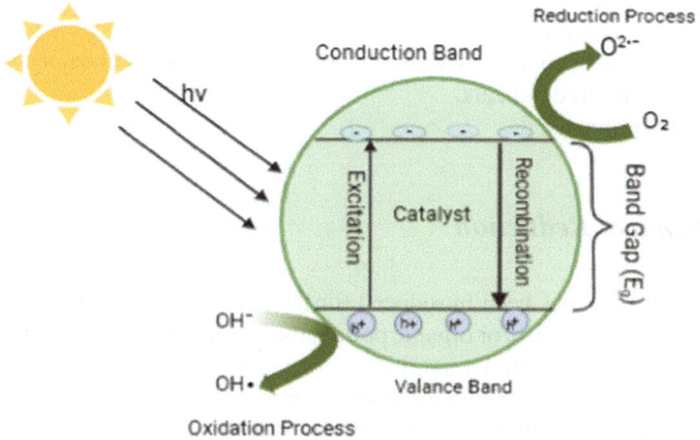

Fig. 4 Schematic representation of the photocatalytic mechanism

them to transition into the conduction band, allowing the semiconductor to conduct electricity.

Holes created in the valence band of a semiconductor play a crucial role in initiating reactions that lead to pollutant degradation. These holes can oxidize donor molecules and interact with water, producing powerful hydroxyl radicals (strong oxidizing power for the degradation of pollutants).

The electrons in the conduction band react with dissolved oxygen species to form superoxide ions, which then induce redox reactions.

To produce the needed products, these holes and electrons may be involved in a series of oxidation and reduction reactions with any species that could be adsorbed on the semiconductor's surface. Abd Elkodous et al. (2021) studied that a layered $Co_{0.5}Ni_{0.5}Fe_2O_4/SiO_2/TiO_2$ nanocomposite matrix integrated with 10% carbon dots achieved 65% UV-assisted photocatalytic degradation of chloramine-T from hazardous wastewater. Ghosh et al (2020) explained the photocatalytic activity of TiO_2-Ag nanocomposites for the degradation of dyes and significant antibacterial activity. Rhodamine B, methylene blue, and methyl orange degraded 95%, 76%, and 23%, respectively, under visible light irradiation for 330min.

A few advantages and disadvantages of the photocatalytic method studied by Saravanan et al. (2017) and Thakur et al. (2024), respectively, are the following:

(1) Photocatalysis serves as an efficient alternative to energy-intensive conventional treatments (like adsorption on activated carbon, ultrafiltration, or chemical coagulation) by utilizing renewable, pollution-free solar energy.
(2) Photocatalysis produces harmless byproducts, unlike conventional treatments that merely shift pollutants between phases.
(3) The photocatalytic process can destroy a variety of hazardous compounds in different wastewater streams.
(4) Minimal secondary waste generation.
(5) The conditions for photocatalysis are mild, requiring modest reaction times and less chemical input.
(6) Unintentional secondary pollution from residual nano-photocatalysts.
(7) Ion leaching, free radicals.
(8) Toxic by-products.

3.3 *Chemical Reduction*

Chemical reduction has been introduced for the detoxification of harmful contaminants, such as heavy metals or organic pollutants, into less toxic or benign substances using nanocomposites that act as catalysts or reactive agents. Generally, chemical reduction is performed by using a reducing agent. For some compounds, reduction with a reducing agent alone is thermodynamically feasible, but it is a kinetically non-feasible process (Rostami-Vartooni et al. 2016). Therefore, metal nanoparticles are used as a catalyst to make it kinetically possible (Khalil et al. 2020).

Nanocomposites, due to their high surface area, unique reactivity, and stability, enable an efficient way to detoxify polluted environments and restore ecological balance. For example, composites like metal oxide nanoparticles supported on carbon materials can reduce toxic heavy metals (e.g., chromium or lead) to less harmful oxidation states, while nanoscale zero-valent iron (nZVI) can break down chlorinated organic compounds in groundwater. The chemical reduction process with nanocomposites provides an efficient, cost-effective, and eco-friendly solution for treating polluted air, water, and soil.

For instance, metal-based nanocomposites like those containing silver, iron, or palladium nanoparticles can facilitate the reduction of toxic pollutants like heavy metals (e.g., chromium, lead) and organic pollutants (e.g., dyes, pesticides) in water. These nanocomposites enhance the reaction rate of chemical reduction, breaking down pollutants at a molecular level and reducing them to safer compounds. Although the catalytic potential of nanomaterials depends on the surface area of nanoparticles, still stabilizing medium plays a key role in finalizing the surface area of NPs (Naz et al. 2021; Begum et al. 2017; Varadavenkatesan et al. 2016; Wu et al. 2014).

Chemical reduction of several dyes by sodium borohydride ($NaBH_4$) as the reducing agent is an economic, safe, highly efficient, and green alternative approach. Hence, catalytic degradation by $NaBH_4$ is superior to other traditional methods in terms of being less time-consuming, simple design, easy operation, and their recyclability (Quiton et al. 2021; Depren et al. 2020). However, this process is thermodynamically favourable and kinetically unfavourable in the absence of a catalyst (Amir et al. 2015; Zheng et al. 2015). So, different catalytic systems have received considerable attention for a variety of applications in several fields as they exhibited tunable physicochemical properties (Wang et al. 2020a, b; Soltani and Zabihi 2020). Bimetallic catalysts possess better catalytic potential as compared to their individual metal (monometallic analogues) owing to synergistic effects between two metals and the degree of freedom that can be adjusted to increase the catalytic efficiency of bimetallic nanomaterials (Wang et al. 2020a, b; Gu et al. 2019; Naz et al. 2021).

3.4 Membrane Filtration

Membrane filtration is a physicochemical separation technique that uses thin (<1 mm) mostly synthetic polymeric, semi-permeable membranes (Gray 2014). Membranes are generally divided into isotropic or anisotropic. Isotropic membranes exhibit uniformity in both composition and physical properties throughout their entire cross-section. In contrast, anisotropic membranes are characterized by non-uniformity, featuring distinct layers that differ in structure and chemical composition. This variability allows for specialized functions and applications in various fields (Cevallos-Mendoza et al. 2022). The chemical, biotechnological, pharmaceutical, food, and metallurgical areas, as well as other separation techniques, all use membrane filtration technology to purify home and commercial water supplies (Li et al. 2011). Membrane separation is a clean process that requires less energy than traditional

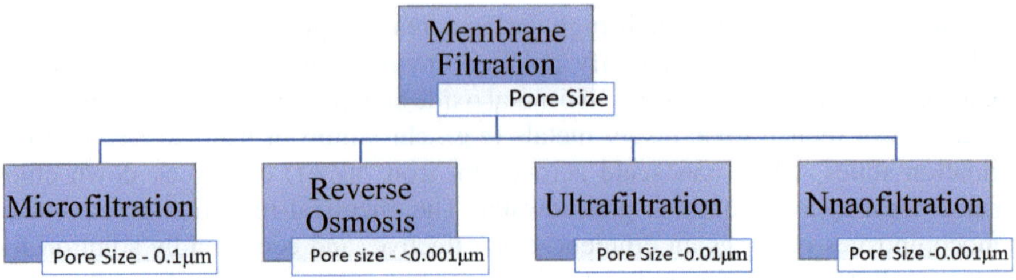

Fig. 5 Types of membrane filtration

processes like filtration, distillation, ion exchange, and chemical treatment methods. It has a wide range of industrial and environmental applications (Siddique et al. 2021).

Depending on pore size and operating pressure, there are four types of membrane filtration (Gray 2014) (Fig. 5).

3.4.1 Role of Nanocomposites in Membrane Filtration

Nanocomposite membranes are made by incorporating inorganic or organic nanoparticles (NPs) into polymeric membrane matrix. When included in the membrane, they improve filtration (Jhaveri and Murthy 2016) because of various factors, such as better adsorption and particle trapping on a larger surface area. Improving water affinity (hydrophilicity), reducing fouling (anti-fouling), and increasing water flux. It enhanced the mechanical rigidity and other suitable characteristics, and also provided anti-microbial characteristics in order to avoid bio-fouling. Kamari and Shahbazi (2021) reported that a high-performance nanofiltration membrane (blending $Fe_3O_4@SiO_2$-CS biocompatible and green bionanocomposite) NFM-3 exhibited 98.8% and 98.2% removal of heavy metals (Cd^{2+}) and dyes, respectively, with stable efficiency during long-term filtration. Liu et al. (2020) studied that the simultaneous incorporation of graphitic carbon nitride (g-C_3N_4) and halloysite nanotubes (HNTs) into a polyamide (PA) matrix significantly enhanced the thin film nanocomposite membrane, achieving nearly double the water permeance.

The general advantages and disadvantages of membrane filtration are the following:

(1) Helpful in water purification.
(2) Improve mechanical, thermal, and chemical properties of the membranes.
(3) Improve permeability and water flux.
(4) High membrane cost.
(5) Deposition of materials on the membrane and requires regular cleaning.
(6) Fouling can reduce performance.

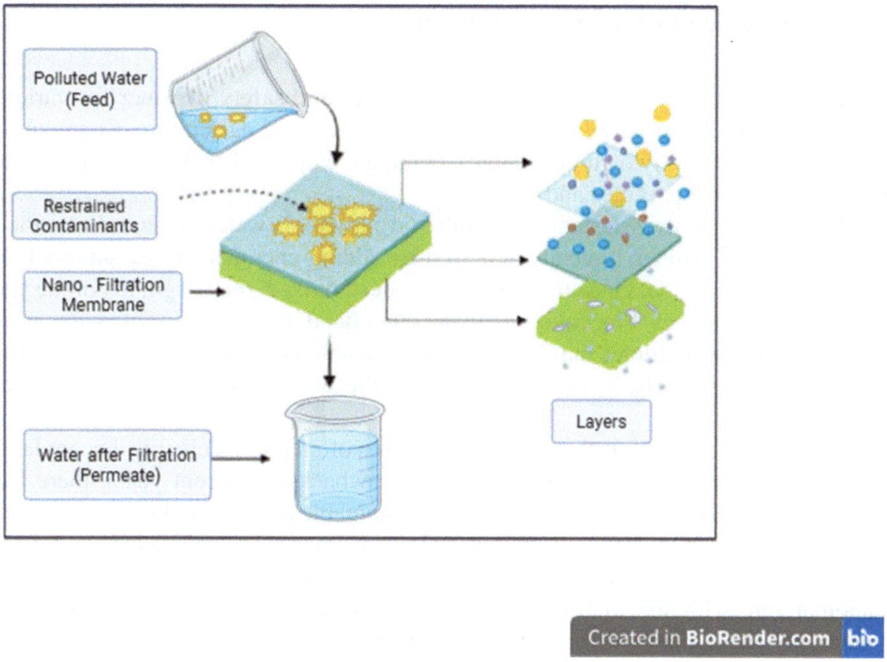

Fig. 6 Mechanism showing nanomembrane filtration

Mechanism of Membrane Filtration

Steps involved in the membrane filtration process (Rahman et al. 2023) (Fig. 6):

(1) **Feed Solution**: The mixture or solution that is first introduced into the membrane system is known as feed. This might be liquid with other contaminants or dissolved solids (solutes).

(2) **Membrane with nanocomposites layer**: The nanocomposite-enhanced membrane selectively allows water or smaller molecules to pass while retaining unwanted substances on the feed side. The purpose of nanocomposite-enhanced membrane is to retain certain ions or molecules while permitting others to flow through. The adsorption of pollutants and solutes onto the membrane may take place when the feed comes into contact with the membrane surface. For the targeted removal of contaminants such as organic pollutants or heavy metals, nanoparticles in the membrane can provide greater adsorption sites. Nanoparticles often enhance transport by providing hydrophilic paths that attract and facilitate the movement of water molecules.

(3) **Permeate and Retentate**: Permeate means the purified solution that has passed through the membrane is called the permeate, and it is collected on the other side. It is clean now and also known as filtrate and water product. Whereas retentate refers to the rejected particles, which may consist of salts, proteins, microorganisms, and other impurities.

3.5 Coagulation and Flocculation

Water treatment is critical to ensuring the quality and safety of water for various applications, including drinking, industrial processes, and agriculture. Coagulation and flocculation are two fundamental techniques employed in water treatment to remove suspended particles and impurities.

Coagulation involves destabilizing colloidal particles in water by neutralizing their surface charges. This is typically achieved by adding chemical coagulants such as alum or ferric chloride. Once the particles are destabilized, they begin to collide and form small aggregates. Flocculation, on the other hand, is the process of aggregating these destabilized particles into larger, more settleable flocs (Mittal et al. 2024). The larger flocs can then be easily removed from the water through sedimentation or filtration.

Traditional coagulation-flocculation processes often rely on chemical coagulants that can have environmental and economic drawbacks. In recent years, there has been a growing interest in developing more sustainable and efficient water treatment methods (Rytwo et al. 2013). Nanomaterials have emerged as promising candidates for enhancing coagulation and flocculation processes. Nanoflocculants, which are nanomaterials with flocculating properties, have shown significant potential in water treatment. These materials can effectively remove a wide range of contaminants, including heavy metals, dyes, and bacteria (Rytwo 2012). The unique properties of nanoflocculants, such as their large surface area and high reactivity, allow them to destabilize and aggregate particles efficiently.

One of the key advantages of nanoflocculants is their ability to improve the efficiency of coagulation-flocculation processes. By rapidly destabilizing particles and forming large, dense flocs, nanoflocculants can reduce the treatment time and chemical dosage required (Oyewo et al. 2020). This can lead to significant cost savings and environmental benefits. However, the application of nanoflocculants in water treatment is still in its early stages, and several challenges need to be addressed. One major challenge is the potential environmental impact of nanomaterials. It is crucial to ensure that the release of nanoflocculants into the environment is minimized and that their long-term effects are well understood (Eslami et al. 2020; Noor et al. 2024).

Another challenge is the development of cost-effective and scalable production methods for nanoflocculants. While nanoflocculants have shown promising results in laboratory studies, their large-scale application requires the development of efficient and affordable production processes. Despite these challenges, nanoflocculants offer a promising avenue for improving water treatment processes. By addressing the limitations of traditional methods and leveraging the unique properties of nanomaterials, nanoflocculants can contribute to the development of more sustainable and efficient water treatment technologies (Unuabonah and Taubert 2014). Eslami et al. (2020) studied the enhancement of the coagulation process by using polytitanium chloride (PTC) and Fe_2O_3-Mn_2O_3 nanocomposite for the removal of Arsenic from polluted water. The highest Arsenic (As(V)) removal efficiency achieved was 99.96% by using PTC+Fe_2O_3-Mn_2O_3 nanocomposite at pH 7, coagulant dose of 20 mg/L, and up to 300 μg/LAs (V) initial concentration. Floc recirculation further decreased coagulant dosage by 25% increasing process efficiency and sustainability. Etemadi et al. (2021) found that polyaluminum chloride (PAC) coagulant could effectively mitigate fouling and improve humic acid (HA) removal in the submerged membrane system. The polyvinyl chloride/alumina (PVC/Al_2O_3) nanocomposite membrane significantly decreased the fouling ratio from 60.7% to 19.4% and showed a high HA removal efficiency of approximately 90.5%.

The general advantages and disadvantages of the coagulation and flocculation methods are the following:

(1) Reduce the time to settle out suspended solids, which decreases the time of wastewater treatment.
(2) Remove organic compounds.
(3) Effective in removing particulate matter.
(4) Operational cost.
(5) Sludge formation.
(6) Coagulation adversely affects brine quality.

3.6 Magnetic Separation

The magnetic separation process includes first attaching contaminants to magnetic carrier substances, for example, magnetite, and then separating the contaminant-laden carrier under a magnetic field (Gong and Zhao 2014). High magnetic content, fast magnetic responsiveness, narrow size distribution, high saturation magnetization, and high capture capacity or selectivity are all desirable characteristics for magnetic separation materials in terms of separation efficiency. Magnetic nanoparticles are characteristically alloys, oxides, or composite structures based on iron, cobalt, and nickel (Yang et al. 2020). In addition to possessing specific magnetic properties, magnetic nanoparticles have the four primary effects of ordinary nanoparticles volume effects, surface effects, quantum size effects, and quantum tunneling of magnetization. The magnetism of nanoparticles can be classified into three types. These are diamagnetism, ferromagnetism, and paramagnetis. When placed in an external magnetic field in the opposite direction to the applied field, diamagnetic substances get weakly magnetized. Ferromagnetic materials are substances that get strongly magnetized in an external magnetic field in a direction that aligns with the direction of the externally applied field. Paramagnetic substances are materials that become weakly magnetized when they are exposed to an external magnetic field aligned in the same direction as the applied field (Kaufman et al. 2008). The most common nanoparticle used is Fe_3O_4 due to its availability and superparamagnetism.

Fatimah et al. (2022) studied magnetically separable nanocomposite γ-Fe_2O_3/montmorillonite (γ-Fe_2O_3/Mt) for removal of dyes. γ-Fe_2O_3/Mt nanocomposite exhibited high photocatalytic activity for the degradation of RhB, incorporating adsorption and photodegradation mechanisms, with easy magnetic separation (22 emu/g) and also reusable without any chemical change up to five cycles. Zhang et al. (2021) reported that magnetic graphene-based (Fe_3O_4/RGO) nanocomposites exhibited 99.9% Chromium (Cr(VI)) removal efficiency with rapid magnetic separation of less than 5 s under optimal experimental conditions and kept 98.13% adsorption rate after 10 cycles.

The key advantages and disadvantages of magnetic separation are the following:

(1) Effective separation from nonmagnetic substances.
(2) Energy and cost savings.
(3) Easy maintenance.
(4) Limited applications.
(5) Need regular maintenance.

Mechanism

Magnetic separation devices generally include magnetic substances, flow channels, and external magnetic fields, permitting separation during continuous flow (Yang et al. 2020).

The steps involved in the magnetic separation mechanism are the following:

(1) Feed material introduction: The mixture contains both magnetic and non-magnetic contaminants that are introduced into the magnetic separation devices. The mixture may be in the form of a slurry, powder, or solution
(2) Exposure to magnetic field: The contaminants that pass through the separation devices are exposed to magnetic fields. A magnet or electromagnet can generate the external magnetic fields.
(3) Magnetic Force Interactions: The magnetic force affects only those charges that are already in motion. It moves throughout by the magnetic field. The magnetic contaminants are adsorbed onto the magnetic source due to the strong magnetic interaction of particles.
(4) Separation: The last step is the separation of the magnetic contaminants or materials either physically or via a field gradient. The magnetic materials are separated from the non-magnetic material and can be used for further processing (Fig. 7).

3.7 *Ion Exchange Method*

Ion exchange refers to a specific chemical process where undesirable dissolved ions, which include nitrate, fluoride, sulfate, and arsenic, in water and wastewater are exchanged for other ions that carry similar charges. An ion is defined as an atom or combination of atoms or molecules that carries a net positive (cation) or net negative (anion) electrical charge. Usually, ion exchangers are ion exchange resins, zeolites, clay, etc. Ion exchangers are either cations or anions (Slater 2013). Cation ion exchanger—These exchange positively charged ions(cations) in solution. Anion ion exchanger—These exchange negatively charged ions(anions) in solution (Nachod 2012).

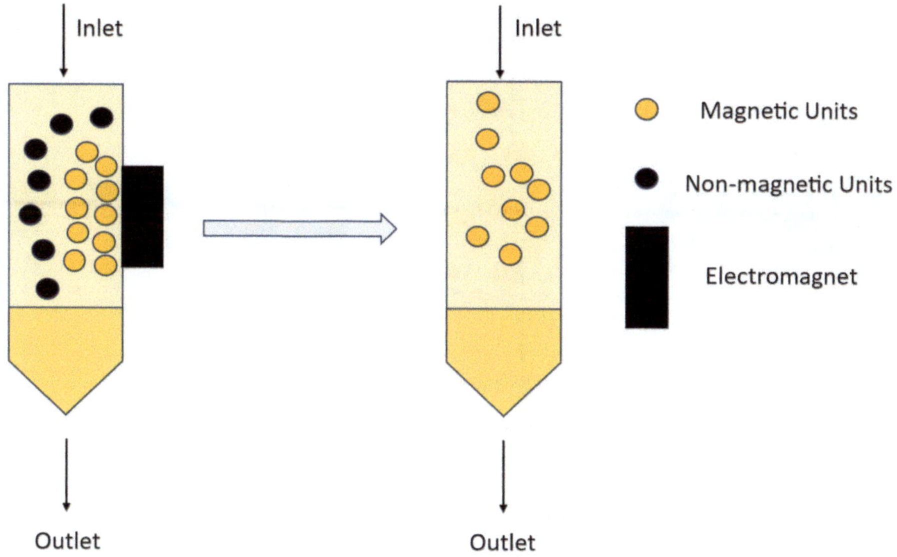

Fig. 7 A simplified diagram of the magnetic separation process

Role of nanocomposites in ion exchange

The unique structural, chemical, and physical properties of nanocomposite materials have led to their increased use in enhancing ion exchange processes. The integration of nanomaterials in ion exchange matrices creates a more efficient nanocomposite-enhanced ion exchange process.

(a) Increased surface area of nanocomposites provides more active sites for ion exchange.
(b) Nanocomposites are a porous structure that facilitates faster ion exchange.

Mechanism of ion exchange method

When nanoparticles are introduced to ion exchange membranes (IEMs), the membrane structure can go through various morphological modifications which result in the emergence of different transport mechanisms, such as the Donnan exclusion, and hopping mechanisms (Allah et al. 2023).

Hopping Mechanism

The most prominent method of ion transport in solid ion exchange membranes (IEMs) is the hopping process. In an ideal, tightly packed crystal, ions are unable to move. Solid IEMs consist of various defects that enhance the ionic mobility of ions during transport. Ions have to pass through the coordination polyhedron edge, which is characterized by its minimal size. Ion energy is much higher, and interionic distances are much shorter in this configuration. So, it is expected that the hopping mechanism is responsible for the ion transport (Peighambardoust et al. 2010).

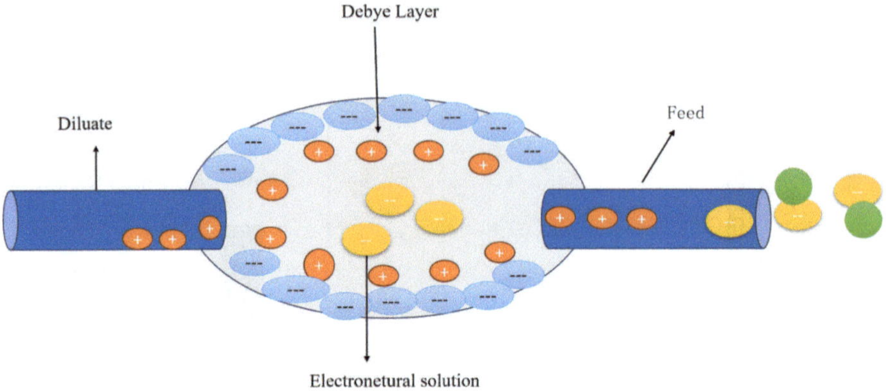

Fig. 8 Transport of ions through an ion exchange membrane

Donnan Exclusion Mechanism

Another method by which ion transport happens in the IEMs is the Donnan exclusion mechanism. Since counterions mostly occur in a thin Debye layer near the pore walls, the ion distribution in membranes is not uniform. Their selective transport is thus provided by this layer, raising the permeable selectivity (Hosseini et al. 2014). An electrically neutral solution occurs inside the membrane matrix's pore centers. When the solution enters and makes contact with the membrane, it is in equilibrium (Fig. 8).

Co-ions and counter-ions may exist in this layer in nearly equal amounts. The conductivity and selectivity of the membrane can both be increased and reduced by nonselective transfer. The electro-neutral layer enables the nonselective transport of ions, including cations and counterions. Water molecules are also transported by the non-selective vehicle. Hence, there is an increase in water uptake (WU), which enhances conductivity even more. However, due to the collision of the ions, adding additional water to the membrane pores decreases conductivity and raises the membrane's ionic resistance. The modified IEMs' sieving mechanism provides the membrane's selectivity (Yang et al. 2007). The feature of the membrane and, consequently, the ion transport mechanisms are also influenced by the method's nature. By altering the membrane's solution's viscosity and porosity, the commonly used nanoparticle improves the membrane's transport phenomena. The transfer of co-ions and counter-ions through the IEMs can be explained using the ion cluster dispersion mechanism. This concepts indicate that the ion cluster of nanoparticles in the IEMs is redistributed whenever nanoparticles are introduced to the polymer matrix, which makes it simpler to create interconnected routes. The method reliably predicts that the WU will increase as the ion exchange capacity increases. Co-ions and counter-ions are transferred by combining all of these processes if nanomaterials are functionalized throughout the synthesis of IEM (Rastgar et al. 2017) (Fig. 9).

Fig. 9 Ion cluster dispersion mechanism

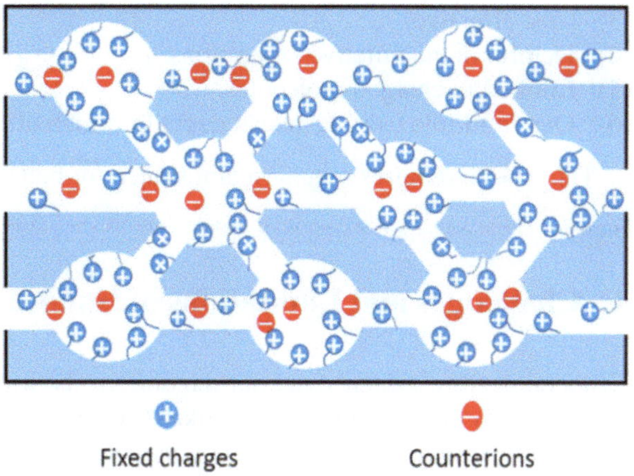

A Case Study on the Use of Carbon Nanotubes to Enhance the Photocatalytic Activity of TiO_2

A carbon nanotube (CNT)/TiO_2 nanocomposite photocatalyst has been developed by a simple impregnation, which is used, for the first time, for gas-phase degradation of benzene. When compared to commercial titania (Degussa P25), the as-prepared CNT/TiO_2 nanocomposite is found to exhibit a greater photocatalytic activity for the decomposition of benzene. The breakdown of methyl orange in the liquid phase has also been found to exhibit a similar behavior. CNT plays two important roles in enhancing the photocatalytic activity of TiO_2, based on the characterization of photocatalysts utilizing a variety of combined techniques, including X-ray diffraction, transmission electron microscopy, ultraviolet/visible (UV/Vis) diffuse reflectance spectra, and photoluminescence spectra. One is to act as an electron reservoir, which helps in trapping electrons released from TiO_2 particles as a result of UV light irradiation, therefore limiting the recombination of electron–hole pairs. The other is utilized as a support or dispersion template to control the shape of the TiO_2 particles in the CNT/TiO_2 nanocomposite (Xu et al. 2010). Khan (2020) reported that polyaniline-based Zirconium (Zr(IV)) iodate nanocomposite has high selectivity for lead (Pb(II)) ions, making it effective for groundwater purification. The incorporation of Zr(IV) iodate enhanced the properties of the ion exchange method. Additionally, it shows antibacterial activity against Gram-negative and Gram-positive bacteria (Sharma et al. (2017) studied the ion exchange capacity of Sodium dodecyl sulphate-ironsilicophosphate (SDS-FeSP) nanocomposite. SDS-FeSP nanocomposite exhibited high ion exchange capacity (1.45 meq/g) and selectivity towards Zn^{2+} and Mg^{2+} metal ions.

Some general advantages and disadvantages of the ion exchange process are the following:

(1) Can be used to purify a wide range of compounds.
(2) The resin can often be regenerated and reused for multiple cycles.

(3) Cost effective.
(4) Helpful in quantitative separations.
(5) Limited to charged molecules.
(6) Over multiple uses, resin can degrade, which affect the performance.

The system requires regular maintenance for better performance.

4 Challenges and Limitations of Nanocomposites

Nanocomposites, which combine nanomaterials with bulk matrices, have shown exceptional improvements in fields like structural reinforcement, enhanced conductivity, and thermal stability. However, numerous challenges and limitations exist in their development, application, and broader adoption. Understanding these obstacles is essential to address existing gaps and maximise their potential.

4.1 Stability

Achieving long-term stability in nanocomposites, especially in challenging conditions, is a primary concern. Exposure to UV radiation, high temperatures, and chemicals can lead to degradation. For example, polymer-based nanocomposites exposed to UV light can experience accelerated breakdown due to the generation of reactive oxygen species (ROS), compromising properties like mechanical strength and thermal stability (Xu et al. 2021; Singh and Gupta 2019). A study on polyethylene (PE) nanocomposites reinforced with titanium dioxide (TiO_2) nanoparticles demonstrated significant stability loss due to UV-induced degradation. While TiO_2 particles initially enhanced strength, sunlight exposure over time accelerated breakdown in the PE matrix due to ROS production (Lal et al. 2017). This case highlights both the performance benefits and stability challenges under UV exposure.

4.2 Durability

Durability is closely related to stability and can be impacted by material compatibility and nanoparticle dispersion quality. Due to their high surface energy, nanoparticles tend to agglomerate, creating weak points within the structure. These aggregations can lead to uneven stress distribution, reducing resistance to wear and tear. Environmental factors like moisture and temperature can further affect durability, particularly in structural applications (Siddiqui and Chaudhry 2017). Also in automotive applications, researchers found that carbon nanotubes (CNTs) in epoxy composites initially increased strength but led to durability issues due to CNT clustering. This clustering created stress concentration points, resulting in micro-cracking under thermal cycling. The clustered areas weakened the material over time, emphasizing the need for uniform nanoparticle dispersion for long-term durability (Kim et al. 2020).

4.3 Economic Viability

While nanocomposites offer substantial advantages, their high production costs, stemming from nanoparticle synthesis and specialized manufacturing, pose economic challenges. Scaling from laboratory to industrial levels is difficult due to the need for precision techniques, such as chemical vapor deposition, making nanocomposites less economically viable compared to conventional materials (Lee et al. 2021; Patel et al. 2022). A feasibility study in the construction industry examined the use of silica-based nanocomposites in concrete to increase strength. The high production costs of silica nanoparticles and the specialized mixing required made this option cost-prohibitive compared to conventional concrete. Although effective in small-scale applications, the nanocomposite was not widely adopted due to economic limitations (Huang and Zhang 2018).

4.4 Toxicity

Nanocomposite toxicity, especially those containing metal oxides or carbon-based nanomaterials, is a considerable safety concern. Many nanomaterials, because of their small size and high reactivity, can interact with biological systems in unpredictable ways, potentially causing oxidative stress, inflammation, or cell death. For example, studies on silver and titanium dioxide nanoparticles have shown toxic effects on human cells and aquatic organisms, raising concerns about safe disposal and recycling (Huang et al. 2018). Magnetic graphene oxide/MgAl-layered double hydroxide nanocomposite: one-pot solvothermal synthesis, adsorption performance, and mechanisms for Pb^{2+}, Cd^{2+}, and Cu^{2+} (Fiorito et al. 2024).

A study was conducted on silver nanoparticles (AgNPs) used in antibacterial coatings found that AgNPs released into water bodies were toxic to aquatic life, affecting algae and fish. AgNPs inhibited photosynthesis in algae and disrupted fish gill function, highlighting the need for controlled use and disposal of bioactive nanoparticles in nanocomposites (Choi et al. 2021).

4.5 Environmental Impact

The environmental impact of nanocomposites is increasingly important, given global sustainability goals. Nanoparticles can persist in ecosystems, bioaccumulate, and disrupt food chains. Certain nanomaterials leach into soil and water, affecting microbial communities essential for ecosystem functions (Rashid et al. 2018). In agricultural settings, zinc oxide (ZnO) nanoparticles from plastic mulches used for UV resistance were found to leach into soil, altering microbial communities. ZnO nanoparticles affected nitrogen-fixing bacteria crucial for soil health, reducing fertility and crop yield. This case underscores the importance of environmental impact assessment before large-scale nanocomposite applications (Rao et al. 2019). The lifecycle challenges, including recycling and disposal, further add to the environmental footprint of nanocomposites, as these materials degrade and release nanoparticles into the environment (Zhou et al. 2021).

5 Future Perspectives

The field of nanotechnology keeps evolving, and new developments in material synthesis and functionalization will result in the production of polysaccharides (PSA)-based nanocomposites that are more efficient as well as adaptable. These developments could enhance membrane-based techniques, photocatalysis, and adsorption, enabling the removal of a greater range of pollutants. Due to their huge surface areas, high reactivity, and ability to be modified for specific contaminants, nanocomposites are showing great potential as solutions for the removal of pollutants. Further developments in this area will focus on strengthening the selectivity and efficiency of nanocomposites in adsorbing or catalyzing the degradation of various contaminants, such as dyes, heavy metals, and pharmaceutical residues. Simple post-treatment removal, effective sunlight-driven degradation, and rapid pollutant capture may all be made possible by innovations like photocatalytic and magnetic nanocomposites. Further, by using eco-friendly and biodegradable elements in nanocomposites, secondary pollution will be reduced, fostering more environmentally friendly treatment techniques.

Considering all of this, nanocomposites for wastewater and water treatment look to have an optimistic future and the potential to totally transform the sector. Global ecosystem conservation, public health safety, and water pollution mitigation are all possible through continued study and smart implementation. All of these, still, rely on budgeting and administration, which are sometimes a component of political or financial constraints on development. Furthermore, while classifying nanocomposites for their future use, the variety of nanocomposite sources derived from the extraction of plants, vegetables, biowaste, and biomass should be taken into consideration. As a result, bio-based nanoscale reinforcements that are applied to nanocomposites ought to be environmentally friendly and financially viable. For this reason, the proposals' feasibility and lab-scale research should be carefully considered.

6 Conclusions

In conclusion, nanocomposites have proven to be highly effective materials in the removal of pollutants due to their unique physicochemical properties, such as high surface area, reactivity, and mechanical stability. The combination of these properties with traditional composite materials has led to significant advancements in environmental pollution control, particularly in the treatment of wastewater. The mechanisms through which nanocomposites remove pollutants—including adsorption, photocatalysis, chemical reduction, and membrane filtration—have been explored in this chapter, each playing a critical role in the efficient degradation and removal of a wide range of contaminants such as heavy metals, organic dyes, and harmful chemicals.

Adsorption is one of the most common and effective methods used for the removal of pollutants, especially in the case of heavy metals and organic compounds, due to the large surface area and functional groups on the nanocomposites. Photocatalysis, an eco-friendly process, has been shown to effectively degrade organic pollutants under light exposure, making it a promising solution for wastewater treatment. Membrane filtration technologies, particularly those using nanocomposite materials, offer a highly efficient approach for separating contaminants from water, providing excellent permeability and selectivity in the removal of dyes and heavy metals.

Despite their promising applications, the use of nanocomposites in pollutant removal is not without challenges. Issues related to the stability, durability, economic viability, and potential toxicity of nanomaterials must be carefully addressed to ensure their sustainable and safe use. Environmental impacts, such as the potential for nanoparticle leaching and bioaccumulation, also need to be considered to prevent unintended ecological consequences. These limitations underline the need for further research into improving the performance, scalability, and sustainability of nanocomposites.

Looking ahead, nanocomposites hold great promise for addressing global pollution challenges, especially with advances in material design, functionalization, and recovery methods. As research progresses, it is likely that innovations in nanocomposite technologies will lead to more efficient, cost-effective, and environmentally sustainable solutions for pollution remediation. Therefore, while there are hurdles to overcome, the future of nanocomposites in environmental applications appears bright, with their potential to play a key role in the development of cleaner and more sustainable technologies.

References

Abd Elkodous M, El-Sayyad GS, Maksoud MA, Kumar R, Maegawa K, Kawamura G, Matsuda A et al (2021) Nanocomposite matrix conjugated with carbon nanomaterials for photocatalytic wastewater treatment. J Hazard Mater 410:124657

Abdi J, Abedini H (2020) MOF-based polymeric nanocomposite beads as an efficient adsorbent for wastewater treatment in batch and continuous systems: modelling and experiment. Chem Eng J 400:125862

Ahmad K (2021) Study of different polymer nanocomposites and their pollutant removal efficiency. Polymer 217:123453

Allah SM, AlMallahi MN, Indira SS, Al-Marzouqi AH, Elgendi M (2023) Recent progress in nanoparticle-based ion exchange membranes for water desalination. Case Stud Chem Environ Eng 100577

Al-Mutairi NH, Mehdi AH, Kadhim BJ (2022) Nanocomposites materials definitions, types and some of their applications: a review. European J Res Dev Sustain 3(2):102–108

Alqadami AA, Naushad M, Abdalla MA, Ahamad T, Alothman ZA, Alshehri SM (2016) Synthesis and characterization of Fe_3O_4@ TSC nanocomposite: highly efficient removal of toxic metal ions from aqueous medium. RSC Adv 6(27):22679–22689

Alqadami AA, Naushad M, Abdalla MA, Ahamad T, ALOthman ZA, Alshehri SM, Ghfar AA (2017) Efficient removal of toxic metal ions from wastewater using a recyclable nanocomposite: a study of adsorption parameters and interaction mechanism. J Clean Prod 156:426–436

Alqadami AA, Naushad M, ALOthman ZA, Alsuhybani M, Algamdi M (2020) Excellent adsorptive performance of a new nanocomposite for removal of toxic Pb (II) from aqueous environment: adsorption mechanism and modeling analysis. J Hazardous Mater 389:121896

Alsuhybani M, Alshahrani A, Algamdi M, Al-Kahtani AA, Alqadami AA (2020) Highly efficient removal of Pb (II) from aqueous systems using a new nanocomposite: adsorption, isotherm, kinetic and mechanism studies. J Mol Liq 301:112393

Amir M, Kurtan U, Baykal A (2015) Rapid color degradation of organic dyes by Fe_3O_4@ His@ Ag recyclable magnetic nanocatalyst. J Ind Eng Chem 27:347–353

Anjum M, Miandad R, Waqas M, Gehany F, Barakat MA (2019) Remediation of wastewater using various nano-materials. Arab J Chem 12(8):4897–4919

Begum R, Farooqi ZH, Ahmed E, Naseem K, Ashraf S, Sharif A, Rehan R (2017) Catalytic reduction of 4-nitrophenol using silver nanoparticles-engineered poly (N-isopropylacrylamide-co-acrylamide) hybrid microgels. Appl Organomet Chem 31(2):e3563

Camargo PHC, Satyanarayana KG, Wypych F (2009) Nanocomposites: synthesis, structure, properties and new application opportunities. Mater Res 12:1–39

Cevallos-Mendoza J, Amorim CG, Rodríguez-Díaz JM, Montenegro MDCB (2022) Removal of contaminants from water by membrane filtration: a review. Membranes 12(6):570

Chen J, Gao L, Liu W (2022) Nanocomposite applications in pollutant removal: a review of recent progress and future directions. Mater Today Chem 24:100758. https://doi.org/10.1016/j.mtchem.2022.100758

Choi O, Hu Z, Sedlak DL (2021) Environmental toxicity of silver nanoparticles in aquatic ecosystems: implications for use in antibacterial coatings. Ecotoxicol Environ Saf 215:112091. https://doi.org/10.1016/j.ecoenv.2021.112091

Dabrowski A (2001) Adsorption—from theory to practice. Adv Colloid Interface Sci 93:135–224

Depren SK, Filiz BC, Kulaklı BN, Figen AK (2020) Ultra-layered sheet CuCo nanoparticles for optimized application in catalytic reduction of organic dye. Mater Charact 160:110116

Dey A, Gogate PR (2021) Nanocomposite photocatalysts-based wastewater treatment. In: Handbook of nanomaterials for wastewater treatment. Elsevier, pp 779–809

Eslami H, Esmaeili A, Ehrampoush MH, Ebrahimi AA, Taghavi M, Khosravi R (2020) Simultaneous presence of poly titanium chloride and Fe_2O_3-Mn_2O_3 nanocomposite in the enhanced coagulation for high rate As (V) removal from contaminated water. J Water Process Eng 36:101342

Etemadi H, Qazvini H, Shokri E (2021) Effect of coagulation treatment on antifouling properties of PVC nanocomposite membrane in a submerged membrane system for water treatment. Water Sci Eng 14(4):295–303

Fatimah I, Purwiandono G, Hidayat A, Sagadevan S, Kamari A (2022) Mechanistic insight into the adsorption and photocatalytic activity of a magnetically separable γ-Fe_2O_3/Montmorillonite nanocomposite for rhodamine B removal. Chem Phys Lett 792:139410

Fiorito S, Silvestri M, Cirignano M, Marini A, Di Stasio F (2024) Controlled growth of large SiO_2 shells onto semiconductor colloidal nanocrystals: a pathway toward photonic integration. ACS Appl Nano Mater 7(4):3724–3733

Frimpong RA, Fraser S, Zach Hilt J (2007) Synthesis and temperature response analysis of magnetic-hydrogel nanocomposites. J Biomed Mater Res Part A: Official J Soc Biomater Japanese Soc Biomater Australian Soc Biomater Korean Soc Biomater 80(1):1–6

Fujishima A, Honda K (1972) Electrochemical photolysis of water at a semiconductor electrode. Nature 238(5358):37–38

Ghosh M, Mondal M, Mandal S, Roy A, Chakrabarty S, Chakrabarti G, Pradhan SK (2020) Enhanced photocatalytic and antibacterial activities of mechanosynthesized TiO_2–Ag nanocomposite in wastewater treatment. J Mol Struct 1211:128076

Gong Y, Zhao D (2014) Physical–chemical processes for phosphorus removal and recovery

Gray NF (2014) Filtration methods. In: Microbiology of waterborne diseases. Academic Press, pp 631–650

Gu Y, Sun S, Liu Y, Dong M, Yang Q (2019) Solvent effect on the solvothermal synthesis of mesoporous NiO catalysts for activation of peroxymonosulfate to degrade organic dyes. ACS Omega 4(18):17672–17683

Gupta RK, Kennel E, Kim KJ (eds) (2009) Polymer nanocomposites handbook. CRC press

Gupta A, Singh R, Verma P (2020) Challenges in conventional wastewater treatment and the emerging role of nanotechnology. J Environ Manage 260:110065

Gupta S, Verma R, Pandey A (2023) Adsorption of heavy metals and organic dyes using nanocomposites: a pathway to sustainable wastewater treatment. Environ Res 221:115104. https://doi.org/10.1016/j.envres.2023.115104

Hosseini SM, Rafiei S, Hamidi AR, Moghadassi AR, Madaeni SS (2014) Preparation and electrochemical characterization of mixed matrix heterogeneous cation exchange membranes filled with zeolite nanoparticles: Ionic transport property in desalination. Desalination 351:138–144

Huang X, Pan M (2016) RETRACTED: the highly efficient adsorption of Pb (II) on graphene oxides: a process combined by batch experiments and modeling techniques

Huang Y, Zhang T (2018) Economic feasibility of silica nanocomposites in construction: a case study on concrete reinforcement. Constr Build Mater 188:768–775. https://doi.org/10.1016/j.conbuildmat.2018.08.072

Huang Q, Chen Y, Yu H, Yan L, Zhang J, Wang B, Xing L et al (2018) Magnetic graphene oxide/MgAl-layered double hydroxide nanocomposite: one-pot solvothermal synthesis, adsorption performance and mechanisms for Pb^{2+}, Cd^{2+}, and Cu^{2+}. Chem Eng J 341:1–9

IPCC (2023) Climate change and water: implications of wastewater treatment. Intergovernmental Panel on Climate Change

Iravani S (2021) Nanomaterials and nanotechnology for water treatment: recent advances. Inorganic Nano-Metal Chem 51(12):1615–1645

Javanbakht V, Ghoreishi SM, Habibi N, Javanbakht M (2016) A novel magnetic chitosan/clinoptilolite/magnetite nanocomposite for highly efficient removal of Pb (II) ions from aqueous solution. Powder Technol 302:372–383

Jhaveri JH, Murthy ZVP (2016) Nanocomposite membranes. Desalin Water Treat 57(55):26803–26819

John MM, Benettayeb A, Belkacem M, Mitchel CR, Brahim MH, Benettayeb I, Hosseini-Bandegharaei A et al (2024) An overview on the key advantages and limitations of batch and dynamic modes of biosorption of metal ions. Chemosphere 142051

Kamari S, Shahbazi A (2021) High–performance nanofiltration membrane blended by Fe_3O_4@ SiO_2–CS bionanocomposite for efficient simultaneous rejection of salts/heavy metals ions/dyes with high permeability, retention increase and fouling decline. Chem Eng J 417:127930

Kaufman AA, Hansen RO, Kleinberg RL (2008) Paramagnetism, diamagnetism, and ferromagnetism. Methods Geochem Geophys 42:207–254

Kecili R, Hussain CM (2018) Mechanism of adsorption on nanomaterials. In: Nanomaterials in chromatography. Elsevier, pp 89–115

Khalil A, Ali N, Khan A, Asiri AM, Kamal T (2020) Catalytic potential of cobalt oxide and agar nanocomposite hydrogel for the chemical reduction of organic pollutants. Int J Biol Macromol 164:2922–2930

Khan MA (2020) Fabrication and applications of polyaniline Zr (IV) iodate nanocomposite ion exchanger for heavy metal sorption with antibacterial activity. Groundw Sustain Dev 11:100427

Khan MM, Adil SF, Al-Mayouf A (2015) Metal oxides as photocatalysts. J Saudi Chem Soc 19(5):462–464

Khan WS, Hamadneh NN, Khan WA (2016) Polymer nanocomposites–synthesis techniques, classification and properties. Sci Appl Tailored Nanostruct 50

Kim J, Choi H, Park S (2020) Durability issues in carbon nanotube-reinforced epoxy composites for automotive applications. Compos Sci Technol 187:107920. https://doi.org/10.1016/j.compscitech.2020.107920

Kumar A, Sharma G, Naushad M, Kumar A, Kalia S, Guo C, Mola GT (2017) Facile hetero-assembly of superparamagnetic Fe_3O_4/$BiVO_4$ stacked on biochar for solar photo-degradation of methyl paraben and pesticide removal from soil. J Photochem Photobiol, A 337:118–131

Kumar P, Sharma N, Patel R (2021) Energy-efficient approaches in wastewater management. Water Res Adv 5:200123

Kumar M, Singh A, Bhardwaj R (2023) Addressing challenges in the application of nanocomposites for environmental cleanup. Environ Adv 6:100264. https://doi.org/10.1016/j.envadv.2023.100264

Kuo MC, Tsai CM, Huang JC, Chen M (2005) PEEK composites reinforced by nano-sized SiO_2 and Al_2O_3 particulates. Mater Chem Phys 90(1):185–195

Lal R, Raghavendra M, Chopra V (2017) Impact of UV-induced degradation on polyethylene nanocomposites reinforced with titanium dioxide. J Appl Polym Sci 134(32):45172. https://doi.org/10.1002/app.45172

Lee SJ, Begildayeva T, Jung HJ, Koutavarapu R, Yu Y, Choi M, Choi MY (2021) Plasmonic ZnO/Au/g-C3N4 nanocomposites as solar light active photocatalysts for degradation of organic contaminants in wastewater. Chemosphere 263:128262

Li F, Zhao D (2022) Ceramic-based nanocomposites for environmental applications: synthesis, properties, and performance evaluation. Ceram Int 48(2):1234–1246. https://doi.org/10.1016/j.ceramint.2021.09.057

Li NN, Fane AG, Ho WW, Matsuura T (eds) (2011) Advanced membrane technology and applications. John Wiley & Sons

Liu Y, Wang X, Gao X, Zheng J, Wang J, Volodin A, Zhu J et al (2020) High-performance thin film nanocomposite membranes enabled by nanomaterials with different dimensions for nanofiltration. J Membr Sci 596:117717

Liu Z, Zhao X, Chen M (2022) Magnetic nanocomposites for pollutant removal and recovery: a sustainable perspective. Sep Purif Technol 285:120394. https://doi.org/10.1016/j.seppur.2022.120394

Mai YW, Yu ZZ (2006) Polymer nanocomposites

Malik S, Singh J, Goyat R, Saharan Y, Chaudhry V, Umar A, Baskoutas S et al. (2023) Nanomaterials-based biosensor and their applications: a review. Heliyon 9(9)

Marcelo LR, de Gois JS, da Silva AA, Cesar DV (2021) Synthesis of iron-based magnetic nanocomposites and applications in adsorption processes for water treatment: a review. Environ Chem Lett 19(2):1229–1274

Mazaheri M, Payandehpeyman J, Hedayatian M (2024) Agglomeration and interphase-influenced effective elastic properties of Metal/Graphene nanocomposites: a developed mean-field model. Compos Struct 329:117762

Melinte V, Stroea L, Chibac-Scutaru AL (2019) Polymer nanocomposites for photocatalytic applications. Catalysts 9(12):986

Mittal M, Tripathi S, Shin DK (2024) Biopolymeric nanocomposites for wastewater remediation: an overview on recent progress and challenges. Polymers 16(2):294

Moges A, Nkambule TT, Fito J (2022) The application of GO Fe_3O_4 nanocomposite for chromium adsorption from tannery industry wastewater. J Environ Manage 305:114369

Nachod FC (ed) (2012) Ion exchange: theory and application. Elsevier

Naz M, Rafiq A, Ikram M, Haider A, Ahmad SOA, Haider J, Naz S (2021) Elimination of dyes by catalytic reduction in the absence of light: a review. J Mater Sci 56(28):15572–15608

Noor A, Ullah R, Tuzen M, Hassan A, Ullah S (2024) Silver-coated activated carbon nanocomposite by Moringa oleifera for dye removal by multivariate optimization approach. Biomass Conver Bioref 14(15):17971–17983

Oyewo OA, Elemike EE, Onwudiwe DC, Onyango MS (2020) Metal oxide-cellulose nanocomposites for the removal of toxic metals and dyes from wastewater. Int J Biol Macromol 164:2477–2496

Panigrahy SK, Nandha A, Chaturvedi M, Mishra PK (2024) Novel nanocomposites with advanced materials and their role in waste water treatment. Next Sustain 4:100042

Patel P, Dhami H, Jain S (2022) Economic and scale-up challenges for nanocomposite materials. Compos Struct 288:115321. https://doi.org/10.1016/j.compstruct.2021.115321

Peighambardoust SJ, Rowshanzamir S, Amjadi M (2010) Review of the proton exchange membranes for fuel cell applications. Int J Hydrogen Energy 35(17):9349–9384

Quiton KGN, Lu MC, Huang YH (2021) Synthesis and catalytic utilization of bimetallic systems for wastewater remediation: a review. Chemosphere 262:128371

Rahman MM, Ahmed L, Anika F, Riya AA, Kali SK, Rauf A, Sharma R (2023) Bioinorganic nanoparticles for the remediation of environmental pollution: critical appraisal and potential avenues. Bioinorg Chem Appl 2023(1):2409642

Rao S, Khanna M, Reddy R (2019) Effects of zinc oxide nanoparticle leaching on soil microbial communities in agricultural settings. Environ Pollut 250:831–839. https://doi.org/10.1016/j.envpol.2019.04.082

Rashid MI et al (2018) Environmental behavior of nanocomposites: leaching and soil impact. Environ Poll 242:1036–1045. https://doi.org/10.1016/j.envpol.2018.07.072

Rastgar M, Shakeri A, Bozorg A, Salehi H, Saadattalab V (2017) Impact of nanoparticles surface characteristics on pore structure and performance of forward osmosis membranes. Desalination 421:179–189

Rehman S, Ullah R, Butt AM, Gohar ND (2009) Strategies of making TiO_2 and ZnO visible light active. J Hazard Mater 170(2–3):560–569

Ricco R, Konstas K, Styles MJ, Richardson JJ, Babarao R, Suzuki K, Falcaro P et al (2015) Lead (II) uptake by aluminium based magnetic framework composites (MFCs) in water. J Mater Chem A 3(39):19822–19831

Rostami-Vartooni A, Nasrollahzadeh M, Alizadeh M (2016) Green synthesis of seashell supported silver nanoparticles using Bunium persicum seeds extract: application of the particles for catalytic reduction of organic dyes. J Colloid Interface Sci 470:268–275

Rytwo G (2012) The use of clay-polymer nanocomposites in wastewater pretreatment. Scientific World J 2012(1):498503

Rytwo G, Lavi R, Rytwo Y, Monchase H, Dultz S, König TN (2013) Clarification of olive mill and winery wastewater by means of clay–polymer nanocomposites. Sci Total Environ 442:134–142

Sadoun AM, Mohammed MM, Fathy A, El-Kady OA (2020) Effect of Al_2O_3 addition on hardness and wear behavior of Cu–Al_2O_3 electro-less coated Ag nanocomposite. J Market Res 9(3):5024–5033

Saravanan R, Gracia F, Stephen A (2017) Basic principles, mechanism, and challenges of photocatalysis. Nanocomposites for visible light-induced photocatalysis, 19–40

Shameem MM, Sasikanth SM, Annamalai R, Raman RG (2021) A brief review on polymer nanocomposites and its applications. Mater Today: Proc 45:2536–2539

Sharma G, Thakur B, Naushad M, Ala'a H, Kumar A, Sillanpaa M, Mola GT (2017) Fabrication and characterization of sodium dodecyl sulphate@ ironsilicophosphate nanocomposite: ion exchange properties and selectivity for binary metal ions. Mater Chem Phys 193:129–139

Siddique T, Dutta NK, Choudhury NR (2021) Mixed-matrix membrane fabrication for water treatment. Membranes 11(8):557

Siddiqui SI, Chaudhry SA (2017) Arsenic removal from water using nanocomposites: a review. Curr Environ Eng 4(2):81–102

Sing KSW, Rouquerol J, Rouquerol F (1998) Adsorption by powders and porous solids. Academic Press, San Diego, USA

Singh B, Gupta R (2019) Challenges in the stability of polymer-based nanocomposites. Polym Degrad Stab 163:45–54. https://doi.org/10.1016/j.polymdegradstab.2019.02.006

Singh R, Kumar N, Joshi H (2022) Mechanistic insights into pollutant removal using nanocomposites: adsorption, photocatalysis, and beyond. Chem Eng J Adv 9:100234. https://doi.org/10.1016/j.ceja.2022.100234

Singh V, Patel R, Kumar S (2023) Advances in nanocomposite technology for dye and heavy metal removal: a review. J Clean Prod 389:136838. https://doi.org/10.1016/j.jclepro.2023.136838

Slater MJ (2013) Principles of ion exchange technology. Butterworth-Heinemann

Smith A, Roberts K, Jenkins T (2021) Versatility in nanocomposite structures: polymer, ceramic, and metal-based systems for environmental remediation. Adv Func Mater 31(6):2100435. https://doi.org/10.1002/adfm.202100435

Soltani M, Zabihi M (2020) Hydrogen generation by catalytic hydrolysis of sodium borohydride using the nano-bimetallic catalysts supported on the core-shell magnetic nanocomposite of activated carbon. Int J Hydrogen Energy 45(22):12331–12346

Thakur S, Ojha A, Kansal SK, Gupta NK, Swart HC, Cho J, Prakash J et al. (2024) Advances in powder nano-photocatalysts as pollutant removal and as emerging contaminants in water: analysis of pros and cons on health and environment. Adv Powder Mater 100233

Thostenson ET, Li C, Chou TW (2005) Nanocomposites in context. Compos Sci Technol 65(3–4):491–516

UNESCO (2021). Retrieved from https://unesdoc.unesco.org/ark:/48223/pf0000375724

Unuabonah EI, Taubert A (2014) Clay–polymer nanocomposites (CPNs): adsorbents of the future for water treatment. Appl Clay Sci 99:83–92

Varadavenkatesan T, Selvaraj R, Vinayagam R (2016) Phyto-synthesis of silver nanoparticles from Mussaenda erythrophylla leaf extract and their application in catalytic degradation of methyl orange dye. J Mol Liq 221:1063–1070

Wang H, Luo C (2024) Nanocomposite membranes in water treatment: development and environmental applications. Membr Sci Technol 15(1):1–19. https://doi.org/10.1016/j.memsci.2024.101573

Wang CG, Xing YH, Li ZP, Li J, Zeng XQ, Ge MF, Niu SY (2009) Synthesis, crystal structures and properties of a series of three-dimensional lanthanide coordination polymers with the rigid and flexible mixed dicarboxylate ligands of 1, 4-benzene dicarboxylic acid and succinic acid. J Mol Struct 921(1–3):126–131

Wang P, Liang YN, Zhong Z, Hu X (2020a) Nano-hybrid bimetallic Au-Pd catalysts for ambient condition-catalytic wet air oxidation (AC-CWAO) of organic dyes. Sep Purif Technol 233:115960

Wang YX, Zhang FP, Luan YX, Ye M (2020b) Ligand-enabled Ni–Al bimetallic catalysis for nonchelated dual C-H annulation of arylformamides and alkynes. Org Lett 22(6):2230–2234

Wei W, Yu D, Huang Q (2020) Preparation of Ag/TiO_2 nanocomposites with controlled crystallization and properties as a multifunctional material for SERS and photocatalytic applications. Spectrochim Acta Part A Mol Biomol Spectrosc 243:118793

World Bank (2022) Tackling water pollution in developing countries. The World Bank Group, Washington, D.C.
Wu YG, Wen M, Wu QS, Fang H (2014) Ni/graphene nanostructure and its electron-enhanced catalytic action for hydrogenation reaction of nitrophenol. J Phys Chem C 118(12):6307–6313
Xu YJ, Zhuang Y, Fu X (2010) New insight for enhanced photocatalytic activity of TiO_2 by doping carbon nanotubes: a case study on degradation of benzene and methyl orange. J Phys Chem C 114(6):2669–2676
Xu L, Huang B, Wang H (2021) Photodegradation effects on polymeric nanocomposites. J Polym Res 28(7):229. https://doi.org/10.1007/s10965-021-02692-3
Yang Y, Zhang H, Wang P, Zheng Q, Li J (2007) The influence of nano-sized TiO_2 fillers on the morphologies and properties of PSF UF membrane. J Membr Sci 288(1–2):231–238
Yang Q, Dong Y, Qiu Y, Yang X, Cao H, Wu Y (2020) Design of functional magnetic nanocomposites for bioseparation. Colloids Surf, B 191:111014
Yang H, Lee C, Chen X (2023a) Innovations in photocatalysis using nanocomposites: a review. J Clean Prod 375:133895
Yang L, Hu X, Zhao Q (2023b) Removal of heavy metals using nanocomposites: mechanisms and environmental applications. Environ Pollut 315:120478. https://doi.org/10.1016/j.envpol.2023.120478
Zhang X, Wang Y (2021) The emerging role of nanocomposites in environmental applications: a review of synthesis, mechanisms, and prospects. Environ Nanotechnol Monitor Manage 16:100549. https://doi.org/10.1016/j.enmm.2021.100549
Zhang H, Wen J, Fang Y, Zhang S, Zeng G (2019) Influence of fulvic acid on Pb (II) removal from water using a post-synthetically modified MIL-100 (Fe). J Colloid Interface Sci 551:155–163
Zhang X, Yi G, Zhang Z, Yu J, Fan H, Li P, Zhang C et al (2021) Magnetic graphene-based nanocomposites as highly efficient absorbents for Cr (VI) removal from wastewater. Environ Sci Pollut Res 28:14671–14680
Zhang Y, Liu P, Huang J (2023) Advances in photocatalytic nanocomposites for environmental remediation: a sustainable approach. J Photochem Photobiol, A 435:114291. https://doi.org/10.1016/j.jphotochem.2023.114291
Zhao R, Wang Y, Li H (2019) Sludge management challenges in conventional wastewater treatment systems. Water Environ Res 91(11):1421–1432
Zheng LQ, Yu XD, Xu JJ, Chen HY (2015) Reversible catalysis for the reaction between methyl orange and $NaBH_4$ by silver nanoparticles. Chem Commun 51(6):1050–1053
Zhou S, Ni X, Zhou H, Meng X, Sun H, Wang J, Yin X (2021) Effect of nZVI/biochar nanocomposites on Cd transport in clay mineral-coated quartz sand: facilitation and rerelease. Ecotoxicol Environ Saf 228:112971

Biodegradable Nanocomposites for Organic Pollutant Removal

Dinesh Kumar Chelike, Ram Krishna Rathore, Ahmed Mohsin Alsayah, Nitish Kumar Singh, Rupak Kumar Deb, Avinash Kumar Namdeo, and Agnivesh Kumar Sinha

Abstract The emergence of contemporary environmental issues has catalysed the development of innovative and eco-friendly water purification techniques. Biodegradable nanocomposites have the potential to be efficient materials for eliminating organic contaminants from water. The nanocomposites get rid of pollutants well and are good for the environment because they use the large surface area and reactivity of nanoparticles with biodegradable matrices. This study synthesizes, characterizes, and uses biodegradable nanocomposites to adsorb and degrade dyes, pharmaceuticals, and pesticides. We integrated biodegradable polymers like polylactic acid (PLA) and polycaprolactone (PCL) with functionalized nanoparticles like titanium dioxide TiO_2, GO and zinc oxide ZnO, and a hybrid material containing inorganic–organic to formulate the eco-friendly nanocomposites. Microparticles and a biodegradable matrix work together to make nanocomposites more stable, easier to spread. Nanocomposites undergo biodegradation, which results in the formation of non-toxic byproducts. Utilizing these biodegradable nanocomposites in extensive water treatment protocols presents a viable and environmentally friendly alternative to existing methodologies. This chapter will demonstrate that the utilization of biodegradable nanocomposites has the potential to effectively eliminate organic contaminants from water and the surrounding environment while adhering to environmentally sustainable practices. The subsequent phase involves enhancing fabrication techniques and scaling up production for industrial applications.

D. K. Chelike · R. K. Rathore (✉) · A. K. Sinha
Rungta International Skills University, Bhilai, India
e-mail: krish.rathore@gmail.com

A. M. Alsayah
Refrigeration and Air-Condition Department, Technical Engineering College, The Islamic University, Najaf, Iraq

N. K. Singh (✉)
School of Engineering & Technology, Vivek University, Bijnor, India
e-mail: hknitish@gmail.com

Avaaya Infotech Private Limited, Patna, India

R. K. Deb · A. K. Namdeo
Department of Mechanical Engineering, Lingaya's Vidyapeeth, Faridabad, India

S. Roy et al. (eds.), *Nanocomposites for Sustainable Wastewater Treatment*, Composites Science and Technology, https://doi.org/10.1007/978-981-95-1369-7_4

Keywords Organic contamination · Biodegradable nanocomposite · Wastewater treatment · Nanomaterial

1 Introduction

The study represents a distinctive hybrid mesoporous compound designed for the degradation of pharmaceutical contaminants in water. The hybrid materials composed of metal–organic frameworks and chitosan, when applied over nanosilica, demonstrate superior drug adsorption via hydrogen bonding (H-Bond) interactions and effective antibiotic photo-degradation. The hybrid compounds reduced band gap and increased conductivity, significantly enhancing pollution reduction by minimizing electron–hole recombination. This improved light absorption and charge transfer efficiency, thereby augmenting the material's ability to degrade contaminants. The illustration of structure and morphology was examined by multiple techniques, such as transmission electron microscopy (TEM), X-ray photoelectron spectroscopy (XPS), Fourier transform infrared spectroscopy (FTIR), X-ray diffraction (XRD), scanning electron microscopy (SEM), Brunauer–Emmett–Teller (BET) analysis and thermogravimetric analysis (TGA). It was essential to examine pH levels, catalyst dosage, and radiation duration to improvement of the efficiency of the adsorption photodegradation process. The model, tetracycline, antibiotics and doxycycline, at a concentration (400 mg/L), were noticed to exhibit an efficiency higher than 85% through non-linear optimization techniques. The optimal values for maximum removal were determined to be a hybrid mesosphere dosage of 4 mg/mL, a radiation period of 10 min, and a pH of 4.3. Pseudo-second-order diffusion models were favored compared to pseudo-first-order models in kinetic experiments. The hybrid material, mesosphere, effectively eliminated over 80% of each antibiotic in real aqueous samples and exhibited sustained efficiency after three cycles. This research illustrates the capability of hybrid mesoporous materials in the elimination of pharmaceutical contaminants from aquatic environments (Deymeh et al. 2024). It is well recognized that organic pollutants have a detrimental impact on the ecology. Significant volumes of organic-contaminated effluent, mostly from sewage treatment plants, industrial effluents, and agricultural runoff, are released into marine environments (Awad et al. 2019).

The source and specific molecules present have a significant impact on the quantities of organic impurities. Nonetheless, it is often noted that concentrations usually lie between 0.01 and 10 μg/L (Babu Ponnusami et al. 2023). One or more cyclic rings, which may indicate an aromatic or aliphatic origin, the absence of polar functional groups, and a range of halogen substitutes, most often chlorine, are typical characteristics of organic contaminants. These qualities allow them to survive in the environment (Ali et al. 2012). One of the most prevalent renewable biopolymers in the biosphere, cellulose is a valuable and potentially valuable resource material, as Varghese et al. showed. Because of its extraordinary, unique qualities, such as its high surface area, excellent crystallinity, and biodegradability, its nano equivalent, nanocellulose, has likewise earned widespread appeal and attention among researchers worldwide. According to reports, nanocellulose and its composites are good options for a variety of applications, including the remediation of contaminated water, the breakdown of harmful organic compounds, the detection of pathogens in water, the adsorption of oil spills in water, and the atmospheric sequestration of CO_2. We have attempted to compile and summarize the most recent research conducted and published on nanocellulose-based membranes used for the elimination of various metal ions, oils, dyes, and pharmaceutical components found in aqueous solutions that continue to endanger aquatic and human life in this report (Varghese et al. 2022). Dopamine self-polymerized on the surface of graphene oxide to create Thu et al. With a comparatively high active surface area, graphene oxide serves as a supporting material. With its varied functional groups, the polymer inspired by mussels (PDOP) will provide active sites that are prepared for the adsorption of contaminants from the water. Additionally, the addition of PDOP material is likely to increase the as-prepared sorbents' hydrophilicity (Chen et al. 2022), which will enhance the adsorption process in an aqueous environment. Using methods from Fourier Transform Infrared (FTIR), Raman, and Scanning Electron Microscopy (SEM), the morphological and structural characteristics of the as-prepared biosorbents were investigated. The Brunauer–Emmett–Teller (BET) method was used to calculate the active surface area. With different contact times and adsorbate concentrations, the adsorption kinetics and isotherms were also investigated. We'll also talk about insights into how organic dyes adsorb onto the surface of the biosorbent while it's being manufactured (Thu et al. 2023).

1.1 Origins and Effects of Organic Pollutants

Organic pollutants come from a range of home, industrial, and agricultural sources. They fall into a number of categories:

(i) **Dyes**: A lot of water is polluted by the synthesis of dye and pigments utilized in the textile industry, due to textile colors do not stick to fabric well, they are discharged into aquatic environments as effluent. Consequently, several textile sectors ongoing discharge of unprocessed wastewater has negative effects on the environment and public health. Textile dyes have the potential to poison aquatic environments and ascend the food chain, endangering aquatic-loving creatures. This research will examine the impact of textile dyes on human health, aquatic life, and water bodies. Textile dyes degrade the aesthetic value of water bodies by increasing the biochemical and chemical oxygen demand (BOD/COD), decreasing photosynthesis, stopping plant growth, entering the food chain, causing resilience and bioaccumulation, and possibly promoting toxicity, mutagenicity, and carcinogenicity. Consequently, in order to avoid negative impacts on human health, the environment, and

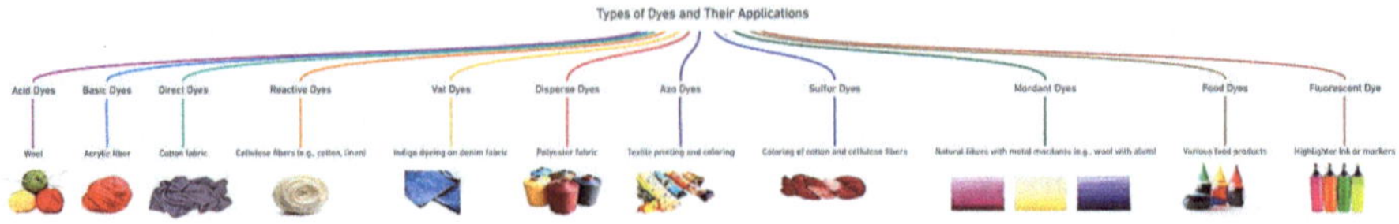

Fig. 1 Top-loaded bottom-pouring stir-casting furnace

natural water resources, wastewater containing dyes must be adequately treated using ecologically friendly technology (Al-Tohamy et al. 2022). Dye is often separated into many varieties according to its source, composition, and use. The textile industries utilize the most synthetic dyes, including sulfide, azo, dispersion, acid, basic, reactive, and mordant dyes as displayed in Fig. 1 (Al-Tohamy et al. 2022).

(ii) **Pesticides**: The many negative effects of chemical pesticides on natural ecosystems and human health have been thoroughly examined by Zhou et al., emphasizing the demanding need for sustainable pest control techniques. The vast utilize of pesticides, such as organochlorine compounds (e.g., Endrin and DDT), which are known for their endurance and bioaccumulation, poses serious risks to water quality, food safety as well as biodiversity due to their accumulation in the food chain process, these constituents risk the highest trophic levels and raise the possibility of contrary health consequences such as cancer, acute poisoning, and neurological disorders. In specific cases, such as atrazine and glyphosate exhibit the prevalence of pesticide pollution is exhibited in a range of environments. Significant discoveries contain the correlation between pesticide exposure and elevated chances of Parkinson's disease and non-Hodgkin lymphoma, as well as the emergence of antibiotic resistance in microbial communities. This overview addresses the variety of approaches to remediation, such as bioremediation strategies based on microbial degradation and physicochemical methods such as photodegradation and improved oxidation processes. Environmental elements like sunshine intensity, soil characteristics, and organic matter concentration affect photodegradation rates. Degrading persistent pesticides and improving soil recovery have been shown benefits of bioremediation using certain microbial consortia. In this study, the necessity of improved regulatory frameworks, user education, and integrated pest management (IPM) is emphasized. A strategy change towards methods that reduce environmental damage and protect human health is recommended. By lowering the need for pesticides and fostering ecological balance, putting these strategies into practice may greatly improve sustainable agriculture (Zhou et al. 2025). Synthetic chemicals are used in agriculture as pesticides, insecticides, herbicides, and fungicides. According to Li et al. (2023), the most often used organophosphate pesticides in the agricultural industry include glyphosate, trichlorfon, malathion, dichlorvos, diazinon, methyl parathion, and dichlorvos. Several nations have outlawed the use of organochlorine insecticides due to

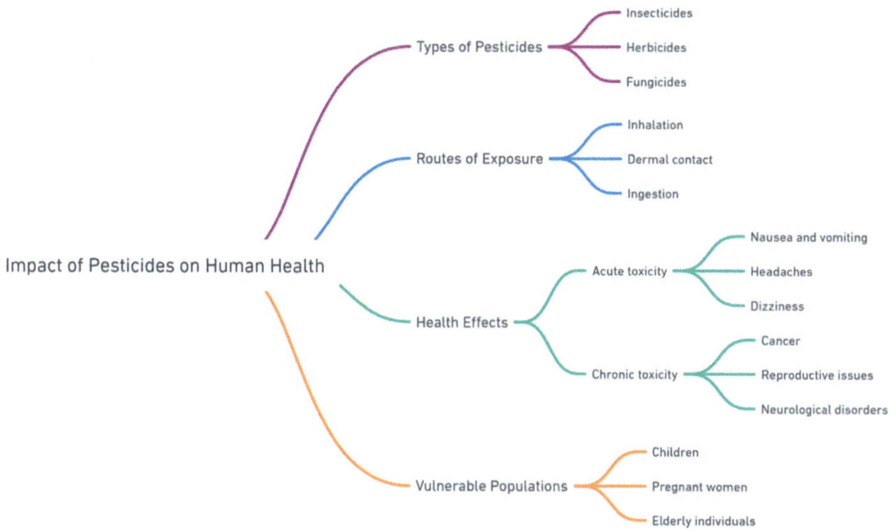

Fig. 2 The effects of pesticides on human health (Ascherio et al. 2006). Reproduced with permission under this license http://creativecommons.org/licenses/by/4.0/

their harmful effects and toxicological characteristics as revealed in Fig. 2 (Li et al. 2023).

(iii) **Pharmaceutical Residues**: Salahshoori et al. revealed that pharmaceutical contaminants are now a major environmental problem when it comes to contaminating natural water supplies. The investigation of pollutant adsorption on different surfaces via typical experimental techniques frequently requires significant labor and resources. Therefore, this work explores the potent integration of in silico techniques, such as quantum mechanics (QM), Monte Carlo simulations (MC), and molecular dynamics (MD), as a comprehensive toolkit to gather extensive perspectives into the adsorption of pharmaceutical contaminants. These computational methods provide for a comprehensive knowledge of the processes involved by spanning many scales, from macroscopic environmental influence to interactions at the molecular level. In this review of pharmaceutical pollutants and their impact on the environment, we stress the need to find effective and long-lasting adsorption solutions. The benefits of each model in simulating the adsorption of pharmaceutical contaminants are then highlighted as the theoretical fundamentals of MC, MD, and QM are examined. Further investigation is the potential for combining these methods to provide a more complete assessment of the adsorption method. Recently, case studies have been found to demonstrate how well in silico methods may be used to predict adsorption behaviors on a range of surfaces and in various environmental conditions. Finally, the environmental effect of pharmaceutical contaminant adsorption is explored, along with how in silico-based methods might be useful in long-term efforts to reduce these effects (Salahshoori et al.

2024). Conventional methods for evaluating the disposal of pharmaceutical waste often depend on experimental techniques, which may be expensive, time-consuming, and have limited applicability. These constraints may be addressed by cutting-edge methods in high-throughput screening and computational modeling, which provide a comprehensive knowledge of the removal of pharmaceutical pollutants at the molecular level (Veclani et al. 2020; Asghari et al. 2024) (Table 1).

2 Biodegradable Polymers for Nanocomposites

Biodegradable polymers, often known as eco-friendly or bio-based polymers, hold great promise in addressing the issue of plastic waste (Dananjaya et al. 2022a, b, c, 2024). In the environment, these polymers spontaneously break down. They might be recognized as ecologically friendly substitutes for conventional polymers made from petroleum (Larson et al. 2013; Keskin et al. 2017). A variety of biodegradable polymers that show promise include PLA, PHA, PBS, PCL, PGA, PBAT, PLGA, starch- and cellulose-based polymers (Dhamaniya et al. 2018). They possess many attributes, including robust mechanical strength, advantageous thermal properties, and barrier properties (Tan et al. 2016; Post et al. 2021). Applications for biodegradable polymers may be found in consumer products, packaging, agriculture, and the biomedical industry. They provide greener alternatives for reducing plastic waste and promoting sustainability for a cleaner planet.

2.1 Polylactic Acid (PLA)

PLA is one of the most well-known biopolymers; its properties, including biocompatibility, elasticity, thermoplastic behavior, excellent molding ability, and stiffness, provide the most promising uses. The biodegradable thermoplastic PLA ($C_3H_4O_2$) is made from renewable resources such as sugarcane bagasse, maize, starch, and food waste. It contains an aliphatic polyester. One of the most extensively manufactured bioplastics worldwide is polylactic acid (PLA), which has a minimal carbon footprint and is environmentally friendly. The number of efforts to use PLA in manufacturing to partly replace petrochemical plastics is increasing annually. This polymer is usually employed in high-end applications, but its usage won't grow until it can be made as cheaply as possible. Consequently, food waste high in carbs may serve as the main source of raw materials for PLA manufacturing. Although biological fermentation is the usual method for producing lactic acid (LA), a good downstream separation process that produces high-quality products at a reasonable cost of production is also necessary (Swetha et al. 2023) (Table 2).

Table 1 Qualitative analysis regarding the environmental and health effects of pesticide toxicity (11)

S. No.	Effect	Specific impact	Quantitative data	References
1	Human-health	Deaths from acute poisoning Parkinson's disease risk Non-Hodgkin Lymphoma risk	Every year, there are over 385 million instances of severe pesticide poisoning. Through almost 11,000 deaths globally 70% greater frequency among those who were exposed Increased risk of 41% among water samples	Joseph Otorkpa et al. (2024) Ascherio et al. (2006) Eriksson et al. (2008)
2	Environment-water	Flow contamination	Pesticides have been identified in 50% of universal water quality assessments	Syafrudin et al. (2021)
3	Environment-soil	Microbial diversity reduction	Reductions of up to 30% in the variety of soil microbes	Zhang et al. (2017)
4	Environment-biodiversity	Bee population falls	In some locations, bee populations have declined by up to 30%	Goulson et al. (2015)
5	Environment-bioaccumulation	The amount of pesticides in marine mammals	DDT and polychlorinated biphenyls were found at lipid weight values of up to 3819 ng/g and 466 ng/g, respectively	Li et al. (2015)
6	Environment-air	Pesticide transport in the atmosphere	Up to 100 km away from the spraying areas, pesticides were found	Lavin and Hageman (2013)

Table 2 Types of biodegradable nanocomposites and their properties (Koolivand et al. 2024)

Type of NP	NP concentration	Additive	EOR (%)	Remarkable findings
SiO_2	1 wt%	Polyacrylamide	10	Increased solution viscosity and thermal stability of the polymer solution
SiO_2, Fe $(OH)_3$, TiO_2, Al_2O_3	0.2 wt%	Xanthan gum	1–9	Improved solution viscosity
Al_2O_3	0.05 wt%	Anionic surfactant	12.5	Wettability alteration
SiO_2	0.01–3 wt%	–	9–19	Improved oil displacement in porous media
Clay	0.9 wt%	Polyacrylamide	5–6	Improved rheological properties
ZrO_2	0.1 g/cc	Cetrimonium bromide surfactant	40	Wettability alteration
Pyroxene	1 wt%	–	10.5	Reduced IFT and altered contact angle
Carbon	1 wt%	–	24.5	Reduced oil viscosity
Fumed silica	2 wt%	Sodium dodecyl sulfate surfactant	13	Increased solution viscosity
CeO_2, ZrO_2, Al_2O_3, TiO_2, CNT, MgO, $CaCO_3$, SiO_2	5 wt%	–	8–9	Increased solution viscosity
TiO_2	1.9–2.5 wt%	Sodium dodecyl sulfate surfactant	4	Improved rheological properties
SnO_2	2 wt%	–	22	Reduced IFT
ZnO	1.5 wt%	–	11	Change in rock wettability, reduced IFT, oil viscosity, mobility ratio and permeability alterations
Carbon	0.1 wt%	–	32	Wettability alteration
Fluorescent carbon	0.05 wt%	–	22	Reduced IFT
SiO_2	0.3 wt%	Xanthan gum polymer–surfactant	27	Increased solution viscosity
SiO_2, TiO_2, Al_2O_3	0.2–0.4 wt%	HPAM	7	Improved rheological properties
Nickel	0.005–0.02 wt%	Xanthan gum	6	Increased solution viscosity
SiO_2	1 wt%	Polyethylene glycol	8	Improved rheological properties

2.2 *Polycaprolactone (PCL) and Chitosan*

According to Schueren et al., the electrospinning method could be used to effectively load the pH-sensitive dye onto biocompatible PCL and PCL containing chitosan nanofibers. It was advised to apply a complexing agent to reduce dye leaching. The functionalized nanofibers' morphology resembled that of the PCL/chitosan and blank PCL nanofibers (Van der Schueren et al. 2013). The generation of stimuli-responsive materials might be significantly aided by the use of pH-sensitive dye incorporated into nanofibers. But before they can be used in biomedical applications, biocompatible polymers must be applied. Therefore, the generation and characterization of pH-sensitive PCL nanofibrous structures and PCL/chitosan nanofibrous blends containing 20% chitosan are the primary concerns of this investigation (Van der Schueren et al. 2013). There are several advantages of using biodegradable polymers in nanocomposites. These polymers guarantee that the components employed for pollutant removal do not linger in the environment by lowering the danger of secondary contamination and environmental pollution. Moreover, biopolymers improve the nanocomposite's general biocompatibility, allowing for its use in delicate ecosystems. The classification of biopolymers and natural polymers was depicted in Fig. 3.

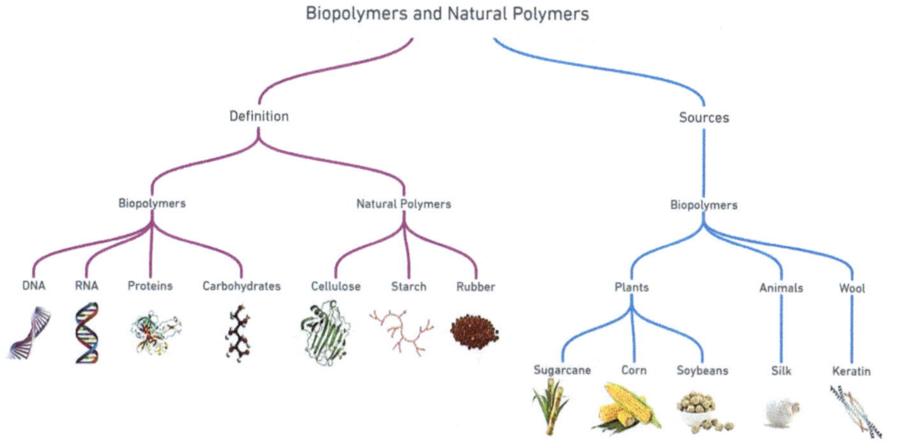

Fig. 3 Categorization of biopolymers according to their source (Ewis et al. 2022)

3 Types of Biodegradable Nanocomposites for Pollutant Removal

A range of biodegradable nanocomposites has been designed to efficiently remove organic contaminants from water. Among them are polymer–clay nanocomposites, which effectively remove dyes and heavy metals by encasing clays like montmorillonite in biodegradable polymers to increase surface area and adsorption capacity (Ewis et al. 2022; Al-Gethami et al. 2024; Hussain et al. 2024). Polymer-carbon-based nanocomposites incorporate carbon materials such as carbon nanotubes and graphene, improving both electrical and adsorption properties, which allows for the effective removal of pharmaceutical residues and pesticides. Biopolymer-metal oxide nanocomposites utilize metal oxides like TiO_2 and ZnO within biopolymer matrices to facilitate photocatalytic degradation of pollutants under light, making them useful for wastewater treatment (Majumder et al. 2020). Moreover, green synthesis approaches are gaining popularity, using plant extracts or microorganisms in environmentally friendly processes to form nanoparticles within biodegradable polymers, aligning with sustainability goals (Díez-Pascual 2020; Panigrahy et al. 2024).

4 The Process by Which Biodegradable Nanocomposites Remove Pollutants

According to the particular kind of nanocomposite used, biodegradable nanocomposites could be able to remove organic pollutants via a number of different methods. Several typical techniques are mentioned, including adsorption procedures that include the physical or chemisorption of organic contaminants onto the nanocomposite surface. The great effectiveness of nanocomposites in absorbing pollutants could be attributed to their porous structure and large surface area. Organic pollutants may also be broken down by redox processes via the use of catalytic degradation nanocomposites that include metal oxides or other catalytic nanoparticles. Complex organic molecules are broken down into smaller, less dangerous chemical compounds by this process. In addition, metal oxide nanocomposites, such as ZnO or TiO_2, may promote photocatalytic reactions when exposed to UV or visible light, which degrade organic pollutants in water. Additionally, the biopolymers incorporated in the nanocomposites may aid in the biodegradation of organic pollutants by creating an environment that fosters the growth of microorganisms capable of breaking down these pollutants under certain conditions (Mayet et al. 2023; Dehghani et al. 2023; Annu et al. 2024) as shown in Fig. 4.

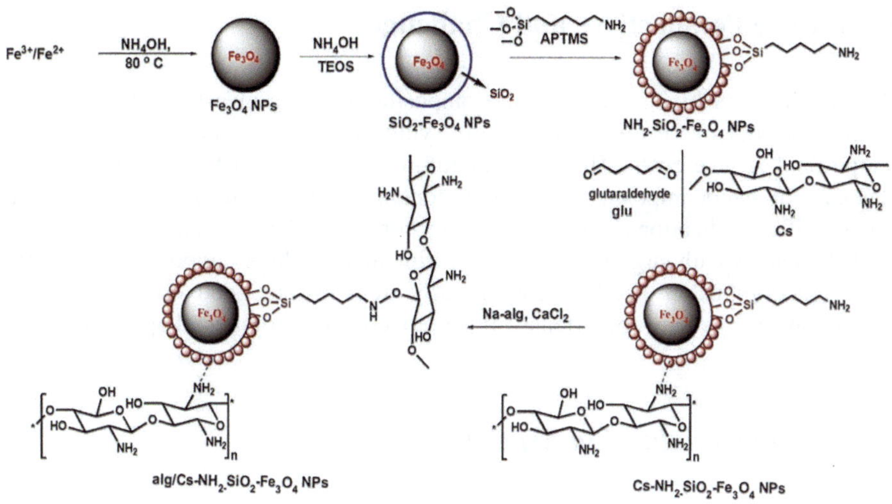

Fig. 4 Synthesis route of alg/Cs-NH_2-SiO_2/Fe_3O_4 adsorbent

5 Advantages and Limitations

When it comes to environmental applications, particularly in the removal of pollutants, the use of biodegradable nanocomposites offers a number of significant advantages. One of the most significant benefits is that they are environmentally friendly; these materials are manufactured from renewable resources and degrade on their own, which reduces the impact that they have on the environment. Furthermore, the unique characteristics of biodegradable nanocomposites, such as their large surface area and high reactivity, make it possible for them to effectively remove pollutants, which contributes to the high efficiency of these materials. The biocompatibility of these materials is an additional key advantage. This feature ensures that they are safe to use in sensitive areas and lowers the likelihood of secondary contamination that may be caused by the use of conventional materials. Although there are many benefits associated with expanding the production of biodegradable nanocomposites, there are also a number of challenges that must be overcome. There is a possibility that the high manufacturing costs of biodegradable polymers and nanoparticles will make it challenging for them to find widespread use. Furthermore, despite the fact that these materials are designed to degrade over the course of time, there are limitations placed on the ways in which they may be recycled and reused, which may reduce their long-term sustainability. The widespread use of biodegradable nanocomposites in environmental restoration projects will be contingent on the successful overcoming of these challenges via advancements in materials science and the development of production procedures that are also economical.

6 Future Perspectives and Challenges

The synthesis of biodegradable nanocomposites is being investigated using novel biopolymer discoveries and eco-friendly synthesis methods in an attempt to boost output and reduce costs. The focus of future research must be on improving the capacity of nanocomposites to remove contaminants, modifying their properties for specific applications, and overcoming commercialization barriers. There are still challenges with large-scale manufacturing and regulatory approval. Biodegradable nanocomposites need to continuously demonstrate good quality and high performance in order to be extensively employed in environmental applications.

6.1 Commercialization Aspects

As more people become aware of the dangers that pollution poses to their health and the environment, there has been an increase in the demand for biodegradable nanocomposites. The reduction of pollution and the promotion of environmentally responsible activities are the goals of regulatory frameworks and environmental initiatives. There is a correlation between increased research funding from both the public and private sectors and an increase in the commercialization of biodegradable nanocomposites. The economic feasibility of biodegradable nanocomposites is of utmost importance, since they have the potential to lower operating costs and reduce their contribution to the environment. The use of fossil fuels may be reduced by the use of renewable resources. Scalability and production methods are essential to commercialization, and breakthroughs in manufacturing processes such as electrospinning and 3D printing are essential to successful commercialization. When it comes to gaining regulatory clearance and commercial acceptability, standardization processes are very necessary. It is possible to speed up the introduction of biodegradable nanocomposites to the market via the formation of partnerships between academic institutions, industrial actors, and government agencies. Research, field applications, pilot studies, integration, and public involvement are some of the future paths and directions that will be taken. (Ghosh and Jones 2021; Zhang et al. 2024).

7 Conclusion

In this chapter, biodegradable nanocomposites were noted to exhibit great promise as a sustainable and effective method of environmental remediation for the removal of organic pollutants. Through the incorporation of nanotechnology's benefits with biodegradable materials such as PCL and PLA. These nanocomposites provide a feasible solution to the world's pollution problems. Their capacity to overcome current obstacles and improve their efficiency for large-scale applications will determine their future contribution to environmental sustainability.

References

Al-Gethami W, Qamar MA, Shariq M, Alaghaz A-NMA, Farhan A, Areshi AA, Alnasir MH (2024) Emerging environmentally friendly bio-based nanocomposites for the efficient removal of dyes and micropollutants from wastewater by adsorption: a comprehensive review. RSC Adv 14(4):2804–2834. https://doi.org/10.1039/D3RA06501D

Ali I, Mohd A, Khan TA (2012) Low cost adsorbents for the removal of organic pollutants from wastewater. J Environ Manage 113:170–183. https://doi.org/10.1016/j.jenvman.2012.08.028

Al-Tohamy R, Ali SS, Li F, Okasha KM, Mahmoud YA-G, Elsamahy T, Jiao H, Fu Y, Sun J (2022) A critical review on the treatment of dye-containing wastewater: ecotoxicological and health concerns of textile dyes and possible remediation approaches for environmental safety. Ecotoxicol Environ Saf 231:113160. https://doi.org/10.1016/j.ecoenv.2021.113160

Annu A, Mittal M, Tripathi S, Shin DK (2024) Biopolymeric nanocomposites for wastewater remediation: an overview on recent progress and challenges. Polymers (Basel) 16(2):294. https://doi.org/10.3390/polym16020294

Ascherio A, Chen H, Weisskopf MG, O'Reilly E, McCullough ML, Calle EE et al (2006) Pesticide exposure and risk for Parkinson's disease. Ann Neurol 60(2):197–203

Asghari M, Salahshoori I, Salmani S, Namayandeh Jorabchi M, Moghaddam A, Ali Khonakdar H (2024) A comprehensive study on transport behaviour and physicochemical characteristics of PU/based 3-phase mixed matrix membranes: effect of [HNMP][HSO_4] ionic liquid and ZnO nanoparticles. Sep Purif Technol 335:126099. https://doi.org/10.1016/j.seppur.2023.126099

Awad AM, Shaikh SMR, Jalab R, Gulied MH, Nasser MS, Benamor A, Adham S (2019) Adsorption of organic pollutants by natural and modified clays: a comprehensive review. Sep Purif Technol 228:115719. https://doi.org/10.1016/j.seppur.2019.115719

Babu Ponnusami A, Sinha S, Ashokan H, V Paul M, Hariharan SP, Arun J, Gopinath KP, Hoang Le Q, Pugazhendhi A (2023) Advanced oxidation process (AOP) combined biological process for wastewater treatment: a review on advancements, feasibility and practicability of combined techniques. Environ Res 237:116944. https://doi.org/10.1016/j.envres.2023.116944

Chen P-Y, Wang B, Chen W-G, Zhuang S, Chen Y, Liu H-J (2022) Carbon-dot-modified polyacrylonitrile fiber as recyclable adsorbent for removing anionic, cationic, and zwitterionic perfluorooctane sulfonates from water. Colloids Surf A Physicochem Eng Asp 651:129763

Dananjaya SAV, Priyanka UMS, Somarathna YR, Karunanayake L, Siriwardena S (2022a) A comparative study on mica waste-filled natural rubber foam composites made out of creamed and centrifuged latex. J Vinyl Add Tech 28(4):907–919. https://doi.org/10.1002/vnl.21937

Dananjaya SAV, Somarathna YR, Karunanayake L, Siriwardena S (2022b) Waste mica as filler for natural rubber latex foam composites. J Polym Res 29(3):71. https://doi.org/10.1007/s10965-022-02930-w

Dananjaya SAV, Somarathna YR, Karunanayake L, Siriwardena S (2022c) Physical properties of natural rubber latex foams produced with processed mica waste powder and creamed natural rubber latex. J Rubber Res Inst Sri Lanka 102(1):1–10. https://doi.org/10.4038/jrrisl.v102i1.1908

Dananjaya V, Somarathna Y, Siriwardena S, Sirimuthu N, Karunanayake L, Abeykoon C (2024) Effects of latex type and processed-mica waste loading on the structural and thermo-physical properties of natural rubber latex foam composites. Int J Lightweight Mater Manuf 7(3):450–466. https://doi.org/10.1016/j.ijlmm.2023.12.002

Dehghani MH, Ahmadi S, Ghosh S, Othmani A, Osagie C, Meskini M, AlKafaas SS, Malloum A, Khanday WA, Jacob AO, Gökkuş Ö, Oroke A, Martins Chineme O, Karri RR, Lima EC (2023)

Recent advances on sustainable adsorbents for the remediation of noxious pollutants from water and wastewater: a critical review. Arab J Chem 16(12):105303. https://doi.org/10.1016/j.arabjc.2023.105303

Deymeh F, Ahmadpour A, Allahresani A, Arami-Niya A (2024) Collaborative adsorption and photocatalytic degradation of high concentration pharmaceutical pollutants in water using a novel dendritic fibrous nano-silica modified with chitosan and UiO-66. Int J Biol Macromol 275:133534. https://doi.org/10.1016/j.ijbiomac.2024.133534

Dhamaniya S, Gupta V, Kakatkar R (2018) Recent advances in biodegradable polymers. J Res Updates Polym Sci 7(2). https://doi.org/10.6000/1929-5995.2018.07.02.3

Díez-Pascual AM (2020) Carbon-based polymer nanocomposites for high-performance applications. Polymers (Basel) 12(4):872. https://doi.org/10.3390/polym12040872

Eriksson M, Hardell L, Carlberg M, Åkerman M (2008) Pesticide exposure as risk factor for non-Hodgkin lymphoma including histopathological subgroup analysis. Int J Cancer 123(7):1657–1663. https://doi.org/10.1002/ijc.23589

Ewis D, Ba-Abbad MM, Benamor A, El-Naas MH (2002) Adsorption of organic water pollutants by clays and clay minerals composites: a comprehensive review. Appl Clay Sci 229:106686

Ghosh K, Jones BH (2021) Roadmap to biodegradable plastics—current state and research needs. ACS Sustain Chem Eng 9(18):6170–6187. https://doi.org/10.1021/acssuschemeng.1c00801

Goulson D, Nicholls E, Botías C, Rotheray EL (2015) Bee declines driven by combined stress from parasites, pesticides, and lack of flowers. Science (1979) 347(6229). https://doi.org/10.1126/science.1255957

Hussain E, Ahtesham A, Shahadat M, Ibrahim MNM, Ismail S (2024) Recent advances of Clay/polymer-based nanomaterials for the treatment of environmental contaminants in wastewater: a review. J Environ Chem Eng 12(3):112401. https://doi.org/10.1016/j.jece.2024.112401

Joseph Otorkpa O, Oche Otorkpa C, Shnada Auta H, Anyalewachi Ukah A (2024) Pattern and impact of pesticide poisoning: a review of published case reports. Texila Int J Public Health 12(2):187–200. https://doi.org/10.21522/TIJPH.2013.12.02.Art016

Keskin G, Kızıl G, Bechelany M, Pochat-Bohatier C, Öner M (2017) Potential of polyhydroxyalkanoate (PHA) polymers family as substitutes of petroleum based polymers for packaging applications and solutions brought by their composites to form barrier materials. Pure Appl Chem 89(12):1841–1848. https://doi.org/10.1515/pac-2017-0401

Koolivand M, Shokouhi M, Esfandyari M, Koolivand salooki M, Sadeghi M (2024) A review of fabrication methods for biodegradable pH-Responsive nanocomposite microgels and their performance in enhanced oil recovery. S Afr J Chem Eng 50:340–352. https://doi.org/10.1016/j.sajce.2024.09.004

Larson S, Newman JK, O Connor G, Griggs C, Martin A, Nijak G, Lord E, Duggar R, Leeson A (2013) Biopolymer as an alternative to petroleum-based polymers to control soil erosion: Iowa army ammunition plant

Lavin KS, Hageman KJ (2013) Contributions of long-range and regional atmospheric transport on pesticide concentrations along a transect crossing a mountain divide. Environ Sci Technol:130109121249000. https://doi.org/10.1021/es304497e

Li H, Dong Z, Weng Q, Chang C, Liu B (2015) Emerging pollutants—part I: occurrence, fate and transport. Water Environ Res 87(10):1849–1872. https://doi.org/10.2175/106143015X14338845156425

Li H, Jiao Y, Li L, Jiao X (2023) Research progress and trend of effects of organophosphorus pesticides on aquatic organisms in the past decade. Comp Biochem Physiol C: Toxicol Pharmacol 271:109673. https://doi.org/10.1016/j.cbpc.2023.109673

Majumder S, Chatterjee S, Basnet P, Mukherjee J (2020) ZnO based nanomaterials for photocatalytic degradation of aqueous pharmaceutical waste solutions—a contemporary review. Environ Nanotechnol Monit Manag 14:100386. https://doi.org/10.1016/j.enmm.2020.100386

Mayet A, Hijji M, Saleh E, Reza A, Kadhim S, Abdullaev S, Alsalamy A, Hassan Z, Gomez C, Tene T (2023) The role of biocomposites and nanocomposites in eliminating organic contaminants from effluents. Water (Basel) 15(17):3093. https://doi.org/10.3390/w15173093

Panigrahy SK, Nandha A, Chaturvedi M, Mishra PK (2024) Novel nanocomposites with advanced materials and their role in waste water treatment. Next Sustain 4:100042. https://doi.org/10.1016/j.nxsust.2024.100042

Post W, Kuijpers LJ, Zijlstra M, van der Zee M, Molenveld K (2021) Effect of mineral fillers on the mechanical properties of commercially available biodegradable polymers. Polymers (Basel) 13(3):394. https://doi.org/10.3390/polym13030394

Salahshoori I, Mahdavi S, Moradi Z, Otadi M, Zare Kazemabadi F, Nobre MAL, Ali Khonakdar H, Baghban A, Wang Q, Mohammadi AH (2024) Advancements in molecular simulation for understanding pharmaceutical pollutant adsorption: a state-of-the-art review. J Mol Liq 410:125513. https://doi.org/10.1016/j.molliq.2024.125513

Swetha TA, Ananthi V, Bora A, Sengottuvelan N, Ponnuchamy K, Muthusamy G, Arun A (2023) A review on biodegradable polylactic acid (PLA) production from fermentative food waste—its applications and degradation. Int J Biol Macromol 234:123703. https://doi.org/10.1016/j.ijbiomac.2023.123703

Syafrudin M, Kristanti RA, Yuniarto A, Hadibarata T, Rhee J, Al-onazi WA, Algarni TS, Almarri AH, Al-Mohaimeed AM (2021) Pesticides in drinking water—a review. Int J Environ Res Public Health 18(2):468. https://doi.org/10.3390/ijerph18020468

Tan Z, Yi Y, Wang H, Zhou W, Yang Y, Wang C (2016) Physical and degradable properties of mulching films prepared from natural fibers and biodegradable polymers. Appl Sci 6(5):147. https://doi.org/10.3390/app6050147

Thu VT, Trieu MH, An NHT, Dat NT, Linh ND, Manh NB (2023) Mussel–inspired biosorbent combined with graphene oxide for removal of organic pollutants from aqueous solutions. Ecotoxicol Environ Saf 255:114793. https://doi.org/10.1016/j.ecoenv.2023.114793

Van der Schueren L, De Meyer T, Steyaert I, Ceylan Ö, Hemelsoet K, Van Speybroeck V, De Clerck K (2013) Polycaprolactone and polycaprolactone/chitosan nanofibres functionalised with the pH-sensitive dye Nitrazine Yellow. Carbohydr Polym 91(1):284–293. https://doi.org/10.1016/j.carbpol.2012.08.003

Varghese RT, Cherian RM, Antony T, Tharayil A, Das H, Kargarzadeh H, Chirayil CJ, Thomas S (2022) A review on the best bioadsorbent membrane- nanocellulose for effective removal of pollutants from aqueous solutions. Carbohyd Polym Technol Appl 3:100209. https://doi.org/10.1016/j.carpta.2022.100209

Veclani D, Tolazzi M, Melchior A (2020) Molecular interpretation of pharmaceuticals' adsorption on carbon nanomaterials: theory meets experiments. Processes 8(6):642. https://doi.org/10.3390/pr8060642

Zhang M, Liang Y, Song A, Yu B, Zeng X, Chen M-S, Yin H, Zhang X, Sun B, Fan F (2017) Loss of soil microbial diversity may increase insecticide uptake by crop. Agric Ecosyst Environ 240:84–91. https://doi.org/10.1016/j.agee.2017.02.010

Zhang X, Yin Z, Xiang S, Yan H, Tian H (2024) Degradation of polymer materials in the environment and its impact on the health of experimental animals: a review. Polymers (Basel) 16(19):2807. https://doi.org/10.3390/polym16192807

Zhou W, Li M, Achal V (2025) A comprehensive review on environmental and human health impacts of chemical pesticide usage. Emerg Contam 11(1):100410. https://doi.org/10.1016/j.emcon.2024.100410

Nanocomposites-Enabled Membrane Technologies for Water Purification

Pappu Kumar

Abstract The increasing demand for clean water, compounded by pollution and scarcity, underscores the need for advanced purification technologies. Membrane-based separation methods are widely adopted for their scalability and pollutant removal efficiency, but traditional membranes face challenges like fouling, low permeability, and limited selectivity. Recent advancements in nanocomposite-enabled membranes, incorporating nanomaterials such as metal oxide nanoparticles, carbon-based materials, and metal–organic frameworks (MOFs), offer promising solutions. This chapter highlights advancements in nanocomposite membranes, focusing on improvements in water flux, fouling resistance, and pollutant selectivity. Methods for embedding nanomaterials into polymer matrices are detailed, highlighting their impact on membrane characteristics. Quantitative findings show that incorporating metal oxide nanoparticles can enhance water flux by 50–80% and fouling resistance by up to 60%, while MOFs improve selectivity for heavy metals by 70–90%. The environmental and health implications of using nanocomposites are critically evaluated, addressing concerns about nanomaterial leaching and toxicity. Challenges such as large-scale production, cost-efficiency, and long-term stability are also discussed, with a focus on scalable fabrication techniques like phase inversion and electrospinning. In conclusion, nanocomposite-enabled membranes represent a transformative approach to water purification. While promising, further research is needed to optimize their sustainability, economic feasibility, and long-term performance, paving the way for more effective water treatment technologies.

Keywords Water purification · Membrane · Nanoparticles · Metal–organic frameworks · Stability

---

P. Kumar (✉)
Department of Civil Engineering, Sandip University, Madhubani, Bihar, India
e-mail: pappukumar.ce16@nitp.ac.in

S. Roy et al. (eds.), *Nanocomposites for Sustainable Wastewater Treatment*, Composites Science and Technology, https://doi.org/10.1007/978-981-95-1369-7_5

1 Introduction

Water scarcity is an increasing global challenge, with nearly four billion people, two-thirds of the world's population, experiencing severe water shortages for at least one month annually, and over two billion people lacking access to adequate water supplies. By 2029, half of the global population is expected to live in water-stressed areas, and by 2040, one in four children could face extreme water scarcity. Population expansion is the primary cause of this problem, urbanization, and climate change place immense pressure on water resources, with global demand projected to rise by 20–25% by 2050. The Middle East and North Africa are already under extreme stress, consuming over 80% of renewable water resources, while Sub-Saharan Africa anticipates a 163% increase in demand by 2050, underscoring the urgent need for sustainable water management strategies (Xiao et al. 2023). The World Health Organization projects that half of all people on Earth will reside in water-stressed areas by 2025 (Kausar 2024). Water is a critical resource for human survival, and as the world population grows, the demand for clean and safe water also increases. Traditional water purification techniques such as coagulation, flocculation, and conventional filtration have limitations in terms of efficiency, operational costs, and the ability to remove specific contaminants. Traditional water purification techniques, such as coagulation, flocculation, and conventional filtration, face significant limitations, including suboptimal efficiency, high operational costs, and restricted ability to remove emerging contaminants such as pharmaceuticals, pesticides, and microplastics. While these methods have been the cornerstone of water treatment for decades, they struggle to meet the demands of modern water quality standards and the increasingly complex nature of water pollution. Recent advancements in membrane technologies, particularly the development of nanocomposite-enhanced membranes, provide innovative solutions by improving contaminant removal efficiency, reducing fouling, and offering scalability. However, despite these advancements, there is a notable gap in understanding their long-term performance, cost-effectiveness at scale, and the environmental implications of nanocomposite materials, highlighting the need for further research and optimization.

Nanocomposite-enabled membrane technologies involve embedding nanoparticles into polymeric or ceramic membranes to enhance their functionality. Figure 1 shows the nanoclays as unique two-dimensional inorganic nanostructures. Nanoclays, due to their layered silicate structure, offer a high surface area and intercalation potential, making them promising materials in various applications, such as drug delivery, catalysis, and environmental remediation (Wu et al. 2023). These nanocomposites are capable of improving the permeability, selectivity, and durability of membranes, making them ideal for applications in water treatment.

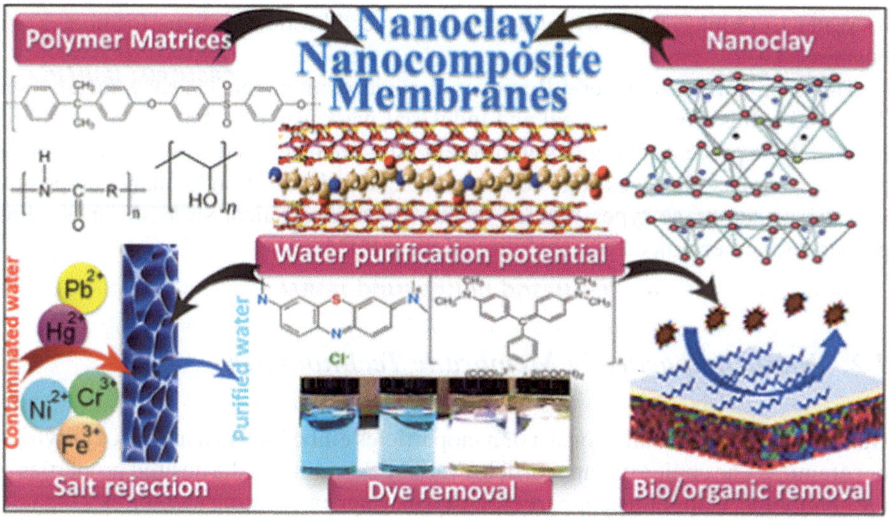

Fig. 1 Nanoclays, exclusive inorganic two-dimensional nanostructures (Ng et al. 2020)

1.1 Fundamentals of Membrane Technologies

Membrane technology is based on the principle of selective separation, where a membrane acts as a physical barrier allowing certain substances to pass while rejecting others. There are various types of membrane processes used in water treatment, including (Xiao et al. 2023).

Ultrafiltration (UF)—A membrane filtration technique called ultrafiltration (UF) is used to extract high molecular weight particles and solutes from liquids. Water and smaller dissolved compounds can pass through, but suspended particles, bacteria, viruses, and macromolecules are retained by the process's use of a semi-permeable membrane with extremely tiny pores, usually between 0.01 and 0.1 microns.

Nanofiltration (NF)—Using a semi-permeable membrane with pore diameters usually between 0.001 and 0.01 microns, nanofiltration (NF) is a membrane filtration technique that eliminates impurities from liquids. Because it fills the gap between reverse osmosis (RO) and ultrafiltration (UF), it can be used to remove organic molecules, certain dissolved salts, and divalent and larger monovalent ions.

Reverse Osmosis (RO)—By pushing water through a semi-permeable membrane at high pressure, reverse osmosis (RO), a very successful water purification method, eliminates dissolved salts, pollutants, and impurities. Desalination, wastewater treatment, and the production of high-purity water for domestic, commercial, and medicinal use are among its many applications. Desalination: produces drinkable water from brackish or saltwater, which is crucial for areas with scarce freshwater supplies. Industrial Procedures: produces ultra-pure water for use in power plants, medicines, and electronics manufacturing.

Water to Drink—Extensively utilised in water filtration systems for homes and businesses.

Treatment of Wastewater—Removes organic contaminants, heavy metals, and dissolved solids from municipal and industrial wastewater so that it can be reused. Each of these processes operates at different pressure levels and is suited to removing particles of specific sizes and contaminants. The integration of nanocomposites into these membrane types has led to improved mechanical strength, antifouling properties, and enhanced permeability (Kausar 2024).

1.2 Nanocomposites in Membrane Technologies

Nanocomposites typically consist of nanoparticles embedded in a matrix of polymers or ceramics. The choice of nanoparticles varies depending on the desired properties of the membrane. Common nanoparticles used in water purification membranes include: (1) Metallic nanoparticles (silver, copper): Known for their antimicrobial properties, these nanoparticles prevent biofouling, which is a common issue in membrane systems. (2) Carbon-based nanoparticles (graphene oxide, carbon nanotubes): These nanoparticles improve membrane hydrophilicity, leading to increased water permeability while maintaining selectivity (Dang et al. 2022). (3) Zeolites and silica nanoparticles: These enhance ion exchange and adsorption capacities, improving the removal of heavy metals and other contaminants.

1.2.1 Graphene-Based Nanocomposite Membranes

Graphene oxide (GO) and reduced graphene oxide (rGO) have shown significant promise in enhancing membrane performance. GO membranes exhibit high water flux and selective permeability due to their unique two-dimensional structure. In addition, GO membranes demonstrate strong antimicrobial activity, which is crucial for preventing biofouling (Ng et al. 2020). Graphene is a nanocomposite that is used for filtration purposes.

1.2.2 Metal Oxide Nanocomposites

Membranes integrated with metal oxide nanoparticles such as titanium dioxide (TiO_2) have gained attention due to their photocatalytic properties (Harito et al. 2019). TiO_2-based membranes can degrade organic contaminants under UV light, making them suitable for applications in areas where both purification and disinfection are required (Wu et al. 2023). It is a metal oxide of a nanocomposite material.

2 Methodology

2.1 Material Selection and Nanocomposite Design

Nanomaterial is based on desired properties such as antimicrobial activity, adsorption capacity, and catalytic ability. Common nanomaterials include silver nanoparticles, carbon nanotubes (CNTs), titanium dioxide (TiO_2), graphene oxide, and metal–organic frameworks (MOFs). Select a suitable polymer matrix to form the membrane base. Examples include polysulfone (PSf), polyether sulfone (PES), polyvinylidene fluoride (PVDF), and polyacrylonitrile (PAN). Incorporate the chosen nanoparticles into the polymer matrix through blending, coating, or in situ synthesis to form the nanocomposite. The incorporation method affects the distribution and performance of the nanomaterials in the membrane. The fabrication of nanocomposite membranes begins with selecting nanomaterials based on desired functionalities such as antimicrobial activity (silver nanoparticles, zinc oxide), adsorption capacity (graphene oxide, CNTs), or catalytic ability (TiO_2, MOFs). These nanomaterials are incorporated into polymer matrices like polysulfone (PSf), polyether sulfone (PES), polyvinylidene fluoride (PVDF), or polyacrylonitrile (PAN) through methods such as blending (using ultrasonic dispersion at 30 kHz for 1 h), dip-coating (10–15 dips with 30-s intervals), or in-situ synthesis (nanoparticle formation at 80 °C for 2 h). The casting solution, prepared by dissolving 15% by weight polymer in solvents like DMF, is processed through phase inversion and dried at 80 °C for 24 h to ensure membrane stability. Quality control measures include evaluating water flux, solute rejection, and fouling resistance, supported by structural analyses like SEM for nanoparticle distribution, pore size determination, and mechanical testing for durability. These steps ensure the development of high-performance, reproducible nanocomposite membranes. High-performance thermoplastic polysulfone (PSf) is renowned for its superior mechanical strength, chemical resistance, and thermal stability. Its chemical structure contains a sulfone group, which identifies it as a member of the sulfone polymer family. Because of its special qualities, PSf is widely employed in many different industries. A member of the fluoropolymer family, polyvinylidene fluoride (PVDF) is a thermoplastic polymer with exceptional performance and versatility. It is well known for having remarkable qualities that make it appropriate for a wide range of applications in different sectors (Yao 2018).

Figure 2 illustrates the workflow of nanocomposite-enabled membrane technology for water purification, beginning with the selection of nanomaterials such as silver nanoparticles, graphene oxide, or titanium dioxide, chosen for their antimicrobial, adsorptive, or catalytic properties. These materials are integrated into polymer matrices like poly sulfone (PSf), polyether sulfone (PES), or polyvinylidene fluoride (PVDF) using methods such as blending (via ultrasonication for uniform dispersion), dip-coating, or in-situ synthesis (forming nanoparticles directly within the polymer). The polymer-nanomaterial mixture is then cast and processed using phase inversion to form porous membranes, followed by drying and post-treatment to enhance mechanical and chemical stability. The final membranes are evaluated through performance testing for water flux, solute rejection, fouling resistance, and structural analyses, ensuring efficient contaminant removal and long-term operational reliability. This systematic approach demonstrates how nanocomposite technology addresses limitations in conventional water purification methods by improving membrane functionality and scalability.

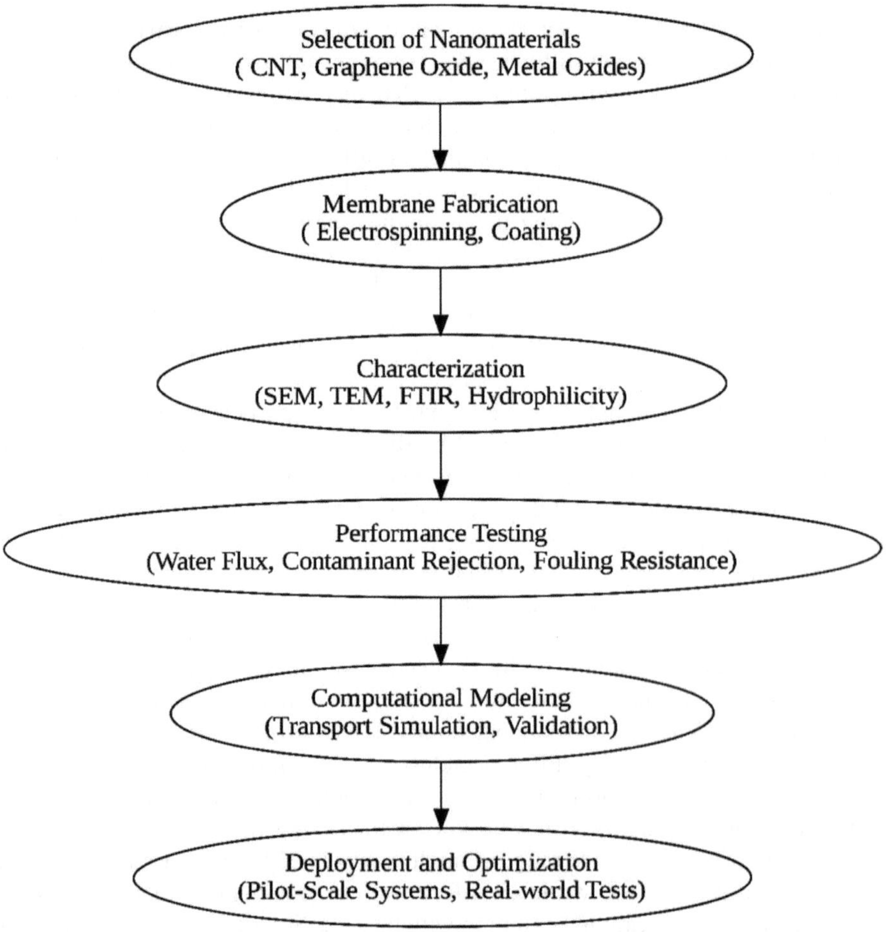

Fig. 2 Methodology of nanocomposite-enabled member technology for water purification

2.2 Membrane Fabrication Techniques

The most common technique, where a polymer solution is cast on a support, then submerged in a non-solvent bath, leading to the formation of the membrane by phase separation. Used to create fibrous nanocomposite membranes with high porosity and surface area. Interfacial Polymerization: Often used for thin-film nanocomposite (TFN) membranes, where two reactive monomers form a thin film on the membrane surface, embedding nanomaterials. Layers of nanoparticles and polymers are applied alternately to build up the membrane structure, allowing for precise control over membrane properties (Moses and Basu 2022). A sophisticated family of membranes called thin-film nanocomposite (TFN) membranes was created to improve performance and get around the drawbacks of conventional thin-film composite (TFC) membranes. Nanomaterials are included in the polyamide (PA) selective layer or other membrane structural elements in TFN membranes. TFN membranes are perfect for a variety of separation processes, such as desalination, wastewater treatment, and gas separation, because of the exceptional qualities that these nanoparticles offer, such as enhanced permeability, selectivity, antifouling resistance, and mechanical strength.

2.3 Characterization of Nanocomposite Membranes

Morphological Analysis: Techniques like Scanning Electron Microscopy (SEM) and Atomic Force Microscopy (AFM) provide information on surface morphology, pore size, and thickness. X-ray Diffraction (XRD) and Fourier Transform Infrared Spectroscopy (FTIR) assess the structural composition of the nanocomposite. Thermal Gravimetric Analysis (TGA) and tensile testing help evaluate the thermal stability and mechanical properties of the membrane. An analytical method called thermogravimetric analysis (TGA) is used to quantify how a material's mass varies with temperature or time. By heating a sample in a controlled environment, researchers can examine its composition, thermal stability, and decomposition behaviour.

2.4 Performance Evaluation in Water Purification

Permeability and Selectivity Testing: Measure water flux and rejection rates for contaminants like salts, heavy metals, and organic compounds to determine the efficiency of the membrane. Evaluate the membrane's resistance to biofouling, scaling, and organic fouling by introducing foulants and monitoring performance degradation (Ng et al. 2013). Microbial tests assess the membranes with antimicrobial nanomaterials, silver, and copper to inhibit bacterial growth. Nanostructured materials that restrict or eliminate microorganisms like bacteria, viruses, and fungi are known as antimicrobial nanomaterials. They are very effective at fighting microbial diseases because of their nanoscale dimensions, which improve their physical, chemical, and biological characteristics. The usage of these materials in consumer goods, food preservation, water treatment, and health care is growing.

3 Application-Specific Optimization and Testing

Desalination: Evaluate salt rejection efficiency and energy consumption for reverse osmosis (RO) membranes. Test adsorption capacity and degradation rates for nanocomposites with photocatalytic properties (TiO_2-based membranes). Determine selectivity for metals such as arsenic, lead, and mercury using adsorption or catalytic reduction properties. Materials' adsorption and catalytic reduction qualities are essential for their use in material science, chemical reactions, and environmental remediation. An outline of these attributes and their importance is provided below:

- **Properties of Adsorption**

Adsorption is the process by which molecules (adsorbates) stick to an adsorbent's surface. The surface properties of the material, including area, porosity, and chemical functionality, control this feature.

3.1 Scale-Up and Real-World Application Testing

Pilot Testing: Run the membrane technology on a larger scale to test performance under real-world conditions. Analyse the cost-effectiveness and sustainability of the nanocomposite membrane, focusing on energy consumption, material longevity, and potential environmental impacts (Liu et al. 2018). The durability and lifespan of a material under a variety of circumstances, such as environmental, chemical, and physical stressors, is referred to as material longevity. It plays a crucial role in establishing the material's dependability, sustainability, and affordability in a variety of applications, from manufacturing and construction to consumer goods and medical equipment.

3.2 Applications in Water Purification

Nanocomposite membranes are increasingly used in desalination processes, particularly in reverse osmosis systems (Wu et al. 2024). Incorporating nanomaterials into RO membranes enhances water flux, reduces fouling, and lowers energy requirements for desalination. Because of its remarkable qualities, unusual structure, and numerous possibilities, graphene has sparked a lot of attention. Graphene-based materials have a lot of potential for use as adsorbents in decontaminating water due to its enormous surface area, variety of functions, ease of manufacture, and inexpensive treatment costs (Ahmaruzzaman et al. 2024).

3.3 Removal of Organic Contaminants

Nanocomposites, such as those using metal oxide nanoparticles, have proven effective in removing organic pollutants, including pesticides, pharmaceuticals, and industrial chemicals. The photocatalytic properties of certain nanoparticles enable the breakdown of complex organic molecules into harmless by-products (Murugesan and Sundaram 2020).

3.4 Heavy Metal Removal

Nanocomposite membranes are highly effective in removing heavy metals such as lead, arsenic, and mercury from water (Ng and Carbery 2020). The high surface area and adsorption capacity of nanoparticles make them ideal for capturing and isolating heavy metal ions, ensuring cleaner water.

3.5 Challenges and Future Prospects

Despite the numerous advantages of nanocomposite-enabled membrane technologies, there are challenges that need to be addressed. These include: The production of nanocomposites at a commercial scale can be expensive, and large-scale manufacturing processes need optimization. The long-term environmental impact of nanoparticles released into water systems must be carefully evaluated to prevent unintended contamination. While nanocomposites enhance the properties of membranes, their long-term durability and resistance to fouling still require further research. The future of nanocomposite-enabled membranes is promising, with ongoing research focused on developing more cost-effective, efficient, and environmentally friendly materials. Innovations such as self-cleaning membranes and the integration of multifunctional nanoparticles are expected to drive the next generation of water purification technologies.

4 Conclusion

Inorganic nanofiller membranes have greatly improved polymeric nanocomposite membranes for treating contaminated water. Nanocomposites have revolutionized the field of membrane technology, offering enhanced water purification capabilities. By integrating nanoparticles into conventional membranes, researchers have been able to improve permeability, selectivity, and fouling resistance. With continued research and development, nanocomposite-enabled membrane technologies are poised to play a crucial role in addressing global water challenges. The reusability and regeneration of nanocomposite membranes are vital for enhancing their economic viability and scalability. Prioritizing these factors during the design process ensures efficient use of resources, prolongs the functional lifespan of the membranes, and offers significant cost-saving opportunities, making them more practical and sustainable for long-term applications.

References

Ahmaruzzaman M, Roy S, Kumar L, Gupta K (2024) Magnetic nanocomposites as an emerging paradigm for mitigation of arsenic from aqueous sources. J Inorg Organomet Polym Mater. https://doi.org/10.1007/s10904-024-03422-8

Dang Y, Wang J, He J, Feng X, Tobin Z, Achola L, Zhao W, Wen L, Suib S (2022) Self-standing ruthenium oxide/nickel oxide nanocomposites enabled reliable and regeneratable seawater splitting. SSRN Electron J. https://doi.org/10.2139/ssrn.4111909

Harito C, Bavykin D, Yuliarto B, Dicks H (2019) Polymer nanocomposites having a high filler content: synthesis, structures, properties, and applications. Nanoscale. https://doi.org/10.1039/C9NR00117D

Kausar A (2024) Nanoclay reinforced thermoplastic polymeric nanocomposite membranes: State-of-the-art and progresses. Polymer-Plastics Technol Mater 63(13):1819–1841. https://doi.org/10.1080/25740881.2024.2360169

Liu P, Cottrill A, Kozawa D, Koman V, Pinar D (2018) Emerging trends in 2D nanotechnology that are redefining our understanding of "nanocomposites.". Nano Today. https://doi.org/10.1016/j.nantod.2017.12.004

Moses J, Basu M (2022) Mesoporous silk-bioactive glass nanocomposites as drug-eluting multifunctional conformal coatings for improving osseointegration and bactericidal properties. ACS Appl Mater Interfaces. https://doi.org/10.1021/acsami.2c00093

Murugesan S, Sundaram T (2020) Copolymer/clay nanocomposites for biomedical applications. Adv Func Mater. https://doi.org/10.1002/adfm.201908101

Ng K, Carbery WP (2020) Frenkel excitons in heat-stressed supramolecular nanocomposites enabled by tunable cage-like scaffolding. Nat Chem. https://doi.org/10.1038/s41557-020-00563-4

Ng K, Webster M, Carbery WP, Visaveliya N, Gaikwad P, Jang SJ, Kretzschmar I, Eisele DM (2020) Frenkel excitons in heat-stressed supramolecular nanocomposites enabled by tunable cage-like scaffolding. Nat Chem 12(12):1157–1164. https://doi.org/10.1038/s41557-020-00563-4

Ng L, Mohammad A, Leo C (2013) Polymeric membranes incorporated with metal/metal oxide nanoparticles: a comprehensive review. Desalination. https://doi.org/10.1016/j.desal.2010.07.020

Wu B, Meng Y, Peng C, Yang J, Xing S, Wu N, Yin C, Yang J, Bai S, Tiwari MK (2023) Porous bulk superhydrophobic nanocomposites for extreme environments. Matter 6(6):1992–2004. https://doi.org/10.1016/j.matt.2023.03.033

Wu D, Gao T, Zhang Q, Wang Y (2024) Simultaneous improvements in energy density and thermal conductivity of PAN polymer nanocomposites enabled by core-shell structure BZCT@BNNS nanofibers. Compos Sci Technol. https://doi.org/10.1016/j.cst.2024.01.002

Xiao R, Zhou G, Wen Y, Ye J, Li X, Wang X (2023) Recent advances on stimuli-responsive biopolymer-based nanocomposites for drug delivery. Compos B Eng 266:111018. https://doi.org/10.1016/j.compositesb.2023.111018

Yao N (2018) Nanocomposites for smart textiles. Wearable Nanotechnol. https://doi.org/10.1002/9781119460367

Nanocomposites for Heavy Metals Removal and Recovery

Deepak Kumar, Meenakshi Suhag, Vijay Kumar, and Vikas Kumar

Abstract The exponential growth in the world population, recent industrialization, and various agricultural and household activities lead to higher water pollution levels in terms of toxic metal ion contaminants. Toxic metals such as mercury, lead, cadmium, arsenic, and many more have poisonous properties, making the contamination of heavy metals dangerous for the environment and human health. Researchers have developed various treatment techniques and processes to remove these pollutants, such as chemical precipitation, ion exchange, flotation, adsorption, and chemical deposition. However, these traditional methods are associated with high capital and operational costs, lower efficiency, low saturation capacity, and the disposal of residual metals. To overcome these problems, nanocomposites have emerged as a versatile and promising solution for eliminating and recovering heavy metals from polluted water. With their unique properties, different nanocomposites, including metal oxide-based, carbon-based, and polymer-based composites, have been studied for their effectiveness in targeting specific heavy metals. This chapter reviews the application of nanocomposites for removing and recovering heavy metals from contaminated water in detail.

Keywords Heavy metals · Nanotechnology · Nanomaterials · Nanocomposites · Wastewater · Adsorption

Abbreviations

CQDs Carbon-based quantum dots

D. Kumar · M. Suhag (✉)
Institute of Environmental Studies, Kurukshetra University, Kurukshetra, Haryana, India
e-mail: msuhag@kuk.ac.in

V. Kumar
Department of Microbiology, Kurukshetra University, Kurukshetra, Haryana, India

V. Kumar
Department of Microbiology, International Medical School (IMS), UIB, Almaty, Kazakhstan

S. Roy et al. (eds.), *Nanocomposites for Sustainable Wastewater Treatment*, Composites Science and Technology, https://doi.org/10.1007/978-981-95-1369-7_6

GO	Graphene oxide
MWCNTs	Multi-walled carbon nanotubes
rGO	Reduced graphene oxide
SWCNTs	Single-walled carbon nanotubes

1 Introduction

Water is considered a fundamental source on our planet that influences virtually every aspect of human existence. Population growth and increased human activity have increased water demand. Two billion people, or 26% of the global population, do not have safe drinking water, while 3.6 billion people do not have access to regulated cleanliness (WHO 2021). The majority of India's river systems are currently significantly polluted. The quality of water resources has been degraded quickly due to contaminants ranging from organic compounds and dyes to inorganic heavy metals (Akhtar et al. 2021). Industrialization and human activities are the primary sources of these water contaminants. Among different pollutants, heavy metal contamination in water bodies is a substantial and consequential concern that adversely affects living organisms and human health (Abdullayev et al. 2024). Heavy metals such as cadmium, arsenic, lead, nickel, chromium, zinc, mercury, and cobalt are very toxic due to their carcinogenic effects (Bilal et al. 2021). Primary sources of these heavy metals are various industries like fertilizer, electroplating, painting, mining, batteries, etc. (Naseer 2024). Various conventional techniques such as coagulation-flocculation, ion exchange, reverse osmosis, photocatalytic, adsorption, and membrane filtration have been employed for a decade to treat wastewater (Teh et al. 2016; Yousaf et al. 2019; Shrestha et al. 2021; Fernandes et al. 2024).

However, these methods are associated with some drawbacks, like energy-intensive, low efficiency, residual toxicity, membrane fouling during wastewater treatment, etc. Among the various types of wastewater treatments existing globally, adsorption through nano-composite materials, such as metal nanocomposite, metal oxide nanocomposite, carbon nanocomposite, and polymer nanocomposite, has attracted much attention nowadays due to their inherent environmentally harmless and degradable properties. There has been a notable shift towards nanotechnology-based, more sustainable, and innovative approaches in recent years. They offer enhanced sorption capacity, profligate dissolution, and high reaction time, making them suitable for wastewater treatment (Smith, 2023)

2 Metal-Oxide-Based Nanocomposites

Metal oxide nanostructures, which react to various external stimuli such as light, electric fields, magnetic fields, and pressure, have attracted significant interest from academic researchers and industrial developers. These materials are considered promising candidates for the creation of efficient and environmentally sustainable devices aimed at remediating water contamination caused by organic compounds (de Jesus et al. 2022; Patnaik et al. 2023; Lu and Didier 2020; Shanaah et al. 2023).

At present, heterogeneous photocatalysis is a prominent technique for the removal of contaminants utilizing metal oxides. The effectiveness of this method is closely linked to the properties of metal oxide nanostructures, which include high chemical and structural stability, a significant surface area-to-volume ratio, enhanced surface interaction with contaminant molecules, and the potential for material reuse in subsequent decontamination cycles (Pereira et al. 2022). Furthermore, metal oxides can combine with other oxides and effective components, like carbon-based materials and conducting polymers, to create nanocomposites. Compared to the results for pure oxides, this interaction leads to a higher surface area-to-volume ratio, less electron–hole recombination, and improved adsorption, photocatalysis, and sensing performance (Pereira et al. 2022; Jamjoum et al. 2021; Patil 2022). These nanostructured materials are now considered vital research topics, facilitating a diverse array of applications.

Recent research has illuminated promising strategies for removing heavy metal contaminants from water through innovative material technologies. Metal oxides and elemental metals have emerged as potential solutions to water purification challenges, though traditional approaches face significant practical limitations. Specifically, the use of nanomaterials has opened new frontiers in environmental remediation. Zero-valent nano-iron (Fe^0) has garnered particular attention among researchers for its multifaceted capabilities in water treatment. This nanomaterial distinguishes itself through several remarkable properties: an expansive surface area, potent reducing capabilities, minimal toxicity, and robust structural stability. Scientific investigations have consistently demonstrated its exceptional performance in extracting toxic metallic substances from aqueous environments. The primary advantage of nano-iron lies in its sophisticated mechanism of contaminant removal. Unlike conventional methods that struggle with material separation and potential environmental redeposition, Fe^0 nanoparticles offer a more sophisticated approach to water purification. Their unique chemical characteristics enable them to not just adsorb but actively transform harmful metal ions, presenting a comprehensive solution to water quality challenges. Researchers continue to explore and refine the application of these nanomaterials, recognizing their potential to address critical environmental remediation needs. The ongoing development of such advanced technological solutions represents a promising pathway toward more effective and sustainable water treatment strategies. For instance, Kanel et al. (2006) reported that Fe^0 was synthesized to achieve a 99% removal rate of pentavalent arsenic from water. In another study, Xiao et al. (2017) demonstrated that FeO, derived from ferric chloride using *Syzygy jambos*, effectively scavenged hexavalent chromium with an adsorption capacity of 983.3 mg/g. Furthermore, the stability of zero-valent nano iron can be enhanced through the incorporation of various stabilizing agents. A notable example is the biodegradable, non-toxic chitosan carboxymethyl β-cyclodextrin complex, which has been utilized to encapsulate Fe0 for the elimination of divalent copper and hexavalent chromium ions. The removal mechanism involves the reduction of Cu^{2+} to Cu^0 and Cr^{6+} to Cr^{3+}, while Fe^0 is oxidized to Fe^{3+} (Sikder et al. 2014). Additionally, the elimination of heavy metals has been achieved using bimetallic nanoparticles, such as kaolinite-embedded Fe/Ni nanoparticles, which exhibited a remarkable 99.8% efficiency in removing divalent copper ions (Cai et al. 2014).

In addition, metal oxides have been shown to effectively treat water contaminated with heavy metal pollutants. Metal oxide nanoparticles present promising solutions for water treatment, particularly in removing heavy metal contaminants. Various non-magnetic metal oxide nanomaterials, including oxides of manganese, copper, iron, aluminum, cerium, and zinc, have shown remarkable potential for environmental remediation. As an illustrative example, researchers have developed advanced nanostructured materials like manganese dioxide (MnO_2) nanofibers enhanced with polymeric coatings. By strategically layering these nanofibers with organic polymers such as polypyrrole and polyacrylonitrile, scientists have significantly improved their heavy metal capture capabilities. In one notable study, these engineered nanocomposites demonstrated exceptional performance in removing divalent lead ions, with an impressive adsorption capacity reaching 251.90 mg per gram of material (Luo et al. 2015). Similarly, CeO_2 nanofibers, treated with 3-mercaptopropyltrimethoxysilane and vinylpyrrolidone, exhibited a greater affinity for lead ions compared to copper ions during the removal of divalent lead and copper ions (Yari et al. 2015). Researchers developed a novel approach by coating zinc oxide (ZnO) nanofibers with polyvinyl alcohol to enhance metal ion adsorption. In their investigation, the composite material demonstrated varying adsorption capabilities for different metal ions, with uranium [U(VI)] showing the highest adsorption potential, while nickel [Ni (II)] exhibited the lowest uptake capacity (Hallaji et al. 2015). The selective adsorption performance was attributed to several key physicochemical properties of the metal ions, including:

- Metallic hydrated ionic radius
- Electronegativity
- Atomic mass.

These fundamental characteristics play a crucial role in determining the interaction between the metal ions and the nanofiber surface, ultimately influencing the adsorption efficiency of the modified ZnO nanofiber material. The specific variation in adsorption capacities highlights the complexity of ion-surface interactions and underscores the importance of material engineering in developing effective metal ion removal technologies.

Alkaline metal oxide NPs have been documented for their effectiveness in the removal of heavy metals. These nanoparticles are characterized by their eco-friendliness and lower toxicity. Among this category, magnesium oxide (MgO) NPs have been extensively researched for their capacity to eliminate heavy metals. MgO has been synthesized not only for the removal of toxic metals but also for the disinfection of bacteria. The developed nano adsorbent demonstrated the ability to inactivate Escherichia coli while simultaneously removing divalent lead and cadmium ions, as noted by Cai et al. (2017).

Alumina (Al_2O_3) serves as a significant metal oxide nano adsorbent for the removal of heavy metals. It exists in various structural forms, including χ, θ, γ, β, and α, which are naturally found in soils (Chu et al. 2019). The ordinary α-Al_2O_3 adsorbent exhibits high stability when utilized in orthodox approaches (Pham et al. 2019). Alumina possesses several advantageous properties, like excellent Insulation against electricity, high compression strength, water and corrosion obstruction, and efficient heat conductivity, along with robust interatomic bonding, making it an outstanding adsorbent (Hojamberdiev et al. 2018). The adsorption capacity of Al_2O_3 can be enhanced through the incorporation of metallic NPs. In a recent investigation, Kumari et al. (Kumari and Tripathi 2020) employed green-synthesized silver nanoparticles stabilized on alumina nanoparticles to facilitate the removal of heavy metals from pharmaceutical seepages. Researchers developed an innovative green synthesis approach using bio-based reductants derived from Bacillus cereus cell extracts. In a groundbreaking study, this nanocomposite demonstrated exceptional environmental remediation capabilities, achieving remarkable metal removal efficiencies from pharmaceutical wastewater. Specifically, the process effectively eliminated 99.5% of lead and 98.44% of chromium contaminants (Kumari and Tripathi 2020). Additionally, researchers have explored surface modification techniques to enhance metal oxide adsorption capacities. One promising method involves utilizing surfactants like sodium dodecyl sulfate (SDS) to improve adsorbent performance. For instance, SDS-modified alumina surfaces showed significant potential in managing manganese contamination. By creating admicelles on the alumina surface, the modified nano-adsorbent enabled a sophisticated adsorption mechanism that successfully removed 92% of manganese (II) ions from both industrial and spiked wastewater samples (Khobragade and Pal 2016). Table 1 represents the recent studies on metal oxide nanocomposites used for the removal of heavy metals:

2.1 *Mechanism of Metal-Oxide Nanocomposites for the Removal of Heavy Metals*

Adsorption

Metal oxide nanocomposites (various forms of iron oxide-based nanocomposites like siderite, magnetite, limonite, hematite, zirconium-oxide, aluminum-oxide, etc.) have a high surface area and numerous active sites, making them excellent adsorbents for heavy metal ions (various forms of arsenic like As (III), As (V), etc.) (Nassar 2012; Gupta et al. 2021). The process involves the binding of heavy metal ions to the surface of the nanocomposites through physical or chemical interactions, primarily:

Table 1 Metal-oxide nanocomposites used for the removal of heavy metals

Nanocomposites	Heavy metal removal	Removal efficiency	Mechanism	References
NC@Co_3O_4	Pb	86.89%	Adsorption	El Mouden et al. (2023)
	Cd	82.06%		
Fe_3O_4 NPs	Pb, Cd, As	High (specific Value varies)	Ion-exchange, surface complexation	Bichave et al. (2023)
TiO_2	Cr	90%	Photocatalysis, Adsorption	Ani and Egbosiuba (2023)
ZnO–Fe_3O_4 NCs	Pb, Cd	High (reusable Via magnetic recovery	Adsorption, magnetic separation	Goyal et al. (2018)
Bio-functionalized Graphene-magnetite	Cr	222.2 mg/g	Adsorption, reduction	Ramalingam et al. (2018)

- **Surface Complexation**: Metal ions can form chemical bonds with surface functional groups (e.g., hydroxyl groups) on the nanocomposite.
- **Electrostatic Interactions**: Oppositely charged metal ions are attracted to the charged surface of the metal oxide.
- **Van der Waals Forces**: These weak intermolecular forces also contribute to adsorption, especially for larger metal ions.

Reduction and Oxidation Reactions

Some metal oxide nanocomposites can catalyze redox reactions, which reduce the solubility of heavy metals and facilitate their removal (Gupta et al. 2021).

- **Reduction**: For example, copper oxide (CuO) and iron oxide (Fe_3O_4) can reduce toxic metal ions like Cr^{6+} (chromium) into less toxic forms (e.g., Cr^{3+}), which then precipitate out of solution.
- **Oxidation**: Oxidizing agents like TiO_2 (titanium dioxide) can convert metal ions into less soluble forms, promoting precipitation.

Coagulation and Precipitation

Some metal oxide nanocomposites can promote the coagulation of metal ions into larger aggregates, which can then be removed by simple filtration or sedimentation (Gupta et al. 2021).

Surface Modification and Functionalization

Metal oxide nanocomposites can be functionalized to enhance their ability to remove specific metal ions. Common surface modifications include (Gupta et al. 2021):

- **Binding Sites**: The introduction of specific functional groups (e.g., amino, thiol, or carboxyl groups) enhances the affinity for heavy metal ions.

- **Composite Formation**: The combination of metal oxides with organic polymers or other nanoparticles can improve the adsorption capacity and selectivity for specific metals.

3 Carbon-Based Nanocomposites

Due to their unique physicochemical characteristics, different CNCs have been widely used in various industries in recent years, including the food industry, energy storage, pharmaceuticals, and wastewater treatment industries, as seen in Fig. 1. Various types of nanocomposites are widely used to remove the organic or inorganic contaminants from wastewater using conventional nano adsorbents (Barhoum et al. 2019). Due to their chemical and morphological characteristics, high adsorption potential, and inoffensiveness, different kinds of carbon-based nanocomposites, such as SWCNTs, MWCNTs, graphene, graphene oxide (GO), reduced graphene oxide (rGO), and fullerene are extensively used for eliminating impurities from wastewater in recent times (Madima et al. 2020; Singh et al. 2021; Gul et al. 2021).

The properties of CNCs, including ease of use, high thermal permanency, high surface area to volume ratio, and an adjustable surface functional group, are crucial for the accumulation of heavy metals and the surface chemistry of carbon particles (Xie et al. 2017). Table 2 represents the carbon-based nanomaterials used for heavy metal removal, and the main mechanism of heavy metal removal is shown in Fig. 2.

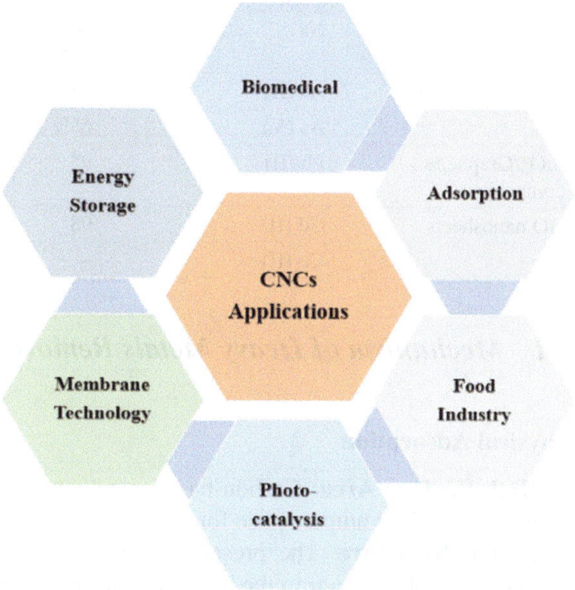

Fig. 1 Applications of carbon-based nanocomposites in different fields

Table 2 Carbon-based nanomaterials for the removal of heavy metals

Carbon-based nanomaterials	Heavy metal removed	Removal rate (% or mg/g)	References
Carbon dot-bentonite clay	Pb (II)	95%	Jlassi et al. (2023)
NC cobalt oxide	Pb (II)	55.24 mg/g	El Mouden et al. (2023)
	Cd (II)	52.91 mg/g	
GO/SiO_2	Cr (VI)	181.81 mg/g	Dan et al. (2023)
MWCNTs	As (V)	33.02 mg/g	Sadeghi et al. (2020)
	Hg (II)	155.61 mg/g	
Vanadium oxide/boron Co-doped carbon nitride(g-C_3N_4)	Diclofenac	100%	Oliveros et al. (2021)
Fe_3O_4 nanoribbon	Cr (VI)	134 mg/g	Song et al. (2022)
Chitosan and GO	U (VI)	320 mg/g	Hu et al. (2020)
Boron nitride NRs	Cd (II)	649 mg/g	Liu et al. (2019)
	Cu (II)	341 mg/g	
Amide CNT/ Polysulfone	Cr (VI)	94.2%	Shah and Murthy (2013)
	Cd (II)	78.2%	
Polypyrrole-rGO	Hg (II)	93.2%	Chandra and Kim (2011)
Magnetic chitosan/GO	Pb (II)	90.1%	Fan et al. (2013)
GO aerogels	Cu (II)	97%	Mi et al. (2012)
Functionalized	Na	66%	Mishra and Ramaprabhu (2011)
Graphene sheets	As (III)	54%	
	As (V)	65%	
SiO_2/Graphene composite	Pb (II)	95	Hao et al. (2012)
GO nanosheets	Cd (II)	98	Zhao et al. (2011)
	Co (II)	–	

3.1 *Mechanism of Heavy Metals Removal*

Physical Adsorption

- **High Surface Area**: Carbon-based nanocomposites have a large surface area, which provides ample space for heavy metal ions to absorb.
- **Porous Structure**: The presence of micropores and mesopores enhances the material's ability to trap the heavy metal ions via van der Waals forces.

Chemical Adsorption

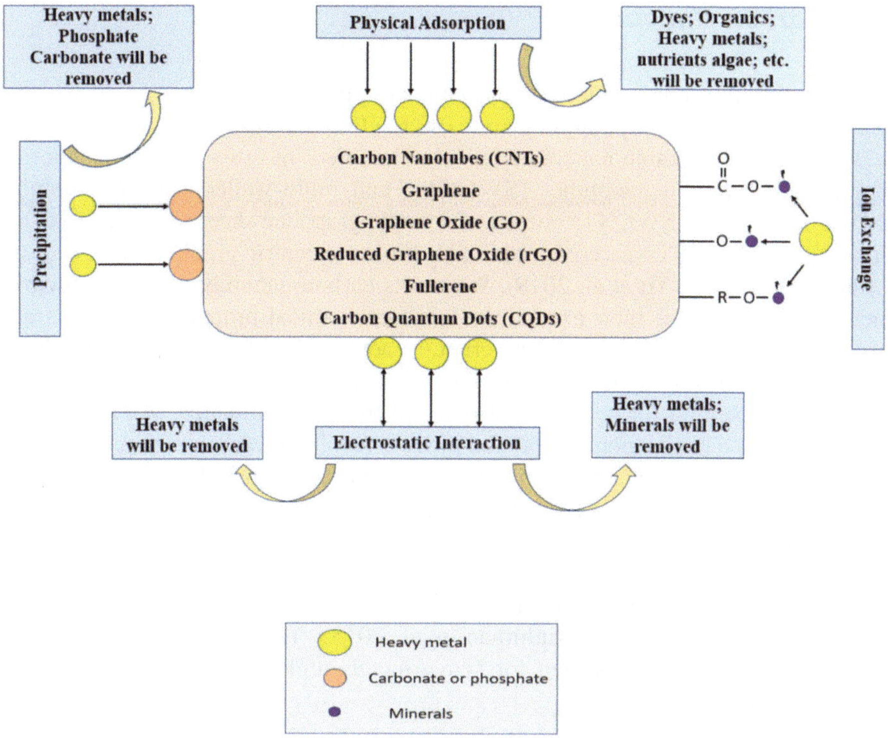

Fig. 2 Mechanism of carbon-based nanomaterials to remove heavy metals

- **Functional Groups**: Oxygen-containing functional groups on the surface of carbon materials can bind to heavy metal ions through ion exchange or complexation.
- **Electrostatic Interaction**: Depending on the pH, the surface charge of the carbon materials can attract positively charged heavy metal ions.
- **Surface Modification**: Incorporating other materials such as metal oxides or polymers introduces additional functional groups or enhances selectivity toward specific heavy metals (Li et al. 2015; Liu et al. 2022).

Ion-Exchange

- Carbon materials modified with ion-exchange resins or certain functional groups can replace loosely bound ions on the composite with heavy metal ions in solution (Liu et al. 2022).

Precipitation

- Carbon-based nanocomposites containing metal oxides can promote the precipitation of heavy metals by altering the pH or releasing hydroxide ions (Liu et al. 2022).

3.2 Carbon Nanotubes

Carbon nanotubes are distinctive nanoscale structures characterized by their cylindrical geometry, typically ranging from 100 to 1000 nm in length and 1 to 3 nm in width. These remarkable nanomaterials are classified into two primary categories: single-walled carbon nanotubes (SWCNTs) and multi-walled carbon nanotubes (MWCNTs). While SWCNTs consist of a single graphene sheet rolled into a cylindrical form, MWCNTs are composed of multiple concentric graphene layers nested within one another (Yu et al. 2018). What sets carbon nanotubes apart in environmental remediation is their exceptional physicochemical properties. Unlike traditional adsorbents like powdered or granular activated carbon, which suffer from limitations in surface area and adsorption efficiency, carbon nanotubes demonstrate remarkable potential for heavy metal removal from contaminated water (Santhosh et al. 2016). The heavy metal extraction process involving carbon nanotubes is intricate and involves multiple interaction mechanisms. These include ion binding, electrostatic attractions, and the formation of complex compounds between metal ions and the functional groups present on the carbon nanotube surface. The unique structural characteristics of carbon nanotubes, such as their precise pore dimensions and extraordinary surface area-to-volume ratio, contribute significantly to their superior adsorption capabilities (Ruthiraan et al. 2015). Table 3 shows the different CNTs-based nanomaterials used for heavy metals removal and their mechanisms are discussed further:

3.3 Mechanisms of CNTs for Removal of Heavy Metals

Adsorption

Adsorption is the primary mechanism for heavy metal removal by CNTs. It involves the interaction between heavy metal ions and the surface of CNTs. CNTs have a high surface area, porous structure, and functional groups, making them excellent adsorbents.

- **Physical Adsorption**: Van der Waals forces help trap heavy metals on the CNT surface.
- **Chemical Adsorption**: Functional groups such as hydroxyl (–OH), carboxyl (–COOH), and amine (–NH_2), either inherent or introduced during functionalization, form chemical bonds with metal ions (Li et al. 2015; Gupta et al. 2021).

Ion Exchange

- Functionalized CNTs with acidic or basic groups can facilitate ion exchange processes. Metal ions in solution are exchanged with hydrogen or other ions attached to the functional groups on CNTs (Gupta et al. 2021).

Table 3 CNT-based nanomaterials used for the removal of heavy metals

CNT-based materials	Heavy metal removal	% efficiency	References
Functionalized CNTs	Pb (II)	98	Akhter et al. (2023)
$MWCNTs_{5-15\ nm}$	Pb (II)	95	Egbosiuba et al. (2021)
	Ni (II)	96	
MWCNTs	Pb (II)	91.2 ± 8.7	Egbosiuba et al. (2022)
KOH@Ni nanoparticles	Cd (II)	80.6 ± 5.8	
	As (V)	88.5 ± 6.5	
Purified CNTs	As	99.99	Bankole et al. (2019)
	Cr	99.80	
	Cd	98.68	
	Ni	78.06	
	Cu	82.91	
	Fe	15.11	
	Zn	21.80	
	Pb	99.44	
Polyhydroxyl butyrate enacted CNTs	As	99.95	Bankole et al. (2019)
	Cr	98.19	
	Cd	99.34	
	Ni	77.95	
	Cu	83.08	
	Fe	15.92	
	Zn	18.34	
	Pb	98.85	
CNT-aided AC	Cr (VI)	72	Atieh (2011)

Electrostatic Interaction

- Positively charged metal ions (e.g., Cu^{2+}, Pb^{2+}) are attracted to negatively charged functionalized CNT surfaces.

Precipitation

- Under certain conditions, CNTs can induce precipitation of heavy metals from water by creating localized environments where metals form insoluble compounds, such as hydroxides or oxides.

3.4 Graphene

Because of advanced characteristics, i.e., high mechanical strength, thermal conductivity, and electrical properties, graphene and graphene-based composites have been widely used for the removal of heavy metals from wastewater has increased in recent years. Additionally, there are some advantages to using them as adsorption materials (Santhosh et al. 2016). Graphene's structural variants, including reduced graphene oxide (rGO) and graphene oxide (GO), have emerged as promising adsorbent materials in environmental remediation and metal ion capture technologies. Graphene oxide is characterized by its unique surface chemistry, featuring various oxygen-containing functional groups that significantly influence its adsorption capabilities. The molecular structure of graphene oxide includes diverse functional groups such as carbonyl and epoxy moieties, which create unique surface properties. When combined with activated carbon (AC), these materials demonstrate enhanced potential for metal ion removal from aqueous solutions. The effectiveness of these adsorbents stems from their complex surface chemistry and structural modifications. Researchers have discovered that strategic functionalization, particularly through EDTA coupling and the introduction of carboxyl (–COOH) and hydroxyl (–OH) groups, can dramatically improve the adsorption performance. These functional groups facilitate stronger electrostatic interactions with metal ions, increasing the material's capacity to capture and remove contaminants from water systems. Composite materials like GO/Fe_3O_4/LA/EDTA represent advanced approaches to maximizing adsorption efficiency, leveraging the unique chemical properties of graphene oxide and its derivatives to address environmental challenges related to metal ion contamination (Wadhawan et al. 2020). Further, these interactions are influenced by surface area and surface charge, i.e. greater the surface area, the greater the removal capacity (Rajabathar et al. 2017). Using pseudo-second-order kinetics, the magnetic dithiocarbamate functionalized reduced graphene oxide (rGO-PDTC/Fe_3O_4) nanocomposites effectively eliminated Cu^{2+}, Cd^{2+}, Pb^{2+}, and Hg^{2+} at a pH range of 5.0 to 6.0 (Singh et al. 2021). The remarkable removal of metal ions, including Se^{4+}, As^{5}, As^{3+}, Cu^{2+}, and Pb^{2+} from the water was achieved by a nanocomposite made from CeO_2 and graphene nanosheet (Baranik et al. 2018). On the other hand, 99%, 97%, and 98% of Pb^{2+}, Hg^{2+}, and Cd^{2+} were removed by SOQDs (stannic oxide quantum dots) capped with a reduced graphene oxide nanocomposite (Dutta et al. 2016). Table 4 represents the different types of graphene-based materials used for the removal of heavy metals.

3.5 Mechanism of Graphene in the Removal of Heavy Metals

Adsorption

- **Physical adsorption**: Heavy metals are attracted to graphene's surface through van der Waals forces or electrostatic interactions. The graphene's high surface area provides abundant adsorption sites (Gupta et al. 2021).
- **Chemical adsorption**: Functional groups on graphene (e.g., hydroxyl, carboxyl, and epoxy groups) chemically bind to heavy metal ions, forming stable complexes (Li et al. 2015).

Table 4 Graphene-based materials for the removal of heavy metals

Graphene-based materials	Heavy metal removed	Removal efficiency	References
PVP/Fe_3O_4/GO	Pb (II)	99.74	Yang et al. (2019)
Fe_3O_4/RGO	Cr (VI)	99.9	Zhang et al. (2021)
Fe_3O_4/PMA-g-PVA	Ag (I)	92.15	Liu et al. (2020)
	Ni (II)	89.2	
	Co (II)	84.52	
	Pb (II)	83.56	
	Cd (II)	90.16	
$W_{18}O_{49}$/graphene material/activated carbon	Cr (III)	96	Mao et al. (2021)
	Pb (II)	92	
	Cd (II)	92	
	Co (II)	94	
	Ni (II)	93	
	Fe (III)	95	
	Cu (II)	98	
Cyclodextrin functionalized 3D-graphene	Cr (VI)	99.8	Wang et al. (2019a, b)
Polyethyleneimine doped graphene oxide (GO-PEI)	Cr (VI)	95	Geng et al. (2019)
Chitosan doped graphene oxide (CS-GO)	Cr (VI)	96	Samuel et al. (2019)
rGO Fe_3O_4@polydopamine (rGO/Fe_3O_4@PDA)	Pb (II)	100	Mehdinia et al. (2020)

Ion Exchange

- Functionalized graphene can exchange cations or anions with heavy metal ions in the solution. This mechanism often involves oxygen-containing functional groups (e.g., –COOH, –OH) or other moieties introduced via doping or coating (Perreault et al. 2015).

Electrostatic Interaction

- Heavy metal ions are attracted to the negatively charged graphene or its derivatives, particularly graphene oxide (GO) and reduced graphene oxide (rGO), which possess negatively charged oxygen-containing groups.

Reduction of Heavy Metals

- Reduced graphene oxide or functionalized graphene can directly reduce heavy metal ions, such as Cr (VI), to Cr (III), making them less toxic and easier to remove. Graphene's electron-donating capacity facilitates this redox process (Yu et al. 2018).

Precipitation

- Graphene-based composites can induce the precipitation of certain heavy metals by altering the pH or through chemical interactions with specific functional groups. This is common in composites doped with elements such as Fe or Mg.

3.6 Carbon-Based Quantum Dots

At present, carbon-based quantum dots (CQDs) are commonly shown as an intriguing form of carbon nanoparticles made up mostly of carbon with a diameter of roughly 10 nm (Abd Rani et al. 2020). Due to their various properties, i.e., high strength, proper biocompatibility, low toxicity (Lim et al. 2018), high-water solubility, less expensive, and far-fetched photo-stability, carbon quantum dots are widely used for wastewater treatment (Das et al. 2018). Because of their optical strength impacts and quantum coherence consequences, carbon quantum dots show rigidity, chemical inactivity, strong fluorescence execution, and reasonable photo-luminescent indications (Shi et al. 2019). In recent years, wheat waste and plant waste have been widely used as a carbon source for manufacturing carbon quantum dots (Nguyen et al. 2021). Table 5 represents the different biomass-based adsorbents used for the heavy metals removal.

Table 5 Carbon-based adsorbents for heavy metal removal

Adsorbent	Heavy metal	References
rCQDs/SPEEK	Pb, Cd	Sgreccia et al. (2024)
Bentonite-CQD	Zn, Cu, Co	Jlassi et al. (2023)
Biomass-derived CQDs	Hg, Pb	Šafranko et al. (2021)
Corn straw (porous carbon)	Cr	Ma et al. (2019)
Discarded mushroom-stick BC	Ni, Cu, Pb, Cd	Wang et al. 2019a, b
Anaerobic digestion sludge BC	Cr, Zn, Cu, Pb, Cd, Ni	Ho et al. (2017), Ni et al. (2019)
Water hyacinth BC	Cd	Zhang et al. (2015)
Rice-husk (Carbon)	Cr	Khan et al. (2016)
Sesame-straw BC	Pb, Zn, Cr, Cd, Zn	Park et al. (2016)

3.7 *Mechanism of Heavy Metal Removal by Carbon-Based Quantum Dots*

Adsorption

- Electrostatic interactions: CQDs typically carry surface functional groups like carboxyl (–COOH), hydroxyl (–OH), and amino ($-NH_2$) groups, which enable the binding of metal ions through electrostatic attraction (Gupta et al. 2021).

Ion Exchange

- Functional groups on CQDs can exchange protons (H^+) with heavy metal cations in water, effectively capturing the metals and releasing protons (Perreault et al. 2015).

Reduction

- Some CQDs have reducing properties due to electron-rich surface groups or dopants, allowing them to reduce toxic heavy metal ions (e.g., Cr^{6+} to Cr^{3+}), making them less toxic and easier to precipitate or remove (Sgreccia et al. 2024).

Precipitation

- CQDs can promote the precipitation of heavy metals through the formation of insoluble complexes with the functional groups on their surface (Sgreccia et al. 2024).

4 Polymer-Based Nanocomposites

In the current landscape of environmental remediation, polymer nanocomposites have emerged as a promising solution for addressing heavy metal contamination in water systems. These advanced materials distinguish themselves through unique characteristics such as their high surface area relative to volume, tailorable surface chemistry, and exceptional adsorption properties (Zeng et al. 2014; Kobielska et al. 2018; Patel et al. 2020; Simeonidis et al. 2018).

Researchers have particularly focused on natural biopolymers as foundational components due to their inherent functional groups. These biopolymers include:

- Cellulose: A glucose-based polymer with hydroxyl groups
- Lignin: A complex phenolic polymer
- Alginate: A linear polysaccharide featuring mannuronic and guluronic acid units
- Chitin: A polysaccharide containing N-acetylglucosamine
- Pectin: A polygalacturonic acid derivative
- Chitosan: A versatile polyaminosaccharide.

The effectiveness of these biopolymers stems from their diverse functional groups, like amines, carboxyls, and hydroxyls, which serve as critical active sites for metal ion

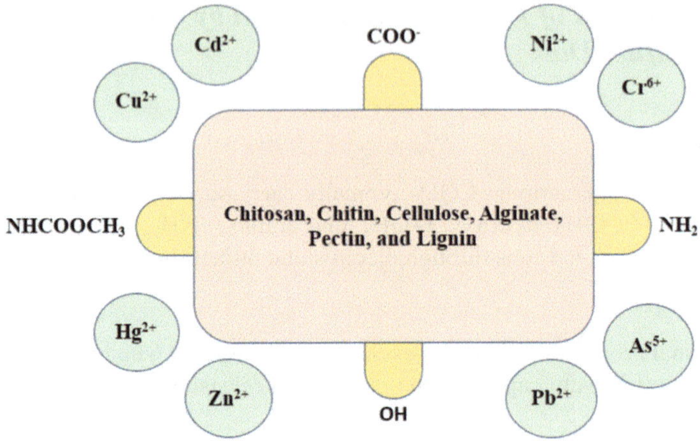

Fig. 3 Active sites of biopolymer composites and removal of various heavy metals

interaction and removal mechanisms. By leveraging these natural polymers' unique structural and chemical properties, researchers can develop sophisticated nanocomposite materials capable of efficiently targeting and extracting heavy metal contaminants from water sources (Kayan and Kayan 2021). In polysaccharide composites, NH_2, COO^-, OH, and $NHCOOCH_3$ act as active sites for heavy metals removal, as shown in Fig. 3.

The strength and distribution of nanoparticles can be improved by adding polymers in addition to offering surface property modification for specific metal binding (Kobielska et al. 2018). Furthermore, in comparison with traditional adsorbents, certain polymer-based nanocomposites have shown better heavy metal uptake capabilities (Patel et al. 2020). Apart from providing the chance for metal recovery and reuse, studies have confirmed that these materials may achieve significant removal efficiency, often surpassing 90% (Patel et al. 2020; Kobielska et al. 2018; Vidu et al. 2020; Simeonidis et al. 2018). The growing problem of heavy metal contamination in water sources requires the development of efficient polymer nanocomposite adsorbents by increasing their surface areas and adding functional groups. These nanocomposites provide other benefits for water treatment applications outside their adsorption features, such as ease of segregation and recyclable qualities (Kobielska et al. 2018). Efficiency, affordability, and ecological responsibility can all be optimized by customizing the polymer formulation and nanoparticle properties (Patel et al. 2020). But there are still some issues, like making sure these nanomaterials are safe and stable in water systems over the long run. Resolving these issues and expanding the application of polymer NCs for successful and efficient heavy metal elimination from water are the main goals of current research and development initiatives (Simeonidis et al. 2018; Lu and Astruc 2018).

4.1 Mechanism of Polymer-Based Nanocomposites for the Removal of Heavy Metals

Adsorption

- **Surface Functionalization**: Polymer-based QDs are functionalized with specific ligands (e.g., carboxyl, amine, or thiol groups) that enhance their affinity for heavy metals.
- **Interaction Mechanism**: Heavy metal ions form coordinate bonds with the functional groups on the QDs' surface, leading to their adsorption (Zhao et al. 2018).

Electrostatic Interactions

- **Surface Charge**: QDs can carry a surface charge, enabling electrostatic attraction to oppositely charged heavy metal ions.
- **Ionic Strength Effects**: The effectiveness of this interaction depends on the ionic strength of the solution, which may influence the repulsion or aggregation of the QDs (Zhao et al. 2018).

Chelation

- **Ligand–Metal Complex Formation**: Functional groups on the polymer bind with metal ions through chelation, forming stable complexes.
- **Selectivity**: Chelation is highly selective for specific heavy metals, depending on the ligand used (e.g., thiol groups for mercury or cadmium) (Zhao et al. 2018).

Reduction/Oxidation

- Redox Active Sites: Some QDs or their functional groups can induce redox reactions, reducing metal ions (e.g., Cr (VI) to Cr (III)) to less toxic forms that can precipitate or be adsorbed (Sgreccia et al. 2024).

5 Clay-Based Nanocomposites

Clay-based nanocomposites have emerged as a promising solution for the exclusion of heavy metals from water due to their high surface area, capacity to exchange cations, adsorption capacity, cost-effectiveness, and environmental friendliness. Nanomaterials have been increasingly studied as adsorbents for heavy metal ions in water, demonstrating their potential for effective environmental remediation (Zeng et al. 2014; Covaliu et al. 2019). One class of nanomaterials that has shown great promise is clay-polymer nanocomposites. These materials combine the unique properties of clay minerals, such as high surface area and cation exchange capacity, with the tolerability and versatility of polymers (Amari et al. 2021). Organic polymer nanomaterials and organic polymer-supported nanocomposites, including those based on cellulose and chitosan, have been found to efficiently adsorb heavy metals like iron, silver, gold, and copper in water (Lu and Astruc 2018). Moreover, clay-polymer nanocomposites have been shown to offer better regeneration capabilities compared to clay materials alone, making them additional cost-effective and sustainable for water treatment applications. (Amari et al. 2021). The minor surface area and sluggish adsorption rate of organic polymer nanomaterials can be overcome by incorporating them into clay-based nanocomposites (Lu and Astruc 2018). Clay-polymer nanocomposites are classified into 3 main classes: (a) exfoliated nanocomposites, (b) intercalated nanocomposites, and (c) phase-separated nanocomposites (Mukhopadhyay and De 2014). The clay sheets are extensively disseminated across the polymer matrix of the exfoliated nanocomposites. They are made up of layers of clay that are around 1 nm apart and enclosed in a matrix of polymer (Amari et al. 2021). Nanocomposites combining clay and polymers offer innovative solutions for water purification, particularly in heavy metal removal. These advanced materials come in two primary structural configurations: intercalated and phase-separated designs. In intercalated nanocomposites, polymeric chains are strategically integrated between silicate crystalline layers, creating a unique molecular arrangement. Phase-separated nanocomposites involve polymer insertion directly within silicate layers, presenting alternative structural characteristics. The versatility of these nanocomposites is demonstrated by the diverse range of polymers that can be employed, including:

- Chitosan
- Polystyrene
- Polypropylene
- Polyesters
- Polyurethanes
- Epoxy resins
- Polyvinyl chloride.

These clay-polymer hybrid materials represent a cutting-edge approach to addressing critical environmental challenges, particularly water contamination. By leveraging nanoscale engineering, researchers can develop sophisticated filtration technologies that potentially offer more efficient and sustainable methods for removing heavy metals from water sources, contributing to global efforts to improve water quality and accessibility (Table 6).

5.1 Mechanism of Heavy Metal Removal by Clay-Based Nanocomposites

Adsorption

- **Surface Adsorption**: Clay-based nanocomposites have a high specific surface area and abundant active sites, which enable them to adsorb heavy metal ions effectively.

Table 6 Removal of contaminants using clay-polymer nanocomposites

Adsorbent	Adsorbate	Removal efficiency	References
CD bentonite	Pb	95%	Jlassi et al. (2023)
Carboxymethylated formic lignin	Pb	107.5 mg/g	Yao et al. (2022)
Polyaniline Montmorillonite Clay nanocomposites	Green 25	100%	Kalotra and Mehta (2021)
Tetraethoxysilane-functionalized Na-bentonite (2 wt%) united into polysulfone/polyethyleneimine sheaths	Methylene Blue	98.9%	Saki et al. (2020)
Carboxymethyl cellulose	Nine pesticides	57–100%	Saki et al. (2020)
Montmorillonite	Methylene Blue	74%	Zhou et al. (2014)
Layer double hydroxide	Methyl orange	88%	Zhou et al. (2014)

- **Ion Exchange**: Many clays (e.g., montmorillonite) naturally possess ion exchange capabilities. Metal cations in solution can exchange with cations present in the clay structure.
- **Physical Adsorption**: Weak van der Waals forces or hydrogen bonding can also contribute to the trapping of heavy metals on the clay surface (Jlassi et al. 2023).

Electrostatic Interaction

- The surface charge of the clay particles plays a significant role. Negatively charged clay layers attract positively charged heavy metal ions from aqueous solutions, facilitating their immobilization.

Complexation and Chelation

- Functional groups (e.g., –OH, –COOH, $–NH_2$) on the clay or the nanocomposite matrix can form stable complexes with heavy metals.
- Modifying the clay surface with organic or inorganic molecules enhances the affinity for specific metals by introducing functional groups that act as chelators (Jlassi et al. 2023).

Precipitation

- Heavy metals can precipitate as insoluble hydroxides, carbonates, or sulfides when interacting with the active sites of clay nanocomposites under suitable pH conditions.

6 Comparative Study of Different Nanocomposites for Heavy Metals Removal

A comparative study of different nanocomposites for heavy metal removal focuses on evaluating their efficiencies, mechanisms, and limitations as shown in Table 7.

Table 7 Comparative study of different nanocomposites for heavy metal removal

Heavy metal	Nanocomposite technique	Treatment mechanism	Removal efficiency	Limitations	References
Pb	GO-TiO_2	Adsorption, photocatalysis	~ 99%	High cost, limited long-term stability in water system	Singh et al. (2024), Covaliu-Mierlă et al. (2023)
Cr	Polyvinyl amine-glutaraldehyde membranes	Nanofiltration	99.9%	Membrane fouling, reduced lifespan due to frequent cleaning	Singh et al. (2024), Covaliu-Mierlă et al. (2023)
Cu	Chitosan-based Bio-nanocomposites	Bioadsorption, ion-exchange	90–95%	Efficiency decreases in mixed-metal solutions, reusability of material requires optimization	Ayach et al. (2024)
Cd	ZnO-graphene nanocomposites	Adsorption, photocatalysis	97–99%	Potential release of nanoparticles, toxicity concern	Singh et al. (2024), Covaliu-Mierlă et al. (2023)
As	Iron-oxide nanoparticles	Adsorption, magnetic removal	95–98%	Limited capacity under high arsenic removal concentrations; requires regeneration for repeated use	Covaliu-Mierlă et al. (2023)
Ni	Polymer–clay hybrid nano composites	Adsorption	80–90%	Performance affected by pH changes, difficult recovery of the adsorbent	Singh et al. (2024), Covaliu-Mierlă et al. (2023)

7 Challenges and Future Directions

Carbon-based nanocomposites hold great promise for the removal of heavy metals from water due to their exceptional properties, such as high surface area, tunable surface chemistry, and strong adsorption capacities. However, several challenges limit their widespread application. One significant issue is their synthesis complexity and cost, which can hinder scalability and economic feasibility. Additionally, the stability of nanocomposites under various environmental conditions, such as changes in pH or the presence of competing ions, can impact their efficiency. Another challenge is the potential release of nanomaterials into the environment during or after use, raising concerns about secondary contamination and ecological impacts. Moreover, the regeneration and reuse of these nanocomposites, while desirable, often lead to reduced efficiency and increased operational costs. Future research directions should focus on developing cost-effective, environmentally friendly synthesis methods that improve the stability and adsorption efficiency of carbon-based nanocomposites. Strategies to enhance their selectivity for specific heavy metals, alongside methods for effective regeneration and safe disposal, are also critical. Integrating nanocomposites into existing water treatment systems and exploring hybrid approaches, such as combining them with biological or photocatalytic processes, may offer sustainable and practical solutions for real-world applications.

8 Conclusion

In conclusion, the creation of new technologies based on nanotechnology and improvements in engineering and environmental research have resulted from the widespread application of nanotechnology. Wastewater treatment is a global issue because of the high levels of undesirable or unknown contaminants that are extremely hazardous to people and the environment. Emerging nano-scale technology has spurred the advanced implementation of new and low-cost techniques, such as adsorption procedures for eliminating environmental pollutants and nano-adsorptive membrane processes that are suitable for filtering. Membrane characteristics, including permeability, hydrophilicity, selectivity, and thermal characteristics, have all improved recently. This can be enhanced by creating more affordable and effective nanocomposite membranes.

References

Abd Rani U, Ng LY, Ng CY, Mahmoudi E (2020) A review of carbon quantum dots and their applications in wastewater treatment. Adv Coll Interface Sci 278:102124

Abdullayev E, Vakili AH, Amr SSA, Karaagaç SU, Alazaiza MY (2024) Navigating heavy metal removal: insights into advanced treatment technologies for wastewater: a review. Global Nest J 260:06247

Akhtar N, Syakir Ishak MI, Bhawani SA, Umar K (2021) Various natural and anthropogenic factors responsible for water quality degradation: a review. Water 13(19):2660

Akhter F, Zoppas FM, Soomro M, Jatoi AS, Noureen F, Akhtar MN, Mehreen F (2023) Carbon-based sorbets for heavy metal removal from aqueous solution, discrepancies, and future prospects: a state-of-the-art review. Biomass Convers Biorefinery 13(12):10343–10359

Amari A, Mohammed Alzahrani F, Mohammedsaleh Katubi K, Salem Alsaiari N, Tahoon MA, Ben Rebah F (2021) Clay-polymer nanocomposites: preparations and utilization for pollutants removal. Materials 14(6):1365

Ani IJ, Egbosiuba TC (2023) Metal oxides and their nanocomposites in wastewater treatment. In: Adsorption through advanced nanoscale materials. Elsevier, pp 185–203

Atieh MA (2011) Removal of chromium (VI) from polluted water using carbon nanotubes supported with activated carbon. Procedia Environ Sci 4:281–293

Ayach J, El Malti W, Duma L, Lalevée J, Al Ajami M, Hamad H, Hijazi A (2024) Comparing conventional and advanced approaches for heavy metal removal in wastewater treatment: an in-depth review emphasizing filter-based strategies. Polymers 16(14):1959

Bankole MT, Abdulkareem AS, Mohammed IA, Ochigbo SS, Tijani JO, Abubakre OK, Roos WD (2019) Selected heavy metals removal from electroplating wastewater by purified and polyhydroxylbutyrate functionalized carbon nanotubes adsorbents. Sci Rep 9(1):4475

Baranik A, Gagor A, Queralt I, Marguí E, Sitko R, Zawisza B (2018) Ceria nanoparticles were deposited on graphene nanosheets for the adsorption of copper (II) and lead (II) ions and of anionic species of arsenic and selenium. Microchim Acta 185:1–9

Barhoum A, Pal K, Rahier H, Uludag H, Kim IS, Bechelany M (2019) Nanofibers as new-generation materials: from spinning and nano-spinning fabrication techniques to emerging applications. Appl Mater Today 17:1–35

Bichave MS, Kature AY, Koranne SV, Shinde RS, Gongle AS, Choudhari VP, Bokil SA et al (2023) Nano-metal oxides-activated carbons for dyes removal: a review. Mater Today: Proc 77:19–30

Bilal M, Ihsanullah I, Younas M, Shah MUH (2021) Recent advances in applications of low-cost adsorbents for the removal of heavy metals from water: a critical review. Sep Purif Technol 278:119510

Cai X, Gao Y, Sun Q, Chen Z, Megharaj M, Naidu R (2014) Removal of co-contaminants Cu (II) and nitrate from aqueous solution using kaolin-Fe/Ni nanoparticles. Chem Eng J 244:19–26

Cai Y, Li C, Wu D, Wang W, Tan F, Wang X, Wong PK, Qiao X (2017) Highly active MgO nanoparticles for simultaneous bacterial inactivation and heavy metal removal from aqueous solution. Chem Eng J 312:158–166

Chandra V, Kim KS (2011) Highly selective adsorption of Hg^{2+} by a polypyrrole–reduced graphene oxide composite. Chem Commun 47(13):3942–3944

Chu TPM, Nguyen NT, Vu TL, Dao TH, Dinh LC, Nguyen HL, Hoang TH, Le TS, Pham TD (2019) Synthesis, characterization, and modification of alumina nanoparticles for cationic dye removal. Materials 12:450

Covaliu CI, Paraschiv G, Stoian O, Vişan A (2019) Nanomaterials applied for heavy metals removal from wastewater. IOP Conf Ser: Mater Sci Eng 572(1):012074

Covaliu-Mierlă CI, Păunescu O, Iovu H (2023) Recent advances in membranes used for nanofiltration to remove heavy metals from wastewater: a review. Membranes 13(7):643

Dan S, Bagheri H, Shahidizadeh A, Hashemipour H (2023) Performance of graphene Oxide/ SiO_2 nanocomposite-based: antibacterial activity, dye and heavy metal removal. Arab J Chem 16(2):104450

Das R, Bandyopadhyay R, Pramanik P (2018) Carbon quantum dots from natural resource: a review. Mater Today Chem 8:96–109

de Jesus RA, de Assis GC, de Oliveira RJ, Bilal M, Bharagava RN, Iqbal HM, Ferreira LFR, Figueiredo RT (2022) Metal oxide nanoparticles for environmental remediation. In: Hafiz

MNI, Bilal M, Nguyen TA, Yasin G (eds) Micro and nano technologies, biodegradation and biodeterioration at the nanoscale. Elsevier, Amsterdam, The Netherlands, pp 529–560
Dutta D, Thiyagarajan S, Bahadur D (2016) SnO_2 quantum dots decorated reduced graphene oxide nanocomposites for efficient water remediation. Chem Eng J 297:55–65
Egbosiuba TC, Abdulkareem AS, Tijani JO, Ani JI, Krikstolaityte V, Srinivasan M, Lisak G et al (2021) Taguchi optimization design of diameter-controlled synthesis of multi-walled carbon nanotubes for the adsorption of Pb (II) and Ni (II) from chemical industry wastewater. Chemosphere 266:128937
Egbosiuba TC, Egwunyenga MC, Tijani JO, Mustapha S, Abdulkareem AS, Kovo AS, Lisak G et al (2022) Activated multi-walled carbon nanotubes decorated with zero-valent nickel nanoparticles for arsenic, cadmium, and lead adsorption from wastewater in batch and continuous flow modes. J Hazard Mater 423:126993
El Mouden A, El Messaoudi N, El Guerraf A, Bouich A, Mehmeti V, Lacherai A, Sher F et al (2023) Multifunctional cobalt oxide nanocomposites for efficient removal of heavy metals from aqueous solutions. Chemosphere 317:137922
Fan L, Luo C, Sun M, Li X, Qiu H (2013) Highly selective adsorption of lead ions by water-dispersible magnetic chitosan/graphene oxide composites. Colloids Surf, B 103:523–529
Fernandes J, Ramísio PJ, Puga H (2024) A comprehensive review on various phases of wastewater technologies: trends and future perspectives. Eng 5(4):2633–2661
Geng J, Yin Y, Liang Q, Zhu Z, Luo H (2019) Polyethyleneimine cross-linked graphene oxide for removing hazardous hexavalent chromium: Adsorption performance and mechanism. Chem Eng J 361:1497–1510
Goyal P, Chakraborty S, Misra SK (2018) Multifunctional Fe_3O_4-ZnO nanocomposites for environmental remediation applications. Environ Nanotechnol, Monit Manage 10:28–35
Gul A, Khaligh NG, Julkapli NM (2021) Surface modification of carbon-based nano adsorbents for advanced wastewater treatment. J Mol Struct 1235:130148
Gupta K, Joshi P, Gusain R, Khatri OP (2021) Recent advances in adsorptive removal of heavy metal and metalloid ions by metal oxide-based nanomaterials. Coord Chem Rev 445:214100
Hallaji H, Keshtkar AR, Moosavian MA (2015) A novel electrospun PVA/ZnO nanofiber adsorbent for U (VI), Cu (II) and Ni (II) removal from aqueous solution. J Taiwan Inst Chem Eng 46:109–118
Hao L, Song H, Zhang L, Wan X, Tang Y, Lv Y (2012) SiO_2/graphene composite for highly selective adsorption of Pb (II) ion. J Colloid Interface Sci 369(1):381–387
Ho SH, Yang ZK, Nagarajan D, Chang JS, Ren NQ (2017) High-efficiency removal of lead from wastewater by biochar derived from anaerobic digestion sludge. Biores Technol 246:142–149
Hojamberdiev M, Daminova SS, Kadirova ZC, Sharipov KT, Mtalo F, Hasegawa M (2018) Ligand-immobilized spent alumina catalyst for effective removal of heavy metal ions from model contaminated water. J Environ Chem Eng 6:4623–4633
Hu X, Wang Y, Yang JO, Li Y, Wu P, Zhang H, Liu Z et al (2020) Synthesis of graphene oxide nanoribbons/chitosan composite membranes for the removal of uranium from aqueous solutions. Front Chem Sci Eng 14:1029–1038
Jamjoum HAA, Umar K, Adnan R, Razali MR, Ibrahim MNM (2021) Synthesis, characterization, and photocatalytic activities of graphene oxide/metal oxides nanocomposites: a review. Front Chem 9:752276
Jlassi K, Al Ejji M, Ahmed AK, Mutahir H, Sliem MH, Abdullah AM, Krupa I et al (2023) A carbon dot-based clay nanocomposite for efficient heavy metal removal. Nanoscale Adv 5(16):4224–4232
Kalotra S, Mehta R (2021) Synthesis of polyaniline/clay nanocomposites by in situ polymerization and its application for the removal of Acid Green 25 dye from wastewater. Polym Bull 78(5):2439–2463
Kanel SR, Greneche J-M, Choi H (2006) Arsenic (V) removal from groundwater using nano scale zero-valent iron as a colloidal reactive barrier material. Environ Sci Technol 40:2045–2050

Kayan GÖ, Kayan A (2021) Composite of natural polymers and their adsorbent properties on the dyes and heavy metal ions. J Polym Environ 29(11):3477–3496
Khan T, Isa MH, Mustafa MRU, Yeek-Chia H, Baloo L, Abd Manan TSB, Saeed MO (2016) Cr (VI) adsorption from aqueous solution by an agricultural waste-based carbon. RSC Adv 6(61):56365–56374
Khobragade M, Pal A (2016) Adsorptive removal of Mn (II) from water and wastewater by surfactant-modified alumina. Desalin Water Treat 57:2775–2786
Kobielska PA, Howarth AJ, Farha OK, Nayak S (2018) Metal-organic frameworks for heavy metal removal from water. Coord Chem Rev 358:92–107
Kumari V, Tripathi A (2020) Remediation of heavy metals in pharmaceutical effluent with the help of Bacillus cereus-based green-synthesized silver nanoparticles supported on alumina. Appl Nanosci
Li L, Wang Z, Ma P, Bai H, Dong W, Chen M (2015) Preparation of polyvinyl alcohol/chitosan hydrogel compounded with graphene oxide to enhance the adsorption properties for Cu (II) in aqueous solution. J Polym Res 22:1–10
Lim H, Liu Y, Kim HY, Son DI (2018) Facile synthesis and characterization of carbon quantum dots and photovoltaic applications. Thin Solid Films 660:672–677
Liu T, Li YL, He JY, Hu Y, Wang CM, Zhang KS, Liu JH et al (2019) Porous boron nitride nanoribbons with large widths as superior adsorbents for rapid removal of cadmium and copper ions from water. New J Chem 43(8):3280–3290
Liu X, Guan J, Lai G, Xu Q, Bai X, Wang Z, Cui S (2020) Stimuli-responsive adsorption behavior toward heavy metal ions based on comb polymer functionalized magnetic nanoparticles. J Clean Prod 253:119915
Liu Z, Ling Q, Cai Y, Xu L, Su J, Yu K, Wang X et al (2022) Synthesis of carbon-based nanomaterials and their application in pollution management. Nanoscale Adv 4(5):1246–1262
Lu F, Astruc D (2018) Nanomaterials for removal of toxic elements from water. Coord Chem Rev 356:147–164
Lu F, Didier A (2020) Nanocatalysts and other nanomaterials for water remediation from organic pollutants. Coord Chem Rev 408:213180
Luo C, Wang J, Jia P, Liu Y, An J, Cao B, Pan K (2015) Hierarchically structured polyacrylonitrile nanofiber mat as highly efficient lead adsorbent for water treatment. Chem Eng J 262:775–784
Ma H, Yang J, Gao X, Liu Z, Liu X, Xu Z (2019) Removal of chromium (VI) from water by porous carbon derived from corn straw: influencing factors, regeneration and mechanism. J Hazard Mater 369:550–560
Madima N, Mishra SB, Inamuddin I, Mishra AK (2020) Carbon-based nanomaterials for remediation of organic and inorganic pollutants from wastewater. A review. Environ Chem Lett 18:1169–1191
Mao M, Yan T, Shen J, Zhang J, Zhang D (2021) Capacitive removal of heavy metal ions from wastewater via an electro-adsorption and electro-reaction coupling process. Environ Sci Technol 55(5):3333–3340
Mehdinia A, Heydari S, Jabbari A (2020) Synthesis and characterization of reduced graphene oxide-Fe_3O_4 @ polydopamine and application for adsorption of lead ions: isotherm and kinetic studies. Mater Chem Phys 239:121964
Mi X, Huang G, Xie W, Wang W, Liu Y, Gao J (2012) Preparation of graphene oxide aerogel and its adsorption for Cu^{2+} ions. Carbon 50(13):4856–4864
Mishra AK, Ramaprabhu S (2011) Functionalized graphene sheets for arsenic removal and desalination of seawater. Desalination 282:39–45
Mukhopadhyay R, De N (2014) Nano clay polymer composite: synthesis, characterization, properties and application in rainfed agriculture. Global J Bio Biotechnol 3(2):133–138
Naseer A (2024) Role of nanocomposites and nano adsorbents for heavy metals removal and dyes. An overview. Desalin Water Treat:100662
Nassar NN (2012) Iron oxide nano adsorbents for removal of various pollutants from wastewater: an overview. In: Application of adsorbents for water pollution control, pp 81–118

Nguyen TTH, Nguyen XC, Nguyen DLT, Nguyen DD, Vo TYB, Vo QN, Van Le Q et al (2021) Converting biomass of agro wastes and invasive plants into alternative materials for water remediation. Biomass Convers Biorefinery:1–16

Ni BJ, Huang QS, Wang C, Ni TY, Sun J, Wei W (2019) Competitive adsorption of heavy metals in aqueous solution onto biochar derived from anaerobically digested sludge. Chemosphere 219:351–357

Oliveros AN, Pimentel JAI, de Luna MDG, Garcia-Segura S, Abarca RRM, Doong RA (2021) Visible-light photocatalytic diclofenac removal by tunable vanadium pentoxide/boron-doped graphitic carbon nitride composite. Chem Eng J 403:126213

Park JH, Ok YS, Kim SH, Cho JS, Heo JS, Delaune RD, Seo DC (2016) Competitive adsorption of heavy metals onto sesame straw biochar in aqueous solutions. Chemosphere 142:77–83

Patel, H. K., Kalaria, R. K., & Khimani, M. R. (2020). Nanotechnology: a promising tool for bioremediation. Removal of toxic pollutants through microbiological and tertiary treatment, 515–547.

Patil SR (2022) Metal oxide-based composites as photocatalysts. In: Delekar SD (ed) Metal oxides, advances in metal oxides and their composites for emerging applications. Elsevier, Amsterdam, The Netherlands, pp 633–672

Patnaik A, Sahu J, Poonia AK, Ghosh P (2023) Current perspective of nano-engineered metal oxide based photocatalysts in advanced oxidation processes for degradation of organic pollutants in wastewater. Chem Eng Res des 190:667–686

Pereira MFG, Nascimento MM, Cardoso PHN, Oliveira CYB, Tavares GF, Araújo ES (2022) Preparation, microstructural characterization and photocatalysis tests of V^{5+}-doped TiO_2/WO_3 nanocomposites supported on electrospun membranes. Inorganics 10:143

Perreault F, De Faria AF, Elimelech M (2015) Environmental applications of graphene-based nanomaterials. Chem Soc Rev 44(16):5861–5896

Pham TD, Tran TT, Pham TT, Dao TH, Le TS (2019) Adsorption characteristics of molecular oxytetracycline onto alumina particles: the role of surface modification with an anionic surfactant. J Mol Liq 287:110900

Rajabathar JR, Shukla AK, Ali A, Al-Lohedan HA (2017) Silver nanoparticle/r-graphene oxide deposited mesoporous-manganese oxide nanocomposite for pollutant removal and supercapacitor applications. Int J Hydrogen Energy 42(24):15679–15688

Ramalingam B, Parandhaman T, Choudhary P, Das SK (2018) Biomaterial functionalized graphene-magnetite nanocomposite: a novel approach for simultaneous removal of anionic dyes and heavy metal ions. ACS Sustain Chem Eng 6(5):6328–6341

Ruthiraan M, Mubarak NM, Thines RK, Abdullah EC, Sahu JN, Jayakumar NS, Ganesan P (2015) Comparative kinetic study of functionalized carbon nanotubes and magnetic biochar for removal of Cd^{2+} ions from wastewater. Korean J Chem Eng 32:446–457

Sadeghi MH, Tofighy MA, Mohammadi T (2020) One-dimensional graphene for efficient aqueous heavy metal adsorption: rapid removal of arsenic and mercury ions by graphene oxide nanoribbons (GONRs). Chemosphere 253:126647

Šafranko S, Goman D, Stanković A, Medvidović-Kosanović M, Moslavac T, Jerković I, Jokić S (2021) An overview of the recent developments in carbon quantum dots—promising nanomaterials for metal ion detection and (bio) molecule sensing. Chemosensors 9(6):138

Saki S, Senol-Arslan D, Uzal N (2020) Fabrication and characterization of silane-functionalized Na-bentonite Polysulfone/polyethyleneimine nanocomposite membranes for dye removal. J Appl Polym Sci 137(36):49057

Samuel MS, Bhattacharya J, Raj S, Santhanam N, Singh H, Singh NP (2019) Efficient removal of Chromium (VI) from aqueous solution using chitosan grafted graphene oxide (CS-GO) nanocomposite. Int J Biol Macromol 121:285–292

Santhosh C, Velmurugan V, Jacob G, Jeong SK, Grace AN, Bhatnagar A (2016) Role of nanomaterials in water treatment applications: a review. Chem Eng J 306:1116–1137

Sgreccia E, Gallardo Gonzalez FS, Prosposito P, Burratti L, Sisani M, Bastianini M, Di Vona ML et al (2024) Heavy metal detection and removal by composite carbon quantum dots/ionomer membranes. Membranes 14(6):134

Shah P, Murthy CN (2013) Studies on the porosity control of MWCNT/Polysulfone composite membrane and its effect on metal removal. J Membr Sci 437:90–98

Shanaah HH, Alzaimoor EFH, Rashdan S, Abdalhafith AA, Kamel AH (2023) Photocatalytic degradation and adsorptive removal of emerging organic pesticides using metal oxide and their composites: recent trends and future perspectives. Sustainability 15:7336

Shi Y, Liu X, Wang M, Huang J, Jiang X, Pang J, Zhang X et al (2019) Synthesis of N-doped carbon quantum dots from bio-waste lignin for selective iron detection and cellular imaging. Int J Biol Macromol 128:537–545

Shrestha R, Ban S, Devkota S, Sharma S, Joshi R, Tiwari AP, Joshi MK et al (2021) Technological trends in heavy metals removal from industrial wastewater: a review. J Environ Chem Eng 9(4):105688

Sikder MT, Mihara Y, Islam MS, Saito T, Tanaka S, Kurasaki M (2014) Preparation and characterization of chitosan–carboxymethyl-β-cyclodextrin entrapped nano zero-valent iron composite for Cu (II) and Cr (IV) removal from wastewater. Chem Eng J 236:378–387

Simeonidis K, Martinez-Boubeta C, Zamora-Perez P, Rivera-Gil P, Kaprara E, Kokkinos E, Mitrakas M (2018) Nanoparticles for heavy metal removal from drinking water. Environ Nanotechnol 1:75–124

Singh S, Kapoor D, Khasnabis S, Singh J, Ramamurthy PC (2021) Mechanism and kinetics of adsorption and removal of heavy metals from wastewater using nanomaterials. Environ Chem Lett 19(3):2351–2381

Singh N, Yadav A, Das S, Debnath N (2024) Recent advances in heavy metal/metalloid ion treatment from wastewater using nanocomposites and bio nanocomposites. Front Nanotechnol 6:1307353

Smith S (2023) Innovative treatment technologies for PFAS-contaminated water: utilizing foam partitioning and exploring electrochemical oxidation. Swedish University of Agricultural Sciences

Song J, Meng Z, Wang X, Zhang G, Bi C, Hou J (2022) One-step microwave method synthesis of Fe_3O_4 nanoribbon@ carbon composite for Cr (VI) removal. Sep Purif Technol 298:121530

Teh CY, Budiman PM, Shak KPY, Wu TY (2016) Recent advancement of coagulation-flocculation and its application in wastewater treatment. Ind Eng Chem Res 55(16):4363–4389

Vidu R, Matei E, Predescu AM, Alhalaili B, Pantilimon C, Tarcea C, Predescu C (2020) Removal of heavy metals from wastewaters: a challenge from current treatment methods to nanotechnology applications. Toxics 8(4):101

Wadhawan S, Jain A, Nayyar J, Mehta SK (2020) Role of nanomaterials as adsorbents in heavy metal ion removal from wastewater: a review. J Water Process Eng 33:101038

Wang X, Li X, Liu G, He Y, Chen C, Liu X, Zhao Y et al (2019a) Mixed heavy metal removal from wastewater by using discarded mushroom-stick biochar: adsorption properties and mechanisms. Environ Sci Process Impacts 21(3):584–592

Wang Z, Lin F, Huang L, Chang Z, Yang B, Liu S, Chen J et al (2019b) Cyclodextrin functionalized 3D-graphene for the removal of Cr (VI) with the easy and rapid separation strategy. Environ Pollut 254:112854

WHO (2021) Progress on household drinking water, sanitation, and hygiene 2000–2020: five years into the SDGs

Xiao Z, Zhang H, Xu Y, Yuan M, Jing X, Huang J, Li Q, Sun D (2017) Ultra-efficient removal of chromium from aqueous medium by biogenic iron-based nanoparticles. Sep Purif Technol 174:466–473

Xie R, Jin Y, Chen Y, Jiang W (2017) The importance of surface functional groups in the adsorption of copper onto walnut shell-derived activated carbon. Water Sci Technol 76(11):3022–3034

Yang B, Wei Y, Liu Q, Luo Y, Qiu S, Shi Z (2019) Polyvinylpyrrolidone functionalized magnetic graphene-based composites for highly efficient removal of lead from wastewater. Colloids Surf, A 582:123927

Yao G, Shao X, Qiu Z, Qiu F, Li Z, Zhang T (2022) Construction of lignin-based nano-adsorbents for efficient and selective recovery of tellurium (IV) from wastewater. Chemosphere 287:132058
Yari S, Abbasizadeh S, Mousavi SE, Moghaddam MS, Moghaddam AZ (2015) Adsorption of Pb (II) and Cu (II) ions from aqueous solution by an electrospun CeO_2 nanofiber adsorbent functionalized with mercapto groups. Process Saf Environ Prot 94:159–171
Yousaf F, Nadeem F, El Zerey-Belaskri A (2019) Comparative analysis of conventional treatment methodologies and advanced processing techniques for reutilization of polluted groundwater–a comprehensive review. Int J Chem Biol Sci
Yu G, Lu Y, Guo J, Patel M, Bafana A, Wang X, Wujcik EK et al (2018) Carbon nanotubes, graphene, and their derivatives for heavy metal removal. Adv Compos Hybrid Mater 1:56–78
Zeng M, Huang Y, Zhang S, Qin S, Li J, Xu J (2014) Removal of uranium (VI) from aqueous solution by magnetic yolk–shell iron oxide@ magnesium silicate microspheres. RSC Adv 4(10):5021–5029
Zhang F, Wang X, Yin D, Peng B, Tan C, Liu Y, Wu S et al (2015) Efficiency and mechanisms of Cd removal from aqueous solution by biochar derived from water hyacinth (Eichhornia crassipes). J Environ Manage 153:68–73
Zhang X, Yi G, Zhang Z, Yu J, Fan H, Li P, Zhang C et al (2021) Magnetic graphene-based nanocomposites as highly efficient absorbents for Cr (VI) removal from wastewater. Environ Sci Pollut Res 28:14671–14680
Zhao G, Li J, Ren X, Chen C, Wang X (2011) Few-layered graphene oxide nanosheets as superior sorbents for heavy metal ion pollution management. Environ Sci Technol 45(24):10454–10462
Zhao G, Huang X, Tang Z, Huang Q, Niu F, Wang X (2018) Polymer-based nanocomposites for heavy metal ions removal from aqueous solution: a review. Polym Chem 9(26):3562–3582
Zhou K, Zhang Q, Wang B, Liu J, Wen P, Gui Z, Hu Y (2014) The integrated utilization of typical clays in the removal of organic dyes and polymer nanocomposites. J Clean Prod 81:281–289

Smart Nanocomposite for Selective Pollutant Removal

Kirti, Monika, Poonam, and Bhawna Dahiya

Abstract The increasing prevalence of environmental pollutants necessitates the development of advanced materials for effective remediation. The nanocomposite integrates functionalized nanoparticles with a responsive matrix that selectively targets and captures pollutants based on their chemical properties. Nanocomposites can remove pollutants as they have superior adsorption properties from the nano-size effect, combination-versatility, and inheritance of the parent constituents. This chapter focuses on eco-friendly nanocomposites such as natural fiber-reinforced nanocomposites, biopolymer-based nanocomposites, polymer nanocomposites, metal nanocomposites, graphene-based nanocomposites, and hybrid nanocomposites that are highly effective in pollutant removal. By leveraging advanced materials science techniques, including surface modification and nanostructuring, the nanocomposite exhibits high selectivity, enhanced adsorption capacity, and rapid response to environmental changes. The present chapter deals with the novel smart nanocomposite designed for the selective removal of specific pollutants from aqueous environments. The smart nanocomposite represents a significant advancement in environmental cleanup technologies, offering a tailored solution for pollutant management, such as heavy metals and organic contaminants, and contributing to sustainable environmental practices by maintaining stability and reusability. The integration of smart nanocomposites with artificial intelligence (AI) represents a transformative approach to the selective removal of pollutants from environmental systems. AI plays a crucial role in optimizing the performance of these nanocomposites. The chapter also focuses on the novel framework that leverages the synergistic potential of advanced nanocomposite materials and AI technologies to achieve targeted pollutant removal with high efficiency. These materials are designed with responsive features that allow them to adapt to varying environmental conditions.

Keywords Artificial intelligence · Contaminant targeting · Environmental remediation · Real-time optimization · Responsive polymer · Smart nanocomposites

Kirti · Monika · Poonam · B. Dahiya (✉)
Institute of Environmental Studies, Kurukshetra University, Kurukshetra, Haryana, India
e-mail: bhawnadahiya@kuk.ac.in

S. Roy et al. (eds.), *Nanocomposites for Sustainable Wastewater Treatment*, Composites Science and Technology, https://doi.org/10.1007/978-981-95-1369-7_7

Abbreviations

ANNs	Artificial neural networks
COFs	Covalent organic frameworks
DDT	Dichlorodiphenyltrichloroethane
MNPs	Magnetic nanoparticles
PHA	Polyhydroxyalkanoates
PLA	Polylactic acid
ROS	Reactive oxygen species

1 Introduction

Environmental pollution has become one of the most pressing global challenges because of the unmanageable discharge of pollutants and contaminants into the environment. Growing population, urbanization, industrialization, and mining are key causes of environmental pollution (Narayanan et al. 2023). The burden of environmental pollution is being shared by both developed and developing countries. Even though global attention is towards pollution, its effects are still being anticipated as a result of its serious long-term effects (Ukaogo et al. 2020). There are many types of pollution, such as air pollution, soil pollution, water pollution, noise pollution, and thermal pollution. Pollutants that cause pollution can be found in air, water, soil, and other sources (Mir et al. 2023). Out of all pollution, water pollution and its remediation by using smart nanocomposites is the main concern in this chapter. Water is an important factor for sustaining life on Earth and maintaining the ecosystem. So, protecting it is mandatory. Human activities such as mining and industrialization have greatly affected the water bodies. The addition of undesired materials to water is called water pollution (Khan et al. 2022). Undesired materials that cause pollution are called pollutants. Pollutants enter water through various sources, such as animal wastes, agricultural, industrial, and residential wastes. Various types of pollutants and their sources and descriptions have been discussed in Table 1. These pollutants can be released from industrial processes, agricultural runoff, household waste, etc. They adversely impact air, water, soil, and living organisms. Heavy metals are toxic even at low concentrations and accumulate in biological tissues, leading to serious health and environmental consequences (Sajeena and Palath 2024; Mohanty and Das 2023; Afshan et al. 2022; Pasqual et al. 2024).

Due to the increase in levels of pollution, efficient and sustainable remediation technologies are essential for mitigating the impacts of pollutants. Traditional methods such as chemical treatments, filtration, and adsorption are often expensive and inefficient in dealing with complex environmental contaminants. Newer approaches, such as the use of advanced materials like nanocomposites, have shown promise in providing highly selective, cost-effective, and environmentally friendly solutions for pollutant removal (Akhtar et al. 2024; Mohammed et al. 2024).

Table 1 Types of pollutants

Types of pollutants	Pollutants	Sources	Description	Reference
Organic pollutants	Pesticides (e.g., atrazine, DDT)	Agricultural runoff, industrial effluent	Chemical compounds designed to kill pests often persist in the environment, toxic to aquatic life	Mishra et al. (2023), Zhou et al. (2024)
	Pharmaceuticals (e.g., antibiotics, painkillers)	Hospitals, laboratories, diagnostic, and research activities, pharmaceutical industry	Pharmaceutical effluents that enter water systems through the release of unmanaged waste	Tyagi et al. (2022), Santos et al. (2013), Okoro et al. (2023)
	Dyes (e.g., methylene blue, rhodamine B)	Textile, printing, tanning, paper, and food industries	Synthetic dyes used in industries like textiles show toxicity towards aquatic life and cause water pollution	Tyagi et al. (2022), Ali et al. (2021), Tkaczyk et al. (2020), Sharma et al. (2021)
	Oil and grease	Ships, urban and river runoff, industrial effluents	Oil and grease released into the ground or surface water can disturb aerobic and anaerobic biological activities	Tyagi et al. (2022), Ali et al. (2021), Hossain et al. (2024), Kanchana et al. (2021)
	Agricultural runoff (e.g., herbicides, fungicides)	Agricultural activities	Chemical uses in agriculture that leach into water bodies can affect water quality and aquatic organisms	Mishra et al. (2023), Zahoor and Mushtaq (2023), Kumar et al. (2021)
Inorganic pollutants	Heavy metals (e.g., Pb^{2+}, Cd^{2+})	Fertilizer, battery, and alloy industries	Heavy metals are one type of harmful substance to the aquatic environment	Masoud et al. (2024)
	Nitrate (NO_3^-) and (NO_2^-)	Agricultural land	Nitrate is found in small amounts in groundwater, but it can be an issue if its concentration increases	Ahamad et al. (2020)
	Fluoride	Aluminum smelters, glass, enamel, textile dyeing, brick and tile works	Fluoride is an important element, but in disproportionate amounts, it can affect various organs and systems	Ahamad et al. (2020), Wu et al. (2022)
	Arsenic	Industrial effluents, agricultural practices	Arsenic bioaccumulation through crop and plant absorption can disrupt the agrarian economy and have harmful impacts on human health.	Jurczynski et al. (2024), Singh et al. (2024)

(continued)

Table 1 (continued)

Types of pollutants	Pollutants	Sources	Description	Reference
	Ammonia	Agricultural runoff, domestic sewage, industrial discharge	Ammonia in drinking water does not pose a direct hazard, but may undesirably affect some water treatment processes' efficiency	Serajuddin et al. (2024)
Other pollutants	Microplastics	Cosmetics, textiles, plastic bottles, and bags	Microplastics are a rising part of worldwide pollution. Microplastics can absorb harmful chemicals from water and can act as vectors for pollutants, which can be ingested by aquatic organisms	Wang et al. (2021), Osman et al. (2023)
	Pathogens (e.g., bacteria, viruses, protozoa)	Sewage, livestock wastewater, and sludge	Intruders that attack host organisms. Numerous can attack a diverse variety of hosts, and some are specific to one host	Yang et al. (2022), Sarmah et al. (2018)

Nanocomposites are engineered materials made by incorporating nanomaterials into a bulk matrix (Okpala 2014). These materials exhibit enhanced properties such as increased surface area, high reactivity, and tunable characteristics, making them ideal candidates for environmental remediation applications (Rando et al. 2022). Nanocomposites can effectively interact with a wide range of pollutants, which improves the performance of adsorption, degradation, and filtration processes (Asghar et al. 2024; Tyagi et al. 2024).

2 Nanocomposites

Nanocomposites are hybrid materials with nanoscale fillers in the matrix, which possess unique properties like high conductivity and strength (Silvestre et al. 2021). Nanocomposites can limit the release of the nanoparticles in water, which helps in the mitigation of negative effects and increases the effectiveness and productivity of nanomaterials (Thondavada et al. 2019).

3 Smart Nanocomposites (Sustainable Nanocomposites)

Smart nanocomposites are materials that combine nanoparticles or nanofibers with a matrix of eco-friendly polymers. These composites offer various advantages such as improved mechanical and thermal properties, better resistance to degradation, and a lower environmental footprint. Figure 1 shows the types and properties of smart nanocomposites, which are discussed in Table 2. (Tripathy et al. 2024).

3.1 Types of Smart Nanocomposites

3.1.1 Natural Fiber-Reinforced Nanocomposites

These composites incorporate natural fibers (e.g., Hemp, flax, jute) into a polymer matrix. Natural fibers are an indiscernible part of composite materials (Mishra et al. 2022). They serve as the reinforcing component, providing unique mechanical,

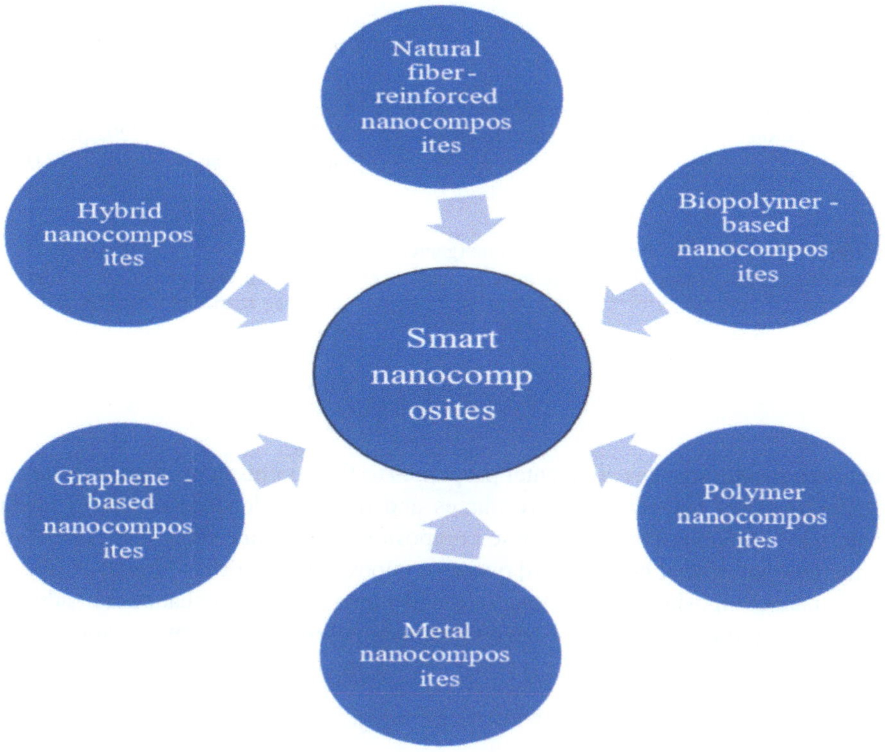

Fig. 1 Types of smart nanocomposites

Table 2 Properties of smart (sustainable) nanocomposites

Smart nanocomposites	Description	Properties	Reference
Natural fiber-reinforced nanocomposites	Natural fibers (e.g., Hemp, flax, jute) combine with nanomaterials in a polymer matrix	Eco-friendly, biodegradable, enhances mechanical properties, is lightweight, and is renewable	Mishra et al. (2022), Bhat et al. (2021)
Biopolymer-based nanocomposites	Biodegradable and renewable biopolymers (e.g., PLA, PHA) are used as the matrix material	Renewable, environmentally friendly, biodegradable, and often used in packaging	Sumrith et al. (2019), Mishra et. al. (2024)
Polymer nanocomposites	Nanoparticles (e.g., clay, silica) dispersed in synthetic or natural polymer	Barrier properties, flame resistance, thermal stability, enhanced strength	Khan et al. (2016), Beryl and Xavier (2023)
Metal nanocomposites	Incorporate metal nanoparticles (e.g., silver, copper) in polymer matrix	Corrosion resistance, high electrical conductivity, antimicrobial properties	Akinniyi et al. (2024), Muresan (2023)
Graphene-based nanocomposites	Use graphene or graphene oxide as the reinforcing agent in the matrix	High strength, lightweight, flexible, and exceptional electrical, and thermal conductivity	Razaq et al. (2022), Zafar et al. (2024), Ji et al. (2016)
Hybrid nanocomposites	Combine organic and inorganic nanofillers to optimize performance	Enhanced mechanical strength, increased barrier properties, biodegradable, thermally stable	Reit et al. (2015), Chen et al. (2015)

thermodynamic, and environmental properties to the composite. Natural reinforced nanocomposites use renewable resources and decrease dependency on synthetic materials (Malkapuram 2023) These composites incorporate inorganic nanofillers to increase interfacial bonding and overall composite properties (Rabbi et al. 2023). These nanocomposites are functionalized with nanoparticles to enhance their adsorption capacity for pollutants and have significant potential to remove various pollutants from wastewater, particularly heavy metals and organic dyes (Castellanos et al. 2024).

3.1.2 Biopolymer-Based Nanocomposites

These nanocomposites utilize natural biopolymers such as chitosan and cellulose, starch, and alginate as a matrix. These offer eco-friendly alternatives to petroleum-based materials. Biopolymers can be combined with inorganic nanoparticles like metal oxides and clay to create eco-friendly and biodegradable materials with excellent pollutant removal capabilities. They are particularly useful in water treatment and soil remediation. Their applications include food packaging and agriculture, where they serve as slow-release carriers for agrochemicals (Wypij et al. 2023). These materials are particularly effective against heavy metals, organic dyes, and nutrients, which make them versatile solutions for water treatment (Peter et al. 2024; Akpınar et al. 2024).

3.1.3 Polymer Nanocomposites

Polymer nanocomposites combine polymers with nanoscale reinforcements like graphene to enhance properties like heat resistance and electrical conductivity (Ray et al. 2024). These nanocomposites are widely used in the automotive and aerospace industries due to their lightweight and durable nature (Akinniyi et al. 2024). Polymer nanocomposites are effective for removing various pollutants, especially organic contaminants such as dyes and phenols. These composites offer improved mechanical, thermal, and chemical properties, which make them suitable for filtration, adsorption, and catalysis in pollutant removal (Bahadi et al. 2024).

3.1.4 Metal Nanocomposites

These nanocomposites incorporate metal nanoparticles (e.g., silver, iron, gold) to improve electrical conductivity and thermal properties, which are suitable for electronic applications. (Akinniyi et al. 2024). Metal-based nanoparticles, such as iron oxide and calcium-based nanoparticles, have been used for phosphorus adsorption selectively from streams of wastewater, which allows the recovery of phosphate-rich materials, and can be used as soil modifications and fertilizers (Ugwuanyi et al. 2024). Metal nanocomposites are effective in removing various pollutants from wastewater, especially heavy metals, dyes, micropollutants, and pathogens. These nanocomposites are particularly effective for catalytic degradation of organic pollutants or adsorbing heavy metals due to the high reactivity and surface area of the metal nanocomposites (Al-gethami et al. 2024).

3.1.5 Graphene-Based Nanocomposites

Graphene and its derivatives (e.g., Reduced graphene oxide) are incorporated into composites because of their outstanding conductivity, surface area, and mechanical properties (Razaq et al. 2022). These composites are essential in sensing applications because of their remarkable strength and conductivity, which enable them to detect a wide range of analytes efficiently. (Sahu et al. 2023) Graphene-based nanocomposites show significant promise in the removal of various pollutants from wastewater, which includes heavy metals, organic dyes, and pharmaceuticals. These nanocomposites are also useful for the adsorption of organic pollutants and toxic ions and for catalytic degradation (Motitswe et al. 2024).

3.1.6 Hybrid Nanocomposites

These nanocomposites combine different fibers and nanoparticles to optimize performance for specific applications. For example, kenaf/nano clay/PLA composites remarkably improve impact resistance and mechanical properties (Moyo 2023). Synthesizing hybrid nanocomposites by deposition of fine particles on larger-sized solid particles has proven an effective method and approach to tackle technical challenges in various applications like gas separation and wastewater treatment (Thondavada et al. 2019). These nanocomposites emerged as effective materials for removing various pollutants from wastewater, which include organic compounds, heavy metals, and pathogens. Various types of nanocomposites can be made by changing their composition, such as surfactants, mobile materials, nanofillers, or polymers (Tesh and Scott 2014).

4 Mechanism of Pollutant Removal

In the removal of pollutants from wastewater, various types of mechanisms are used. Some mechanisms are specific to pollutants and some are applicable to a variety of pollutants. Various mechanisms of pollutants with their efficiency have been discussed in Table 3.

4.1 Adsorption Mechanism

Gendy et al. 2021 demonstrated that covalent organic frameworks (COFs) possess high adsorption capabilities, large surface areas, and structural recyclability, making them suitable for the efficient elimination of heavy metals. The presence of functional groups and the arrangement of substances within COFs further enhance their effectiveness in capturing contaminants.

Kolya et al. (2023), observe that biopolymer-based nanocomposites demonstrate significant potential for effective water treatment through enhanced adsorption mechanisms. The integration of various interactions, such as electrostatic forces, surface complexation, and π-π stacking, along with the synergistic effects of combining

Table 3 Mechanism of smart nanocomposites

Type of smart nanocomposites	Mechanism	Pollutants	Efficiency	Reference
Natural fiber-reinforced nanocomposites	Adsorption, surface interaction, bioadsorption	Heavy metal, organic dye	High adsorption capacity, and environmental stability. It may have lower stability in harsh conditions	Mohammed et al. (2023), Wong et al. (2021), Maia et al. (2023), Zhang ct al. (2022), Rabbi et al. (2024), Castellanos et al. (2024), Haroon et al. (2023)
Biopolymer-based nanocomposites	Adsorption, chemical reaction, electrostatic interaction, coagulation	Heavy metal, organic dye, nutrients, organic pollutants	High-dimensional stability, processibility, water resistance	Xiang et al. (2024), Yaashikaa et al. (2022), Udayakumar et al. (2021), Nangia et al. (2018), Glaskova-Kuzmina et al. (2021)
Polymer nanocomposites	Adsorption, catalysis, ion exchange	Organic contaminants such as dyes and phenols	High efficiency for organic contaminants, but performance can degrade over multiple cycles	Melinte et al. (2019), Li et al. (2022), Intisar et al. (2023), Gusain et al. (2019), Nechifor (2016), Enesca et al. (2022), Amari et al. (2021)
Metal nanocomposites	Adsorption, redox reactions, catalysis	Heavy metals, dyes, micropollutants, and pathogens	Excellent for selective removal of specific contaminants, but long-term stability can be an issue	Boher et al. (2023), Natarajan et al. (2023), Ahmad et al. (2024), Al-Gethami et al. (2024), Hyder and Mir (2022), Castro-Muñoz et al. (2021)

(continued)

Table 3 (continued)

Type of smart nanocomposites	Mechanism	Pollutants	Efficiency	Reference
Graphene-based nanocomposites	Adsorption, filtration, catalysis, redox reactions	Heavy metals, organic dyes, and pharmaceuticals	Extremely high adsorption capacity and recyclability, though cost and scalability remain challenges	Donga et al. (2021), Thakur and Kandasubramanian (2019), Goyat et al. (2022), Tewari et al. (2023), Van Tran et al. (2020)
Hybrid nanocomposites	Adsorption, catalysis	Heavy metals, organic impurities	High performance across a range of pollutants with synergetic effects, but complexity in synthesis and stability	Chauhan et al. (2022), Sadjadi (2020), Yu et al. (2020), Muneeshwaran et al. (2021)

biopolymers with nanofillers. It leads to improved performance in removing contaminants from water. However, challenges related to the scalability, stability, and regeneration of these materials must be addressed to fully realize their potential practical applications.

Xu et al. (2018) focused on iron oxide nanoparticles, specifically Fe_2O_3 and Fe_3O_4, which showed promising adsorptive capacities for arsenic, with values reaching 24.05 mg g^{-1} for trivalent arsenic and 47.41 mg g^{-1} for pentavalent arsenic. The development of GO-$MnFe_2O_4$ nanohybrids significantly improved these capacities to 146 mg g^{-1} and 207 mg g^{-1}, respectively, due to the superior oxidation properties of tetravalent manganese complexes compared to trivalent iron complexes.

The adsorption isotherm models, including Langmuir, Freundlich, Redlich-Peterson, and Dubinin-Radushkevich, provide insights into the nature of adsorption processes. The Langmuir model is typically associated with monolayer adsorption on uniform surfaces, while the Freundlich model accounts for heterogeneous surfaces. The Redlich-Peterson model serves as a hybrid approach, and the Dubinin-Radushkevich model helps differentiate between physical and chemical adsorption processes (Mashile et al. 2020).

Hao et al. (2018) found that the adsorption of arsenic onto magnetic nanoparticles (MNPs) is endothermic, suggesting that higher temperatures can enhance adsorption efficiency. The formation of bidentate binuclear corner-sharing complexes was proposed as a mechanism for As(V) adsorption on MNPs (Farhan et al. 2023).

4.2 Catalytic Degradation

Catalytic degradation processes utilize nanocatalysts like palladium and platinum nanoparticles to break down organic pollutants into less harmful byproducts through oxidation or reduction reactions. These nanocatalysts generate reactive oxygen species (ROS), facilitating efficient pollutant removal under mild conditions. Photocatalytic nanomaterials, such as titanium dioxide and zinc oxide, harness solar energy to degrade organic contaminants and disinfect wastewater (Fu et al. 2022; Lu and Astruc 2020; Zhao et al. 2022).

4.3 Membrane Filtration

Membrane filtration involves the use of semi-permeable membranes to separate contaminants from water based on size and charge. Figure 2 shows the membrane filtration process. There are two main types of membranes used in water treatment:

Symmetric (Homogeneous) Membranes: These membranes consist of a single polymer material with uniform pore sizes throughout the membrane.

Asymmetric (Heterogeneous) Membranes: These membranes have two different layers: a thin outer layer that controls selectivity and a thicker porous inner layer that provides structural support. Due to their thinner structure, asymmetric membranes generally offer higher water flux.

Membrane filtration techniques include:

Microfiltration: It removes macroparticles and some microorganisms.

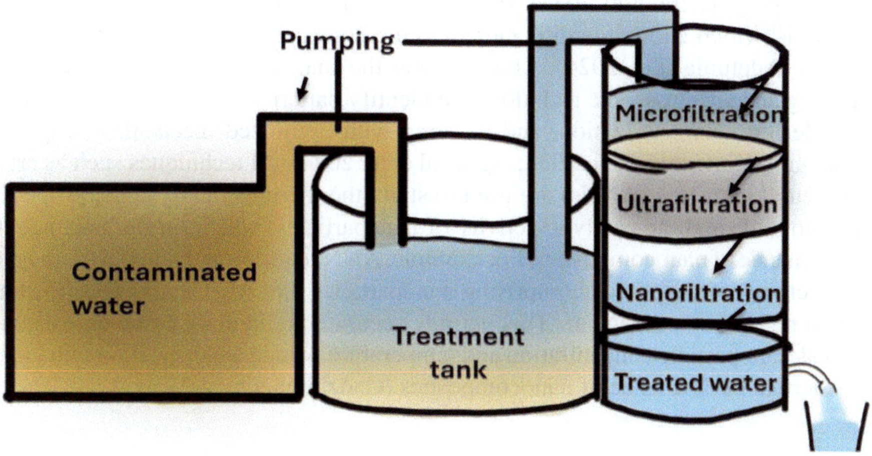

Fig. 2 Membrane filtration process

Ultrafiltration: It targets smaller particles, including organic colloids and some dissolved substances.

Nanofiltration: It operates between ultrafiltration and reverse osmosis and effectively separates dyes and salts.

Reverse osmosis: This process removes dissolved ions and small molecules.

Despite their effectiveness, membrane filtration faces challenges like fouling, which takes place when unwanted substances accumulate on the membrane surface, reducing efficiency, and increasing energy consumption. Fouling can result from the decomposition of particulate matter, scaling, biofilm formation, or the adsorption of organic compounds (Mpala et al. 2023; Kolya et al. 2022; Barman et al. 2021; Baig et al. 2022; Guo et al. 2021; Feng et al. 2022; Hakami et al. 2020; Kotobuki et al. 2021; Cevallos-Mendoza et al. 2022).

5 Integration of Artificial Intelligence

5.1 AI in Optimizing Nanocomposite Performance

AI can optimize the performance of nanocomposites by predicting interactions between nanomaterials and pollutants, recommending material formulations, and optimizing process parameters for maximum pollutant removal efficiency. AI techniques are machine learning models, data-driven insights, and thermal dynamic properties. Mechanical learning techniques are especially used to predict the mechanical performance of polymer nanocomposites. For example, models trained on datasets from tensile tests demonstrate how silver-coated gold nanoparticles (AgC NPs) improve material properties and optimize compositions for desired properties. (Alshoraihy and Ibrahim 2024) Machine learning also enables rapid screening of materials, which allows identification of tailored nanocomposites for specific applications (Adetunla et al. 2024). AI accelerates the analysis of complex datasets from imaging and spectroscopic techniques to identify patterns that inform nanocomposites' design, characterization, and tailoring with optimized mechanical responses (Gadzhimagomedova et al. 2022; Agrawal et al. 2024). AI techniques such as artificial neural networks (ANNs) are used to study the thermal and dynamic properties of nanofluids and the catalytic activity of nanoparticles, which further enhance the performance of nanocomposites. For example, ANNs can correctly model the thermal conductivity of nanofluids comprising nanoparticles like AL_2O_3 and TiO_2 to attain a high regression coefficient. This permits accurate predictions based on variables such as nanoparticle concentration and temperature, which are critical for enhancing the thermal performance of nanocomposites (Gadzhimagomedova et al. 2022).

5.2 *AI Techniques for Pollutant Detection and Analysis*

AI techniques for the detection and analysis of pollutants are nanosensorics, predictive modeling, and data processing. AI methods are applied in nanosensorics to detect pollutants and enable real-time monitoring and analysis of environmental contaminants, which is essential for timely responses to environmental hazards. Nanosensors using nanoparticles can selectively adsorb and detect contaminants like dyes in wastewater. One of the promising applications of nanosensorics is the development of electronic noises in which sensor arrays are combined with AI algorithms. These devices can analyze volatile organic compounds (Gadzhimagomedova et al. 2022). Machine learning algorithms predict the biological and environmental impacts of nanomaterials, aiding in assessing their safety and effectiveness in pollutant remediation. This includes two types of models: the traditional model and the deep learning method. Traditional methods include linear and non-linear regression, decision trees, and support vector machines. They have been widely used to model the biological properties of nanomaterials based on physicochemical descriptors. Deep learning methods include more complex architectures such as convolutional neural networks (CNNs) and generative adversarial networks (GANs), which are capable of automatically generating useful descriptors from raw data that can enhance the predictive power of models without extensive manual input (Winkler 2020). Advanced AI techniques process large datasets generated from environmental monitoring, improving the accuracy and efficiency of pollutant detection (Winkler 2020). While AI significantly enhances the performance of nanocomposites and pollutant detection, challenges remain in fully harnessing its potential, particularly in automating data analysis and ensuring the safety of nanomaterials in environmental applications (Winkler 2020; Chen et al. 2023).

6 Eco-friendly Nanocomposite Materials

Natural and biodegradable materials such as biopolymers play an important role in reducing the environmental footprint of nanocomposites. For example, nanocellulose exhibits high mechanical strength and biodegradability, which make it ideal for sustainable applications. (Norizan et al. 2022) Chitosan enhances antibacterial properties in nanocomposites and is suitable for biomedical applications. Lignin is utilized as a nanofiller; it enhances the properties of composites while being derived from waste materials (Puglia et al. 2022). Eco-friendly materials should decompose naturally to minimize long-term waste. They must also be designed for reuse and to reduce resource depletion. The production process should mitigate energy consumption and emissions (low carbon footprint) and its source materials should be sustainable. These components should be non-toxic (Purnama et al. 2022; Norizan et al. 2022). For nanocomposites, criteria for eco-friendliness should involve biodegradability, low environmental toxicity, sustainable sourcing of raw materials, energy-efficient

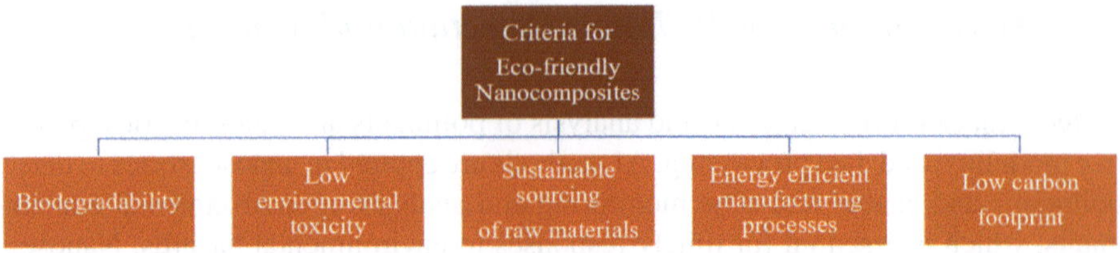

Fig. 3 Criteria for eco-friendly nanocomposites

manufacturing processes, and a low carbon footprint which as shown in Fig. 3. When exposed to environmental conditions, nanocomposite materials should dissociate into non-toxic materials over time. Nanocomposites made from biodegradable polymers or containing biodegradable nanoparticles can decompose naturally and minimize environmental impacts (Matei et al. 2022). Sustainable nanocomposites should utilize materials that are non-toxic and designed to avoid harmful effects when they degrade. Furthermore, low-toxic nanoparticles that depend on natural minerals (e.g., silica, clay) or non-toxic metals (e.g., silver, zinc) are preferable (Banu and Chakraborty 2024) Nanocomposite production should rely on renewable, recyclable sources. For example, using agro-waste (Silva et al. 2022). Manufacturing processes should use less energy, optimize the use of raw materials, and employ renewable energy resources (Das et al. 2022). Reducing the carbon footprint of nano-composite materials includes using renewable energy in production, minimizing transportation, and incorporating energy-efficient material choices (Vafaeva and Zegait 2023). Eco-friendly nanocomposite use ensures that nanocomposites perform effectively in pollutant removal and minimize long-term environmental damage (Bhatt et al. 2022). Additionally, their recyclability and low toxicity contribute to a circular economy in environmental remediation (Shaik and Franchetti 2022).

7 Challenges

Scalability, stability, and durability, cost-effectiveness, and environmental impact of nanocomposites present significant challenges to their widespread use. The production of nanocomposites at a large scale while maintaining their effectiveness and desired properties remains a major hurdle. Ensuring the stability and, over time, effectiveness of nanocomposites, especially in harsh conditions, is very important for practical applications. Furthermore, the synthesis of advanced nanocomposites can be expensive, limiting their widespread adaptability in industrial applications. Moreover, the long-term environmental effects of using nanocomposites include potential toxicity and biodegradability, which need thorough investigation.

8 Future Research and Development Needs

There is a need to develop cost-effective and scalable synthesis methods for nanocomposites with a focus on their durability and reusability. The integration of advanced artificial intelligence (AI) is also essential to improve the performance of nanocomposites; for instance, AI can be used for dynamic system optimization to improve pollutant removal efficiency. Assessment of long-term environmental impacts (e.g., potential toxicity, biodegradability) of nanocomposites is also important to ensure the safe application of nanocomposites.

9 Conclusion

This study concludes that smart nanocomposites hold great promise for effective pollutant removal of pollutants. Their unique properties enable target remediation strategies that can significantly improve water quality and environmental health. However, addressing the outlined challenges and future developments is also important. Continued research and collaboration among scientists, industry, stakeholders, and regulatory bodies will be crucial to harness the full potential of nanocomposites.

References

Adetunla A, Akinlabi E, Jen TC, Ajibade SS (2024, April) Harnessing the power of artificial intelligence in materials science: an overview. In: International conference on science, engineering and business for driving sustainable development goals (SEB4SDG). IEEE, pp 1–6

Afshan N, Bari A, Ashar A, Saleem N (2022) Treatment techniques of industrial effluents and wastewater treatment plants. In: Advanced oxidation processes for wastewater treatment. CRC Press, pp 235–247

Agrawal R, Tilak P, Devand A, Bhatnagar A, Gupta P (2023, June) Artificial intelligence in nanotechnology. In: International conference on data science and big data analysis. Springer Nature Singapore, Singapore, pp 239–248

Ahamad A, Madhav S, Singh AK, Kumar A, Singh P (2020) Types of water pollutants: conventional and emerging. In: Sensors in water pollutants monitoring: role of material. pp 21–41

Ahmad I, Aftab MA, Fatima A, Mekkey SD, Melhi S, Ikram S (2024) A comprehensive review on the advancement of transition metals incorporated on functional magnetic nanocomposites for the catalytic reduction and photocatalytic degradation of organic pollutants. Coord Chem Rev 514:215904

Akhtar MS, Ali S, Zaman W (2024) Innovative adsorbents for pollutant removal: exploring the latest research and applications. Molecules 29(18):4317

Akinniyi JN, Atanda SA, Ariyo DO, Ahmed T, Ayanda IS, Badmos FO et al (2024) Perspective chapter: nanocomposites–unlocking the potentials for diverse applications

Akpınar Ş, Gurbuz F, Özcan S, Odabaşı M (2024) Biopolymeric composite columns for improving water quality in a freshwater stream receiving wastewater treatment plant effluents. J Polym Environ 1–13

Al-Gethami W, Qamar MA, Shariq M, Alaghaz ANM, Farhan A, Areshi AA, Alnasir MH (2024) Emerging environmentally friendly bio-based nanocomposites for the efficient removal of dyes and micropollutants from wastewater by adsorption: a comprehensive review. RSC Adv 14(4):2804–2834

Ali I, Naz I, Peng C, Abd-Elsalam KA, Khan ZM, Islam T et al (2021) Sources, classifications, constituents, and available treatment technologies for various types of wastewater: an overview. Aquananotechnology 11–46

Alshoraihy A, Ibrahim A (2024) Innovative vision: exploring artificial intelligence and machine learning for advancing polymer nanocomposites. In: XXVII International conference on soft computing and measurements (SCM). pp 196–199

Amari A, Mohammed Alzahrani F, Mohammedsaleh Katubi K, Salem Alsaiari N, Tahoon MA, Ben Rebah F (2021) Clay-polymer nanocomposites: preparations and utilization for pollutants removal. Materials 14(6):1365

Asghar N, Hussain A, Nguyen DA, Ali S, Hussain I, Junejo A, Ali A (2024) Advancement in nanomaterials for environmental pollutants remediation: a systematic review on bibliometrics analysis, material types, synthesis pathways, and related mechanisms. J Nanobiotechnol 22(1):26

Bahadi SA, Drmosh QA, Onaizi SA (2024) Adsorptive removal of organic pollutants from aqueous solutions using novel GO/bentonite/MgFeAl-LTH nanocomposite. Environ Res 248:118218

Baig N, Matin A, Faizan M, Anand D, Ahmad I, Khan SA (2022) Antifouling low-pressure highly permeable single step produced loose nanofiltration polysulfone membrane for efficient Erichrome Black T/divalent salts fractionation. J Environ Chem Eng 10(4):108166

Banu KS, Chakraborty P (2024) An overview of bio-assisted nanoparticles: synthesis, application and challenges in nature's toolbox. Nano-Struct Nano-Objects 39:101317

Barman SR, Roy U, Das P, Mukhopadhayay A (2021) Membrane processes for removal of polyaromatic hydrocarbons from wastewater

Beryl JR, Xavier JR (2023) Halloysite for clay–polymer nanocomposites: effects of nanofillers on the anti-corrosion, mechanical, microstructure, and flame-retardant properties—a review. J Mater Sci 58(27):10943–10974

Bhat KM, Rajagopalan J, Mallikarjunaiah R, Rao NN, Sharma A (2021) Eco-friendly and biodegradable green composites. IntechOpen, London, UK

Bhatt P, Pandey SC, Joshi S, Chaudhary P, Pathak VM, Huang Y, Wu X, Zhou Z, Chen S (2022) Nanobioremediation: a sustainable approach for the removal of toxic pollutants from the environment. J Hazard Mater 427:128033

Boher S, Ullah R, Tuzen M, Saleh TA (2023) Metal doped nanocomposites for detection of pesticides and phenolic compounds by colorimetry: trends and challenges. OpenNano 13:100168

Castellanos HG, Aryanfar Y, Mohtaram S, Keçebaş A, Karaca-Dolgun G, Ahmad S et al (2024) The efficacy of nano-cellulose-based composites in heavy metal removal from wastewater: a comprehensive review. J Chem Technol Biotechnol

Castro-Muñoz R, González-Melgoza LL, García-Depraect O (2021) Ongoing progress on novel nanocomposite membranes for the separation of heavy metals from contaminated water. Chemosphere 270:129421

Cevallos-Mendoza J, Amorim CG, Rodríguez-Díaz JM, Montenegro MDCB (2022) Removal of contaminants from water by membrane filtration: a review. Membranes 12(6):570

Chauhan N, Singh V, Kumar S, Vohra A (2022) Hydrothermal synthesis of nitrogen and iron-doped ZnO/MWCNTs nanocomposites for enhanced photocatalytic performance. Arab J Sci Eng 47(6):6989–6999

Chen GG, Qi XM, Li MP, Guan Y, Bian J, Peng F et al (2015) Hemicelluloses/montmorillonite hybrid films with improved mechanical and barrier properties. Sci Rep 5(1):16405

Chen Z, Li Y, Cai Y, Wang S, Hu B, Li B et al (2023) Application of covalent organic frameworks and metal–organic frameworks nanomaterials in organic/inorganic pollutants removal from solutions through sorption-catalysis strategies. Carbon Res 2(1):8

Das O, Babu K, Shanmugam V, Sykam K, Tebyetekerwa M, Neisiany RE et al (2022) Natural and industrial wastes for sustainable and renewable polymer composites. Renew Sustain Energy Rev 158:112054

Donga C, Mishra SB, Abd-El-Aziz AS, Mishra AK (2021) Advances in graphene-based magnetic and graphene-based/TiO_2 nanoparticles in the removal of heavy metals and organic pollutants from industrial wastewater. J Inorg Organomet Polym Mater 31(2):463–480

Enesca A, Cazan C (2022) Polymer composite-based materials with photocatalytic applications in wastewater organic pollutant removal: a mini review. Polymers 14(16):3291

Farhan A, Zulfiqar M, Samiah, Rashid EU, Nawaz S, Iqbal HM et al (2023) Removal of toxic metals from water by nanocomposites through advanced remediation processes and photocatalytic oxidation. Curr Pollut Rep 9(3):338–358

Feng X, Peng D, Zhu J, Wang Y, Zhang Y (2022) Recent advances of loose nanofiltration membranes for dye/salt separation. Sep Purif Technol 285:120228

Fu Y, Yin Z, Qin L, Huang D, Yi H, Liu X et al (2022) Recent progress of noble metals with tailored features in catalytic oxidation for organic pollutants degradation. J Hazard Mater 422:126950

Gadzhimagomedova ZM, Pashkov DM, Kirsanova DY, Soldatov SA, Butakova MA, Chernov AV, Soldatov AV (2022) Artificial intelligence for nanostructured materials. Nanobiotechnol Rep 17(1):1–9

Gendy EA, Ifthikar J, Ali J, Oyekunle DT, Elkhlifia Z, Shahib II et al (2021) Removal of heavy metals by covalent organic frameworks (COFs): a review on its mechanism and adsorption properties. J Environ Chem Eng 9(4):105687

Glaskova-Kuzmina T, Starkova O, Gaidukovs S, Platnieks O, Gaidukova G (2021) Durability of biodegradable polymer nanocomposites. Polymers 13(19):3375

Goyat R, Saharan Y, Singh J, Umar A, Akbar S (2022) Synthesis of graphene-based nanocomposites for environmental remediation applications: a review. Molecules 27(19):6433

Guo S, Wan Y, Chen X, Luo J (2021) Loose nanofiltration membrane custom-tailored for resource recovery. Chem Eng J 409:127376

Gusain R, Gupta K, Joshi P, Khatri OP (2019) Adsorptive removal and photocatalytic degradation of organic pollutants using metal oxides and their composites: a comprehensive review. Adv Coll Interface Sci 272:102009

Hakami MW, Alkhudhiri A, Al-Batty S, Zacharof MP, Maddy J, Hilal N (2020) Ceramic microfiltration membranes in wastewater treatment: filtration behavior, fouling and prevention. Membranes 10(9):248

Hao L, Liu M, Wang N, Li G (2018) A critical review on arsenic removal from water using iron-based adsorbents. RSC Adv 8(69):39545–39560

Haroon K, Kherb J, Jeyaseelan C, Sen M (2023) Recent advances and sustainable approaches towards efficient wastewater treatment using natural waste derived nanocomposites: a review. Nat Environ Pollut Technol 22(3)

Hossain I, Ahmed MK, Chowdhury KA, Moniruzzaman M, Shampa MTA (2024) Ecological risk assessment of oil and grease (OG) and heavy metals in the surface water of Naf River, Bangladesh. Heliyon 10(9)

Hyder MMZ, Mir SH (2022) Performance of Metal-based nanoparticles and nanocomposites for water decontamination. In: Inorganic-organic composites for water and wastewater treatment: volume 2, pp 65–112

Intisar A, Ramzan A, Hafeez S, Hussain N, Irfan M, Shakeel N et al (2023) Adsorptive and photocatalytic degradation potential of porous polymeric materials for removal of pesticides, pharmaceuticals, and dyes-based emerging contaminants from water. Chemosphere 336:139203

Ji L, Meduri P, Agubra V, Xiao X, Alcoutlabi M (2016) Graphene-based nanocomposites for energy storage. Adv Energy Mater 6(16):1502159

Jurczynski Y, Passos R, Campos LC (2024) A review of the most concerning chemical contaminants in drinking water for human health. Sustainability 16(16):7107

Kanchana CM, Chandrasekara NK, Weerasinghe KDN, Pathirana S, Piyadasa RU (2021) Heavy metal, oil and grease pollution of water and sediments in estuarine lagoons in Sri Lanka: a case study in negombo estuarine lagoon. In: Multi-hazard early warning and disaster risks. Springer International Publishing, pp 429–442

Khan WS, Hamadneh NN, Khan WA (2016) Polymer nanocomposites–synthesis techniques, classification and properties. Sci Appl Tailored Nanostruct 50

Khan SA, Ponce P, Yu Z, Golpîra H, Mathew M (2022) Environmental technology and wastewater treatment: strategies to achieve environmental sustainability. Chemosphere 286:131532

Kolya H, Singh VK, Kang CW (2022) Polymeric membranes and hybrid techniques for water purification applications. In: Polymer-based advanced functional materials for energy and environmental applications. Springer Singapore, Singapore, pp 75–91

Kolya H, Kang CW (2023) Next-generation water treatment: exploring the potential of biopolymer-based nanocomposites in adsorption and membrane filtration. Polymers 15(16):3421

Kotobuki M, Gu Q, Zhang L, Wang J (2021) Ceramic-polymer composite membranes for water and wastewater treatment: bridging the big gap between ceramics and polymers. Molecules 26(11):3331

Kumar S, Singh RP, Kalambukattu JG (2021) Modeling daily surface runoff, sediment and nutrient loss at watershed scale employing Arc-APEX model interfaced with GIS: a case study in Lesser Himalayan landscape. Environ Earth Sci 80(15):498

Li H, Cheng B, Gao W, Feng C, Huang C, Liu Y et al (2022) Recent research progress and advanced applications of silica/polymer nanocomposites. Nanotechnol Rev 11(1):2928–2964

Lu F, Astruc D (2020) Nanocatalysts and other nanomaterials for water remediation from organic pollutants. Coord Chem Rev 408:213180

Maia LS, da Silva AIC, Zanini NC, Carvalho LT, Pereira PHF, Medeiros SF et al (2023) Natural fiber-based adsorbents for heavy metals and dyes removal. In: Materials from natural sources. CRC Press, pp 46–92

Malkapuram D (2023) Development of hybrid natural fiber reinforced composite material for automotive applications (No. 2023-28-0131). SAE Technical Paper

Mashile GP, Mpupa A, Nqombolo A, Dimpe KM, Nomngongo PN (2020) Recyclable magnetic waste tyre activated carbon-chitosan composite as an effective adsorbent rapid and simultaneous removal of methylparaben and propylparaben from aqueous solution and wastewater. J Water Process Eng 33:101011

Masoud MS, Zidan AA, El Zokm GM, Elsamra RM, Okbah MA (2024) Humic acid and nano-zeolite NaX as low cost and eco-friendly adsorbents for removal of Pb (II) and Cd (II) from water: characterization, kinetics, isotherms and thermodynamic studies. Biomass Conver Biorefinery 14(3):3615–3632

Matei E, Predescu AM, Râpă M, Ţurcanu AA, Mateş I, Constantin N, Predescu C (2022) Natural polymers and their nanocomposites used for environmental applications. Nanomaterials 12(10):1707

Melinte V, Stroea L, Chibac-Scutaru AL (2019) Polymer nanocomposites for photocatalytic applications. Catalysts 9(12):986

Mir RA, Mantoo AG, Sofi ZA, Bhat DA, Bashir A, Bashir S (2023) Types of environmental pollution and its effects on the environment and society. In: Geospatial analytics for environmental pollution modeling: analysis, control and management. Springer Nature Switzerland, Cham, pp 1–31

Mishra T, Mandal P, Rout AK, Sahoo D (2022) A state-of-the-art review on potential applications of natural fiber-reinforced polymer composite filled with inorganic nanoparticle. Compos Part C: Open Access 9:100298

Mishra AK, Singh P, Raizada P, Vadivel S, Selvasembian R (eds) (2023) Advanced functional materials and methods for photodegradation of toxic pollutants. Elsevier

Mishra B, Panda J, Mishra AK, Nath PC, Nayak PK, Mahapatra U et al (2024) Recent advances in sustainable biopolymer-based nanocomposites for smart food packaging: a review. Int J Biol Macromol 135583

Mohammed M, Jawad AJAM, Mohammed AM, Oleiwi JK, Adam T, Osman AF et al (2023) Challenges and advancement in water absorption of natural fiber-reinforced polymer composites. Polym Testing 124:108083

Mohammed MN, Zaed MA, Saidur R, Tan KH (2024) Nanocomposite-based solar desalination: recent developments and future prospects. J Water Process Eng 64:105733

Mohanty B, Das A (2023) Heavy metals in agricultural cultivated products irrigated with wastewater in India: a review. AQUA—Water Infrastruct Ecosyst Soc 72(6):851–867

Motitswe MG, Badmus KO, Khotseng L (2024) Application of reduced graphene oxide-zinc oxide nanocomposite in the removal of Pb (II) and Cd (II) contaminated wastewater. Appl Nano 5(3):162–189

Moyo M (2023) Performance of nanoclay infused plant fibre-reinforced hybrid biocomposites under impact loading. Doctoral dissertation

Mpala TJ, Etale A, Richards H, Nthunya LN (2023) Biofouling phenomena in membrane distillation: mechanisms and mitigation strategies. Environ Sci Adv 2(1):39–54

Muneeshwaran M, Srinivasan G, Muthukumar P, Wang CC (2021) Role of hybrid-nanofluid in heat transfer enhancement—a review. Int Commun Heat Mass Transfer 125:105341

Muresan LM (2023) Nanocomposite coatings for anti-corrosion properties of metallic substrates. Materials 16(14):5092

Nangia S, Warkar S, Katyal D (2018) A review on environmental applications of chitosan biopolymeric hydrogel based composites. J Macromol Sci Part A 55(11–12):747–763

Narayanan M, Ali SS, El-Sheekh M (2023) A comprehensive review on the potential of microbial enzymes in multipollutant bioremediation: mechanisms, challenges, and future prospects. J Environ Manage 334:117532

Natarajan B, Kannan P, Rather JA, Sheikh RA (2023) Recent developments in metal nanoparticles functionalized nanocomposite adsorbents for heavy metals removal from wastewaters. J Taiwan Inst Chem Eng 147:104942

Nechifor M (2016) Factors influencing the photochemical behavior of multicomponent polymeric materials. In: Photochemical behavior of multicomponent polymeric-based materials, pp 21–65

Norizan MN, Shazleen SS, Alias AH, Sabaruddin FA, Asyraf MRM, Zainudin ES et al (2022) Nanocellulose-based nanocomposites for sustainable applications: a review. Nanomaterials 12(19):3483

Okoro HK, Orosun MM, Oriade FA, Momoh-Salami TM, Ogunkunle CO, Adeniyi AG, Zvinowanda C, Ngila JC (2023) Potentially toxic elements in pharmaceutical industrial effluents: a review on risk assessment, treatment, and management for human health. Sustainability 15(8):6974

Okpala CC (2014) The benefits and applications of nanocomposites. Int J Adv Eng Tech 12:18

Osman AI, Hosny M, Eltaweil AS, Omar S, Elgarahy AM, Farghali M, Yap PS, Wu YS, Nagandran S, Batumalaie K, Gopinath SC (2023) Microplastic sources, formation, toxicity and remediation: a review. Environ Chem Lett 21(4):2129–2169

Pasqual O, Rathnayake A, Thiripuranathar G, Ekanayake S (2024) Main inorganic pollutants and their risk to living beings. In: Nanotechnology-based sensors for detection of environmental pollution. Elsevier, pp 11–38

Peter S, Lyczko N, Thomas S, Leruth D, Germeau A, Fati D, Nzihou A (2024) Fabrication of eco-friendly nanocellulose-chitosan-calcium Phosphate ternary nanocomposite for wastewater remediation. Chemosphere 363:142779

Puglia D, Luzi F, Torre L (2022) Preparation and applications of green thermoplastic and thermosetting nanocomposites based on nanolignin. Polymers 14(24):5470

Purnama P, Samsuri M, Iswaldi I (2022) A review on fully bio-based materials development from polylactide and cellulose nanowhiskers. Polymers 14

Rabbi MS, Das S, Tasneem T, Billah MM, Hasan A (2023) Effect of nano-filler on the manufacturing and properties of natural fiber-based composites: a review. J Eng Adv 4(04):101–115

Rabbi S, Das S, Sharmin S, Al Mamun A (2024) Effect of nanomaterials on the mechanical and morphological properties of jute-GFRP hybrid nanocomposites. Malays J Compos Sci Manuf 14(1):16–33

Rando G, Sfameni S, Galletta M, Drommi D, Cappello S, Plutino MR (2022) Functional nanohybrids and nanocomposites development for the removal of environmental pollutants and bioremediation. Molecules 27(15):4856
Ray SS, Temane LT, Orasugh JT (2024) Polymer nanocomposites. In: Graphene-bearing polymer composites: applications to electromagnetic interference shielding and flame-retardant materials. Springer Nature Switzerland, Cham, pp 1–5
Razaq A, Bibi F, Zheng X, Papadakis R, Jafri SHM, Li H (2022) Review on graphene-, graphene oxide-, reduced graphene oxide-based flexible composites: from fabrication to applications. Materials 15(3):1012
Reit R, Lund B, Voit W (2015) Shape memory polymer–inorganic hybrid nanocomposites. Org-Inorg Hybrid Nanomater 313–350
Sadjadi S (2020) Halloysite-based hybrids/composites in catalysis. Appl Clay Sci 189:105537
Sahu KM, Patra S, Singh SP, Swain SK (2023) Graphene-based polymer nanocomposites for sensing
Sajeena TAM, Pallath N (2024) Industrial waste. In: Handbook of biomass. Springer Nature Singapore, Singapore, pp 231–246
Santos LH, Gros M, Rodriguez-Mozaz S, Delerue-Matos C, Pena A, Barceló D, Montenegro MCB (2013) Contribution of hospital effluents to the load of pharmaceuticals in urban wastewaters: identification of ecologically relevant pharmaceuticals. Sci Total Environ 461:302–316
Sarmah P, Dan MM, Adapa D, Sarangi TK (2018) A review on common pathogenic microorganisms and their impact on human health. Electron J Biol 14(1):50–58
Serajuddin M, Chowdhury MA, Haque MM, Khan TA (2024) The concentration of organic & ammonium pollution and their relationship in River Water: a case study. Sch J Eng Tech 2:82–88
Shaik S, Franchetti MJ (2022) Recycling and sustainability of nanocomposites. In: Applications of nanocomposites. CRC Press, pp 236–259
Sharma J, Sharma S, Soni V (2021) Classification and impact of synthetic textile dyes on Aquatic Flora: a review. Reg Stud Mar Sci 45:101802
Silva LP, Silveira AP, Bonatto CC, Barbosa EF, Medeiros KA, Viol LCDS et al (2022) Sustainable exploitation of agricultural, forestry, and food residues for green nanotechnology applications. In: Biogenic nanomaterials. Apple Academic Press, pp 301–344
Silvestre AJD, Freire CSR, Vilela C (2021) Special issue: advanced biopolymer-based nanocomposites and hybrid materials. Materials 14:493
Singh V (2024) Water pollution. In: Textbook of environment and ecology 2024 Mar 23. Springer Nature Singapore, Singapore, pp 253–266
Sumrith N, Rangappa SM, Dangtungee R, Siengchin S, Jawaid M, Pruncu CI (2019) Biopolymers-based nanocomposites: properties and applications. In: Bio-based polymers and nanocomposites: preparation, processing, properties and performance, pp 255–272
Tesh SJ, Scott TB (2014) Nano-composites for water remediation: a review. Adv Mater 26(35):6056–6068
Tewari C, Tatrari G, Kumar S, Pathak M, Rawat KS, Kim YN et al (2023) Can graphene-based composites and membranes solve current water purification challenges-a comprehensive review. Desalination 116952
Thakur K, Kandasubramanian B (2019) Graphene and graphene oxide-based composites for removal of organic pollutants: a review. J Chem Eng Data 64(3):833–867
Thondavada N, Chokkareddy R, Naidu NV, Redhi GG (2019) Environmental science and engineering applications of polymer and nanocellulose-based nanocomposites. Compos Environ Eng 135–178
Tkaczyk A, Mitrowska K, Posyniak A (2020) Synthetic organic dyes as contaminants of the aquatic environment and their implications for ecosystems: a review. Sci Total Environ 717:137222
Tripathy J, Mishra A, Pandey M, Thakur RR, Chand S, Rout PR, Shahid MK (2024) Advances in nanoparticles and nanocomposites for water and wastewater treatment: a review. Water 16(11):1481
Tyagi I, Goscianska J, Dehghani MH, Karri RR (eds) (2022) Sustainable materials for sensing and remediation of noxious pollutants. Elsevier

Tyagi S, Kapoor RT, Solanki S, Goyal A, Singh R (2024) Nanomaterial mediated wastewater treatment: a new frontier in environmental remediation. In: Microbiome-based decontamination of environmental pollutants. Academic Press, pp 31–49

Udayakumar GP, Muthusamy S, Selvaganesh B, Sivarajasekar N, Rambabu K, Sivamani S et al (2021) Ecofriendly biopolymers and composites: preparation and their applications in water-treatment. Biotechnol Adv 52:107815

Ugwuanyi ED, Nwokediegwu ZQS, Dada MA, Majemite MT, Obaigbena A (2024) The impact of nanomaterials in enhancing wastewater treatment processes: a review. Magna Sci Adv Res Rev 10(1):273–285

Ukaogo PO, Ewuzie U, Onwuka CV (2020) Environmental pollution: causes, effects, and the remedies. In: Microorganisms for sustainable environment and health. Elsevier, pp 419–429

Vafaeva KM, Zegait R (2023) Carbon nanotubes: revolutionizing construction materials for a sustainable future: a review

Van Tran T, Nguyen DTC, Le HT, Vo DVN, Nanda S, Nguyen TD (2020) Optimization, equilibrium, adsorption behavior and role of surface functional groups on graphene oxide-based nanocomposite towards diclofenac drug. J Environ Sci 93:137–150

Wang J, Guo X, Xue J (2021) Biofilm-developed microplastics as vectors of pollutants in aquatic environments. Environ Sci Technol 55(19):12780–12790

Winkler DA (2020) Role of artificial intelligence and machine learning in nanosafety. Small 16(36):2001883

Wong D, Anwar M, Debnath S, Hamid A, Izman S (2021) A review: recent development of natural fiber-reinforced polymer nanocomposites. Jom 73(8):2504–2515

Wu S, Wang Y, Iqbal M, Mehmood K, Li Y, Tang Z, Zhang H (2022) Challenges of fluoride pollution in environment: mechanisms and pathological significance of toxicity—a review. Environ Pollut 304:119241

Wypij M, Trzcińska-Wencel J, Golińska P, Avila-Quezada GD, Ingle AP, Rai M (2023) The strategic applications of natural polymer nanocomposites in food packaging and agriculture: chances, challenges, and consumers' perception. Front Chem 10:1106230

Xiang L, Zheng S, Lei J, Pan D, Li S, Wang J et al (2024) Research progress on preparation and application of sodium alginate aerogel. ES Mater Manufact 25:1215

Xu J, Cao Z, Zhang Y, Yuan Z, Lou Z, Xu X, Wang X (2018) A review of functionalized carbon nanotubes and graphene for heavy metal adsorption from water: preparation, application, and mechanism. Chemosphere 195:351–364

Yaashikaa PR, Kumar PS, Karishma SJER (2022) Review on biopolymers and composites–evolving material as adsorbents in removal of environmental pollutants. Environ Res 212:113114

Yang W, Cai C, Guo Y, Wu H, Guo Y, Dai X (2022) Diversity and fate of human pathogenic bacteria, fungi, protozoa, and viruses in full-scale sludge treatment plants. J Cleaner Prod 380:134990

Yu Z, Lou R, Pan W, Li N, Tang B (2020) Nanoenzymes in disease diagnosis and therapy. Chem Commun 56(99):15513–15524

Zafar M, Imran SM, Iqbal I, Azeem M, Chaudhary S, Ahmad S, Kim WY (2024) Graphene-based polymer nanocomposites for energy applications: recent advancements and future prospects. Res Phys 107655

Zahoor I, Mushtaq A (2023) Water pollution from agricultural activities: a critical global review. Int J Chem Biochem Sci 23(1):164–176

Zhang S, Vanessa C, Khan A, Ali N, Malik S, Shah S et al (2022) Prospecting cellulose fibre-reinforced composite membranes for sustainable remediation and mitigation of emerging contaminants. Chemosphere 305:135291

Zhao W, Adeel M, Zhang P, Zhou P, Huang L, Zhao Y et al (2022) A critical review on surface-modified nano-catalyst application for the photocatalytic degradation of volatile organic compounds. Environ Sci Nano 9(1):61–80

Zhou W, Li M, Achal V (2024) A comprehensive review on environmental and human health impacts of chemical pesticide usage. Emerg Contam 100410

Environmental Footprint and Green Design of Nanocomposites for Wastewater Remediation

Tanaswini Patra, Sankar Prasad Sahu, Jagannath Panda, Sridevi Jadav, Jagadish Kundu, Girija Shankar Behera, Tankadhar Behera, Binay Prakash Panigrahy, Rahul Kumar Singh, Suren Nayak, Saroj Kumar Moharana, and Nirasindhu Desinayak

Abstract Nanocomposites such as metals, metal oxides, carbon-based polymer membranes, and metal–organic framework (MOF) hybrid composites have shown great promise in improving wastewater treatment due to their advanced properties and ability to effectively remove contaminants. However, their environmental impact and sustainability are important considerations as these materials become

---

T. Patra
Iconic-Odisha Adarsha Vidyalaya, Andharua, Bhubaneswar, Odisha, India

S. P. Sahu
Department of Applied Geology, NIT Raipur, Raipur, Chhattisgarh, India

J. Panda (✉)
Department of Industrial Chemistry, Pukyung National University, Busan, South Korea
e-mail: 704720@pknu.ac.kr

Department of Chemistry, NIST University, Institute Park, Berhampur, India

S. Jadav
Department of Applied Geology, IIT (ISM), Dhanbad, Jharkhand, India

J. Kundu
Earth Science Department, Sapienza University of Rome, Rome, Italy

G. S. Behera
Holtec Consulting Pvt Ltd., Gurgaon, Haryana, India

T. Behera · N. Desinayak
Department of Geology, Ravenshaw University, Cuttack, Odisha, India

B. P. Panigrahy
Department of Environmental Science and Engineering, IIT (ISM), Dhanbad, Jharkhand, India

R. K. Singh
Geological Survey of India, Ranchi, Jharkhand, India

S. Nayak
Department of Geology, Utkal University, Bhubaneswar, Odisha, India

S. K. Moharana
Riddhi Siddhi Crusher and Land Transport (Asian Paints), Al Tawyeen, Fujairah, UAE

S. Roy et al. (eds.), *Nanocomposites for Sustainable Wastewater Treatment*, Composites Science and Technology, https://doi.org/10.1007/978-981-95-1369-7_8

more widely used. This chapter examines the potential environmental risks associated with nanocomposites, including the release of nanoparticles into ecosystems, their persistence, and their possible effects on human health and the environment. It also addresses the challenges related to their recycling, disposal, and long-term degradation. In terms of sustainability, the chapter explores energy-efficient production methods, the use of renewable materials, and the potential for green synthesis approaches to minimize environmental harm. Future perspectives focus on the development of biodegradable and eco-friendly nanocomposites, along with strategies for improving their recyclability and reducing their ecological footprint, ensuring that these materials can be both effective in water treatment and sustainable in the long term.

1 Introduction

Water is an indispensable resource, vital for sustaining life on Earth and playing a crucial role in our daily activities and overall survival (Jackson et al. 2001). However, water pollution has become a pressing global challenge, primarily caused by human interventions and natural activities (Kumar et al. 2024). Domestic waste, environmental degradation, and industrial discharges collectively contribute to the increasing levels of water contamination worldwide (Singh et al. 2024a). Industries are significant contributors, with the United Nations reporting that an estimated 300–500 million tons of toxic substances, including solvents, heavy metals, and chemical by-products, are released into water systems annually (Chowdhary et al. 2020). In addition to anthropogenic sources, natural processes also exacerbate water pollution (Häder et al. 2020). Toxic heavy metals such as mercury, zinc, lead, and cadmium are major contributors to water pollution and pose a significant threat to the environment and human health. Both domestic and industrial wastewater are key sources of these contaminants, leading to diseases, environmental degradation, and impacting the quality of life for communities worldwide (Kapoor and Singh 2021; Singh et al. 2023). The increasing toxicity of heavy metals has emphasized the urgent need for effective removal methods to restore and protect marine ecosystems (Oliveira et al. 2024). Notable examples include the Itai-Itai disease in Japan, caused by cadmium contamination, and the Minamata disaster, a result of mercury poisoning (Kaji 2015). Heavy metals accumulate in living tissues because of their non-biodegradable nature and persistence in the environment. Various industries are responsible for discharging heavy metals into water bodies. These include iron and steel production, non-ferrous metal mining, mineral processing, manufacturing, dyeing, painting, photography, electroplating, and metal finishing processes (Kumar et al. 2019). Additionally, heavy metals are released through wastewater from sectors like electrical equipment production, where copper is commonly used, along with the manufacturing of alloys, water pipes, and chemical catalysts. For example, lead is widely used in anti-fouling paints on ship hulls, where it can eventually leach into aquatic systems and accumulate within living organisms (Krishna and Mohan

2014). The presence of heavy metals in wastewater highlights the need for effective treatment and stricter regulations to prevent their harmful impacts on ecosystems and public health (Zhao et al. 2016). A notable example is arsenic contamination, which arises from the natural weathering and erosion of arsenic-bearing rocks. This issue is particularly severe in regions like Nepal, West Bengal, and Bangladesh, where high arsenic levels in groundwater have become a persistent problem. The presence of arsenic in drinking water poses serious health risks, including chronic illnesses such as cancer and cardiovascular diseases, making it a significant public health concern (Raju 2022). Addressing water pollution requires a comprehensive approach that includes regulations of industrial waste, improved water management practices, and innovative solutions to mitigate natural contamination sources like arsenic. The growing demand for clean water has driven advancements in wastewater treatment technologies (Kaur et al. 2017). Among these, nanocomposites considered as a transformative solution because of their superior properties including high surface area, tunable chemical functionality, and enhanced reactivity (Khodakarami and Bagheri 2021). These materials, which include hybrids of metals, metal oxides, carbon-based compounds, polymer membranes, and MOFs as mentioned above, help in Tripathy et al. (2021), demonstrating exceptional efficiency in removing contaminants from water (Vo et al. 2024; Mohammed and Andrzej 2024). However, the widespread adoption of nanocomposites necessitates a thorough examination of their environmental impacts and sustainability. Nanocomposites combine multiple materials at the nanoscale to achieve unique functionalities. For instance, metal nanoparticles exhibit excellent catalytic and antimicrobial properties, while polymers offer flexibility and stability (Dutt et al. 2020). This synergy allows for efficient removal of contaminants, ranging from heavy metals to organic pollutants (Vo et al. 2024; Mohmood et al. 2013). Despite these advantages, the environmental risks associated with the release of nanoparticles and their persistence in ecosystems are significant concerns (Hegde et al. 2016). Furthermore, the production and disposal of nanocomposites often involve energy-intensive processes and hazardous chemicals, challenging their sustainability (Carroccio et al. 2022). This chapter explores the dual aspects of environmental risks and sustainability of nanocomposites in wastewater treatment. It delves into the types of nanocomposites, their environmental impact, challenges in recycling and disposal, and strategies for achieving sustainability. The chapter concludes with future perspectives on designing biodegradable and eco-friendly nanocomposites, emphasizing the importance of balancing performance with ecological responsibility.

2 Types of Nanocomposites in Wastewater Treatment

2.1 Metal and Metal Oxide Nanocomposites

Metal oxide nanocomposites have emerged as highly effective materials for wastewater treatment due to their unique properties, including high surface area, catalytic activity, and tunable chemical functionality. Several metal oxide nanoparticles, such as CuO, TiO_2, Al_2O_3, ZnO, MgO, WO_3, CeO_2, and their composites have been immensely explored for wastewater treatment through adsorptive removal and photocatalytic degradation (Joshi et al. 2020). The size, shape, surface features, and inherent properties of metal oxide NPs and their composites govern the adsorption and photocatalytic processes. TiO_2 is a commonly used nanomaterial, known for its stability and efficiency in photocatalytic degradation of organic pollutants and dyes (Kholief et al. 2024). However, its high bandgap limits its activity to ultraviolet light. Modifications such as doping with metals (e.g., iron) or coupling with reduced graphene oxide enhance its photocatalytic activity under visible light (Deshmukh et al. 2020). ZnO is another widely used nanomaterial, favored for its ability to efficiently separate electron–hole pairs. Its effectiveness in degrading pollutants such as arsenic and organic dyes is further improved when used in composite forms, such as CuO/ZnO (Jeevarathinam and Asharani 2024). Similarly, manganese oxide has been utilized for its high adsorption capacity for heavy metals, including lead and cadmium, especially when combined with carbon-based materials derived from biomass (Wan et al. 2020). Iron oxide (Fe_3O_4) nanocomposites stand out for their magnetic properties, enabling easy recovery and reuse (Hammad et al. 2021). These composites, often combined with graphene or polymers, are highly effective in removing contaminants like arsenic and fluoride (Su et al. 2024). Other materials, such as cerium oxide and zirconium oxide, are used in hybrid nanostructures to address specific pollutants like phosphates and nitrates from agricultural runoff (Su et al. 2015). These advancements demonstrate the versatility and potential of metal oxide nanocomposites in addressing various water contamination challenges effectively and sustainably.

2.2 Carbon-Based Nanocomposites

Carbon-based nanocomposites have emerged as highly effective materials for wastewater treatment as they possess high surface area, tunable porosity, and remarkable adsorption capacities. These materials include graphene oxide, carbon nanotubes, activated carbon hybrids, and fullerene-based composites, which are widely explored for their ability to remove a variety of contaminants such as organic pollutants, heavy metals, dyes, and pathogens from wastewater (Ambika and Singh 2021). Graphene oxide is a functionalized derivative of graphene with oxygen-containing groups (hydroxyl, epoxy, and carboxyl), enhancing its hydrophilicity and surface activity. GO-based composites can efficiently adsorb heavy metals like lead (Pb), cadmium (Cd), and arsenic (As) through electrostatic interactions and chemical bonding (Ahmad et al. 2020). In addition, they exhibit excellent performance in removing dyes such as methylene blue and rhodamine B due to their large surface area and π-π interactions with aromatic pollutants (Xiao et al. 2016). CNTs, both single-walled (SWCNTs) and multi-walled (MWCNTs), are renowned for their tubular nanostructure, high mechanical strength, and electrical conductivity. Functionalized CNTs (e.g., oxidized or amine-modified CNTs) offer enhanced adsorption of contaminants like heavy metal ions and organic compounds. Moreover, their excellent electrical properties allow CNT-based composites to be integrated into electrochemical water treatment systems such as capacitive deionization and electrocoagulation (Chen and Huang 2018). Activated carbon remains a conventional yet highly effective carbon-based material for wastewater treatment. When combined with nanomaterials like metal oxides, polymers, or graphene, activated carbon hybrids demonstrate superior adsorption performance. The high porosity and surface functionality of activated carbon ensure efficient removal of both organic and inorganic pollutants. Functionalization of carbon-based materials is key to enhancing their performance, as it enables selective adsorption and interaction with target contaminants. For instance, surface modification using metal nanoparticles (e.g., Fe, Ag) or polymers can improve pollutant removal efficiency while addressing specific challenges like biofouling or reusability (Agrawal et al. 2021). The integration of carbon-based nanocomposites in advanced treatment technologies, such as photocatalysis, membrane filtration, and electrochemical oxidation, further expands their utility (Nasrollahzadeh et al. 2021). For example, GO-TiO_2 nanocomposites exhibit excellent photocatalytic degradation of organic pollutants under UV or visible light, while CNTs incorporated into membranes enhance filtration efficiency and mechanical strength (Wang et al. 2018) (Fig. 1).

2.3 Polymer-Based Nanocomposites

Polymer-based nanocomposites are emerging as versatile and efficient materials for wastewater treatment due to their enhanced mechanical strength, chemical stability, and functional adaptability (Gayatri et al. 2024). These composites are created by incorporating nanoparticles, such as metal oxides, carbon-based nanomaterials, or functionalized particles, into polymer matrices. The resulting materials exhibit superior performance in removing contaminants such as heavy metals, oil residues, and organic pollutants from wastewater systems (Bhatia et al. 2024). Chitosan-metal hybrids are among the most widely studied polymer-based nanocomposites. Chitosan, a natural biopolymer derived from chitin, offers excellent adsorption properties due to its high density of amino ($-NH_2$) and hydroxyl (–OH) functional groups (Ben Amor et al. 2024). When combined with nanoparticles like Fe_3O_4 or Ag, chitosan-based composites can effectively remove heavy metals like lead (Pb), arsenic (As), and chromium (Cr) via chelation and ion-exchange mechanisms (Xiao

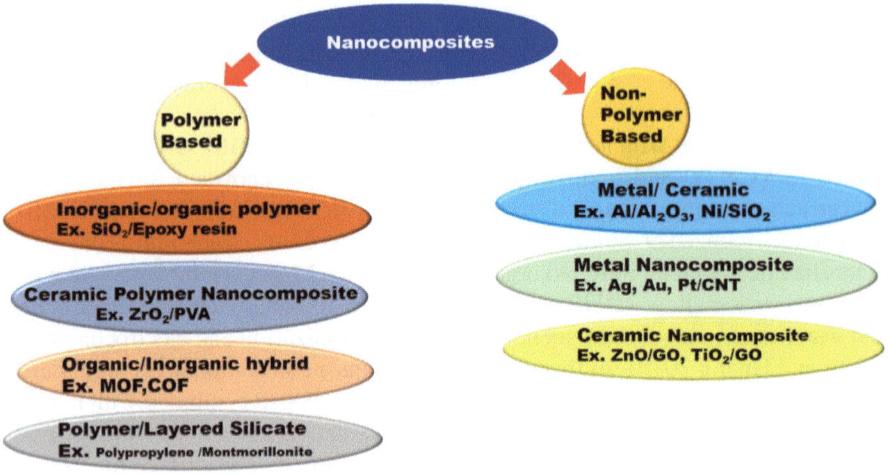

Fig. 1 A schematic diagram of nanocomposite materials

et al. 2021). The addition of metal nanoparticles also introduces antibacterial properties, making these composites particularly valuable for wastewater treatment applications. PVA-based composites are another promising category of polymer-based nanomaterials. PVA is a synthetic, water-soluble polymer that exhibits excellent film-forming ability, mechanical strength, and chemical resistance. Functionalizing PVA with nanoparticles, such as graphene oxide, MOFs, or clay minerals, enhances its adsorption capacity and stability. For instance, PVA-coated composites loaded with ZIF-8 or other MOFs have demonstrated efficient adsorption of heavy metals and dyes while maintaining structural integrity in aqueous environments (Mushtaq et al. 2022; Wang et al. 2023). The flexibility of polymer matrices allows for the incorporation of multiple functionalities into a single material. This includes enhancing pollutant selectivity, improving mechanical properties, and introducing catalytic or antimicrobial activities. For example, polymers combined with TiO_2 nanoparticles create photocatalytic composites that effectively degrade organic pollutants under light irradiation. Similarly, PVA-chitosan blends functionalized with magnetic nanoparticles enable easy separation of contaminants via external magnetic fields, improving operational efficiency. Polymer-based nanocomposites are also utilized in membrane filtration systems. Nanoparticles embedded in polymer membranes enhance filtration performance by increasing hydrophilicity, preventing fouling, and enabling selective rejection of contaminants. These membranes are particularly effective in removing oil residues, heavy metals, and microorganisms from wastewater.

2.4 *Membrane Hybrid Composites*

Membrane-based nanocomposites are widely used in filtration technologies. Incorporating nanoparticles into polymeric membranes enhances their permeability, fouling resistance, and selectivity. Applications include reverse osmosis, ultrafiltration, and nanofiltration (Esfahani et al. 2019). For example, membranes embedded with silver nanoparticles exhibit antimicrobial properties, extending their operational lifespan (Huang et al. 2016).

2.5 *Metal–Organic Framework Hybrid Composites*

MOFs are highly porous crystalline materials composed of metal ions coordinated with organic linkers, offering tunable porosity, high surface area, and functional versatility (Jagannath et al. 2023). These unique properties make MOFs excellent candidates for wastewater treatment applications, particularly in the removal of heavy metals, organic dyes, and pharmaceutical residues (Sahoo et al. 2021). However, challenges such as structural instability in aqueous environments and scalability have driven the development of MOF-based hybrid composites. MOF-carbon hybrids combine the high adsorption capacity of MOFs with the excellent stability and conductivity of carbon materials, such as GO, carbon nanotubes, or activated carbon. For instance, GO-MOF composites have demonstrated enhanced adsorption of heavy metals like Pb(II), Cd(II), and Cr(VI) due to the synergistic interaction between GO's oxygen-containing groups and MOFs' porosity (Wang et al. 2022). Additionally, GO enhances the structural stability of MOFs under aqueous conditions and facilitates the removal of organic dyes (e.g., methylene blue and Congo red) via π-π interactions (Rehman Shah et al. 2021). The integration of carbon materials also enables MOF hybrids to participate in electrochemical water treatment systems, improving pollutant degradation and recovery efficiency. Polymer-MOF hybrids offer further improvements in stability, mechanical strength, and processability. Polymers such as PVA, chitosan, and PANI are commonly used to stabilize MOFs and prevent their structural collapse during water treatment (Ahmadijokani et al. 2022). For instance, PVA-coated UiO-66 composites demonstrate exceptional adsorption of pharmaceutical residues like antibiotics and painkillers, owing to their enhanced structural integrity and functional group interactions (Xing et al. 2022). Similarly, chitosan-MOF hybrids effectively remove heavy metals via chelation and ion exchange, while maintaining biocompatibility and reusability. Photocatalytic MOF hybrids represent another promising category, where MOFs are combined with semiconductor materials like TiO_2 or g-C_3N_4. These composites utilize light energy to degrade organic pollutants through photocatalytic reactions, offering a sustainable solution for wastewater treatment (AbouSeada et al. 2024). For example, TiO_2-MIL-125 composites efficiently degrade dyes and pharmaceutical pollutants under UV or visible light irradiation (Ahmadpour et al. 2020).

3 Environmental Impact of Nanocomposites

3.1 Release of Nanoparticles into Ecosystems

The amplified usage of nanocomposites in various fields, including wastewater treatment, energy, and materials science, raises concerns about their environmental impact, particularly the release of nanoparticles into ecosystems. During manufacturing, usage, and disposal of nanocomposites, nanoparticles can unintentionally enter the environment, posing significant risks to aquatic, terrestrial, and atmospheric systems. In aquatic systems, nanoparticles often leach into water bodies through industrial effluents, wastewater treatment plants, or improper disposal of nanomaterial-based products. Once released, nanoparticles such as metal oxides, carbon-based NPs, or polymer composites can persist in water due to their small size, high reactivity, and resistance to degradation (Singh et al. 2024b). Their ability to adsorb pollutants and interact with natural organic matter can lead to bioaccumulation in aquatic organisms, including algae, fish, and invertebrates. For instance, metal nanoparticles like ZnO or TiO_2 can cause oxidative stress and cellular damage in aquatic species, disrupting normal biological functions (Marisa et al. 2022). The bioaccumulation of nanoparticles in lower trophic-level organisms creates a domino effect in food chains. Nanoparticles consumed by small organisms can be transferred to higher-level predators, leading to biomagnification and long-term ecological damage. Additionally, nanoparticles can alter water quality by influencing pH, turbidity, and microbial activity, which impacts aquatic ecosystems' balance and biodiversity. In terrestrial environments, nanoparticles released during disposal or soil applications may leach into soil and groundwater systems. For example, nanocomposites containing heavy metals like Cd or Pb can contaminate agricultural soil, reducing soil fertility and posing health risks to plants and microorganisms. The toxicity of nanoparticles can also affect microbial communities essential for nutrient cycling, thereby disrupting soil ecosystems (Asghar et al. 2024). Airborne nanoparticles released during the manufacturing process can also pose significant environmental and health risks. Nanoscale particles can remain suspended in the air for long periods, facilitating their inhalation and deposition in terrestrial and aquatic environments through atmospheric deposition. The environmental persistence and potential toxicity of nanoparticles necessitate strategies for monitoring, controlling, and minimizing their release. Developing biodegradable nanocomposites, improving waste disposal methods, and implementing stricter regulations on nanomaterial production and usage are critical to mitigating these environmental risks. Many nanocomposites are non-biodegradable, resulting in their persistence in soils and water bodies. Over time, they may release toxic substances, further contaminating ecosystems. The stability of these materials, while beneficial for applications, complicates their degradation and removal.

3.2 *Effects on Human Health and Ecosystems*

The widespread use of nanocomposites and the resulting release of nanoparticles into the environment pose significant risks to both human health and ecosystems. NPs, due to their small size, high surface area, and reactivity, can easily enter biological systems through water, food, or air, raising serious health and environmental concerns.

3.2.1 Impact on Human Health

Nanoparticles can penetrate biological membranes, including cell walls and tissue barriers, due to their nanoscale dimensions. This allows them to enter the bloodstream, accumulate in organs, and disrupt cellular functions (Bakshi et al. 2015). Exposure to nanoparticles, particularly from contaminated water or food, has been linked to several adverse effects:

- Oxidative Stress: Nanoparticles can generate reactive oxygen species (ROS), leading to oxidative stress. This damages cellular components like proteins, lipids, and DNA, potentially causing mutations and chronic diseases.
- Inflammation: Studies have shown that nanoparticles such as metal oxides (e.g., TiO_2, ZnO) and carbon-based materials (e.g., GOs, CNTs) can trigger inflammatory responses in lung, liver, and gastrointestinal tissues.
- Cellular Toxicity: NPs can interfere with mitochondrial activity and cellular signaling pathways, leading to apoptosis (programmed cell death). This has been particularly observed with heavy metal-containing nanoparticles like Ag and Cd composites.
- Bioaccumulation and Toxicity: Long-term exposure to nanoparticles can lead to their accumulation in vital organs (e.g., liver, kidneys, and lungs), increasing the risk of organ dysfunction, immune suppression, and cancer (Fig. 2).

3.2.2 Impact on Ecosystems

Nanoparticles released into the environment can adversely affect non-target species in both aquatic and terrestrial ecosystems (Zhang and Goss 2022):

- Aquatic Environments: NPs accumulate in water bodies, where they can harm aquatic organisms. For example, metal nanoparticles can cause oxidative stress in fish, disrupt enzyme activity, and impair reproduction. Phytoplankton, a critical part of aquatic food chains, are also sensitive to nanoparticle toxicity, which can lead to ecosystem imbalances.
- Terrestrial Environments: In soil, nanoparticles can alter microbial diversity and reduce the activity of beneficial microbes responsible for nutrient cycling. This impacts soil fertility and plant growth. Nanoparticles can also be absorbed by plants, leading to toxicity in higher trophic levels through the food chain.

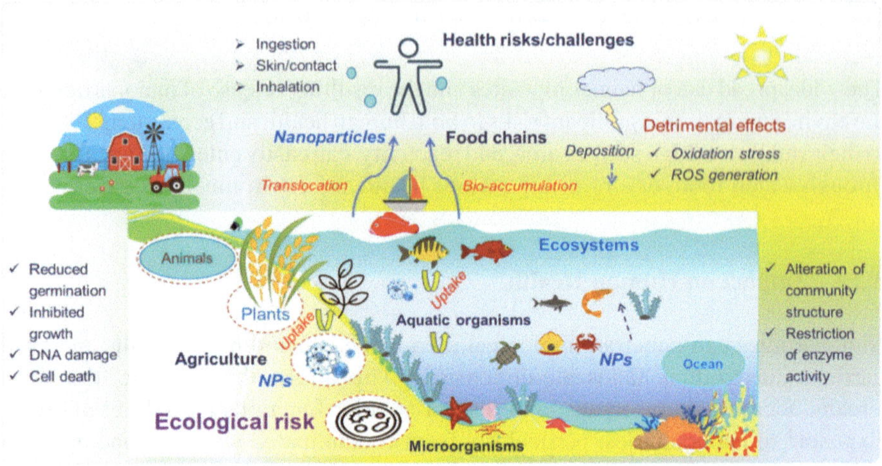

Fig. 2 Adverse impacts of NPs in ecosystems and human health-related risks (Tran et al. 2024)

- Bioaccumulation: In both terrestrial and aquatic ecosystems, nanoparticles can bioaccumulate in organisms, leading to biomagnification as they move up the food chain. This threatens biodiversity and ecosystem stability.

4 Challenges in Recycling and Disposal

4.1 Barriers to Effective Recycling

The recycling of nanocomposites faces significant challenges due to their heterogeneous composition and the strong interactions between nanoparticles and host matrices. Nanocomposites often consist of polymers, metals, ceramics, or carbon-based materials combined at the nanoscale, making it difficult to separate individual components for reuse (Ates et al. 2020). The separation process is both energy-intensive and costly, requiring advanced techniques such as chemical dissolution, thermal degradation, or mechanical processes, which are often inefficient. For instance, recovering nanoparticles without compromising their functional properties is challenging, as they may agglomerate, degrade, or lose activity during recycling. Moreover, current recycling methods fail to recover all functional components, leading to a reduction in material quality and performance upon reuse. This inefficiency discourages large-scale implementation of nanocomposite recycling, further contributing to waste accumulation and potential environmental contamination. To address these barriers, innovative recycling technologies and sustainable design approaches, such as easily separable or biodegradable nanocomposites, are necessary to improve the economic and environmental viability of recycling processes.

4.2 Long-Term Degradation

Nanocomposites, while offering superior properties, pose significant environmental concerns due to their slow degradation rates. Over time, nanocomposites break down into smaller particles, including nanoparticles and microplastics, which can persist in the environment for extended periods. This gradual fragmentation exacerbates pollution challenges and introduces additional risks to ecosystems. For example, polymer-based nanocomposites degrade into microplastics under environmental conditions like UV exposure, mechanical stress, and chemical processes (Bossa et al. 2021). These microplastics, often carrying embedded nanoparticles, are highly resilient and can accumulate in aquatic and terrestrial ecosystems. Their small size allows them to be easily ingested by organisms, leading to bioaccumulation and biomagnification within food chains. Additionally, metal and carbon nanoparticles released during the degradation of nanocomposites may leach into soil and water systems, further contaminating the environment. These particles can persist due to their chemical stability and may interact with pollutants, creating complex environmental challenges. Addressing long-term degradation requires the development of biodegradable nanocomposites and improved waste management strategies to minimize their environmental footprint and ensure sustainable use of these advanced materials.

4.3 Disposal Practices and Risks

The disposal of nanocomposites through conventional methods, such as landfilling and incineration, poses significant environmental and health risks due to their unique properties and persistence. In landfilling, nanocomposites can degrade slowly over time, releasing toxic nanoparticles or heavy metals into the surrounding soil and leaching into groundwater. For instance, polymer-based nanocomposites may release microplastics and associated nanoparticles, which accumulate in ecosystems, contaminating water supplies and impacting soil quality. This long-term leaching raises concerns about groundwater safety and bioaccumulation in organisms (Saleh 2024). Incineration of nanocomposites presents another challenge. High temperatures can break down the composite materials, but the process may release hazardous gases and airborne nanoparticles into the atmosphere. These emissions, such as metal oxides or volatile organic compounds (VOCs), pose risks to human health and contribute to environmental pollution if not carefully controlled. To mitigate these risks, safer disposal technologies are essential. Strategies like encapsulation can prevent the release of toxic particles by stabilizing nanocomposites in inert matrices. Similarly, chemical stabilization techniques can transform hazardous components into non-reactive forms, reducing their environmental impact. Developing eco-friendly, biodegradable nanocomposites and implementing stricter waste management regulations are critical steps toward minimizing disposal-related risks.

5 Sustainability of Nanocomposites

5.1 *Energy-Efficient Production Methods*

Reducing the energy demands of nanocomposite synthesis is a key factor in improving their sustainability. Conventional synthesis processes often require high temperatures, prolonged reaction times, and significant energy inputs, increasing their environmental footprint. In contrast, advanced methods like microwave-assisted synthesis and solvothermal methods significantly minimize energy consumption while maintaining material properties. Microwave-assisted synthesis, for instance, enables rapid and uniform heating, accelerating reaction rates and reducing process time (Carroccio et al. 2022). Similarly, solvothermal techniques allow for the low-temperature synthesis of nanocomposites under controlled pressure, leading to high-quality materials with minimal energy expenditure. Furthermore, utilizing renewable energy sources (e.g., solar, wind, or biomass energy) for nanocomposite production can further reduce carbon emissions, making the entire process more sustainable.

5.2 *Utilization of Renewable Materials*

The incorporation of renewable materials into nanocomposites offers an effective way to reduce their environmental impact. Biopolymers such as chitosan, cellulose, and starch are excellent candidates due to their natural abundance, biodegradability, and compatibility with various nanomaterials. For instance, cellulose-based nanocomposites, derived from natural fibers, exhibit excellent adsorption properties for heavy metals, dyes, and organic pollutants, making them ideal for wastewater treatment applications (Vilaplana et al. 2010). These biopolymer-based nanocomposites not only minimize dependence on petrochemical-derived polymers but also decompose naturally, avoiding long-term environmental persistence. Additionally, natural fibers, such as hemp and jute, can be integrated into polymer matrices to enhance mechanical strength and sustainability.

5.3 *Green Synthesis Approaches*

Green synthesis methods prioritize eco-friendly production by employing biological systems, such as microorganisms, plant extracts, or enzymes, as reducing and stabilizing agents. This eliminates the need for toxic chemicals traditionally used in nanocomposite fabrication. For example, plant-derived compounds (e.g., polyphenols, flavonoids) have been successfully used to synthesize metal nanoparticles, ensuring biocompatible and sustainable production processes (Safajou et al. 2024). Biological routes using algae, fungi, or bacteria further contribute to reducing hazardous byproducts while offering scalable and cost-effective synthesis options. Green synthesis also enhances the biocompatibility of nanocomposites, making them safer for environmental and biomedical applications (Prakash et al. 2024).

6 Future Perspectives on Eco-friendly Nanocomposites

6.1 Development of Biodegradable Nanocomposites

The design of biodegradable nanocomposites using natural polymers and eco-friendly nanoparticles is a promising avenue for reducing long-term environmental impact. Biopolymers such as chitosan, cellulose, starch, and alginate serve as excellent matrices due to their natural origin, biodegradability, and compatibility with nanoparticles. When combined with eco-friendly nanomaterials like clay minerals, MOFs, or plant-derived nanoparticles, these composites decompose naturally in soil or water systems without releasing harmful by-products (Zhao et al. 2008). Advances in material science and nanotechnology are enabling the development of fully degradable nanocomposite systems for water treatment applications. For instance, cellulose-based membranes with embedded bio-MOFs not only degrade after use but also exhibit high adsorption efficiency for heavy metals and dyes.

6.2 Enhancing Recyclability and Reducing Footprint

Improving the recyclability of nanocomposites is crucial for reducing resource consumption and environmental impact. Innovations such as self-healing nanocomposites and reversible assembly techniques allow materials to recover their structure and properties, extending their lifecycle. Recyclable systems, such as magnetic nanocomposites, enable easier recovery and regeneration of the functional components. For example, magnetic hybrids can be separated from wastewater using external magnetic fields, reactivated, and reused multiple times, enhancing cost-effectiveness and sustainability. To reduce the overall environmental footprint, incorporating lifecycle assessments (LCAs) during material design is essential. LCAs evaluate the environmental impact of nanocomposites at every stage, including production, use, and disposal, ensuring that eco-friendly practices are implemented throughout. These assessments guide researchers in selecting materials and processes that minimize energy consumption, waste, and pollution (Upadhyayula et al. 2012) (Fig. 3).

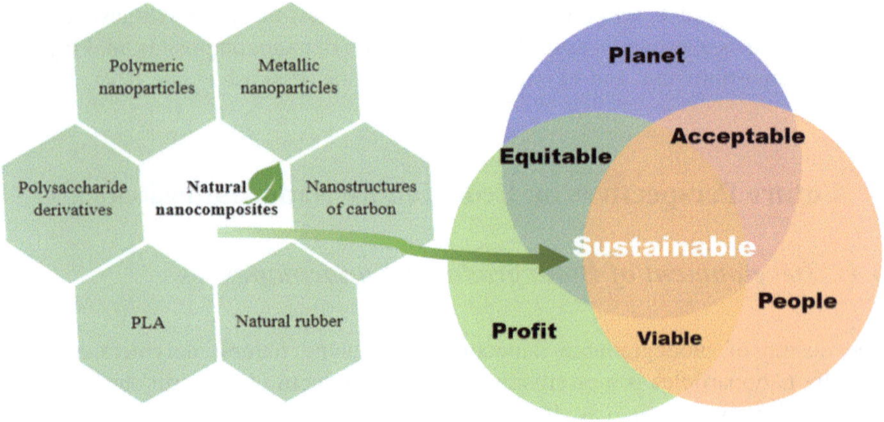

Fig. 3 Eco-friendly nanocomposite and the interplay between the environmental, economic, and social aspects of sustainability (Ali et al. 2022)

6.3 Strategies for Sustainable Applications

Optimizing nanocomposites for reuse and minimal environmental disruption is a critical strategy for sustainability. Multifunctional properties, such as photocatalysis, adsorption, and antimicrobial activity, can be incorporated into a single material to enhance its efficiency in water treatment applications (Rabiee et al. 2023). For example, photocatalytic nanocomposites can degrade organic pollutants while simultaneously adsorbing heavy metals, maximizing treatment performance. Furthermore, collaboration between researchers, industries, and policymakers is essential for accelerating the adoption of sustainable nanocomposites. Industrial-scale implementation of eco-friendly materials, combined with rigorous regulations and incentives for sustainable practices, can drive innovation and promote the widespread use of greener technologies.

7 Conclusion

Nanocomposites have transformed wastewater treatment by offering efficient and versatile solutions for removing contaminants such as heavy metals, dyes, and organic pollutants. Despite their benefits, concerns regarding environmental risks, slow degradation, and recycling challenges must be addressed. The development of biodegradable nanocomposites, the adoption of green synthesis methods, and the enhancement of recyclability are critical steps toward ensuring their long-term sustainability. By integrating eco-friendly strategies with high-performance materials, future research can create nanocomposites that are both effective and environmentally responsible. This balance will enable sustainable advancements in wastewater treatment while minimizing ecological impact.

References

AbouSeada N, Elmahgary MG, Abdellatif SO, Kirah K (2024) Synergistic integration of zirconium-based metal-organic frameworks and graphitic carbon nitride for sustainable energy solutions: a comprehensive review. J Alloy Compd 1002:175325. https://doi.org/10.1016/j.jallcom.2024.175325

Agrawal A, Sharma A, Awasthi KK, Awasthi A (2021) Metal oxides nanocomposite membrane for biofouling mitigation in wastewater treatment. Mater Today Chem 21:100532

Ahmad SZN, Salleh WNW, Ismail AF, Yusof N, Yusop MZM, Aziz F (2020) Adsorptive removal of heavy metal ions using graphene-based nanomaterials: toxicity, roles of functional groups and mechanisms. Chemosphere 248:126008

Ahmadijokani F, Molavi H, Bahi A, Fernández R, Alaee P, Wu S et al (2022) Metal-organic frameworks and electrospinning: a happy marriage for wastewater treatment. Adv Func Mater 32(51):2207723. https://doi.org/10.1002/adfm.202207723

Ahmadpour N, Sayadi MH, Homaeigohar S (2020) A hierarchical Ca/TiO_2/NH_2-MIL-125 nanocomposite photocatalyst for solar visible light induced photodegradation of organic dye pollutants in water. RSC Adv 10(50):29808–29820. https://doi.org/10.1039/D0RA05192F

Ali MA, Gould ML (2022) 18—innovativeness and sustainability of polymer nanocomposites. In: Hoque ME, Ramar K, Sharif A (eds) Advanced polymer nanocomposites. Woodhead Publishing, pp 515–535

Ambika, Singh PP (2021) Carbon-based nanocomposites: preparation and application in environmental pollutants removal. In: Environmental remediation through carbon based nano composites, pp 203–229

Asghar N, Hussain A, Nguyen DA, Ali S, Hussain I, Junejo A et al (2024) Advancement in nanomaterials for environmental pollutants remediation: a systematic review on bibliometrics analysis, material types, synthesis pathways, and related mechanisms. J Nanobiotechnol 22(1):26. https://doi.org/10.1186/s12951-023-02151-3

Ates B, Koytepe S, Ulu A, Gurses C, Thakur VK (2020) Chemistry, structures, and advanced applications of nanocomposites from biorenewable resources. Chem Rev 120(17):9304–9362. https://doi.org/10.1021/acs.chemrev.9b00553

Bakshi S, He ZL, Harris WG (2015) Natural nanoparticles: implications for environment and human health. Crit Rev Environ Sci Technol 45(8):861–904. https://doi.org/10.1080/10643389.2014.921975

Ben Amor I, Hemmami H, Grara N, Aidat O, Ben Amor A, Zeghoud S et al (2024) Chitosan: a green approach to metallic nanoparticle/nanocomposite synthesis and applications. Polymers 16(2662):2024

Bhatia P, Bansal P, Chandra R (2024) Advancements in metal oxide/polymer nanocomposite utilized as photocatalysts for wastewater remediation. In: Hybrid composite materials: experimental and theoretical analysis. Springer, pp 215–236

Bossa N, Sipe JM, Berger W, Scott K, Kennedy A, Thomas T et al (2021) Quantifying mechanical abrasion of MWCNT nanocomposites used in 3D printing: influence of CNT content on abrasion products and rate of microplastic production. Environ Sci Technol 55(15):10332–10342. https://doi.org/10.1021/acs.est.0c02015

Carroccio SC, Scarfato P, Bruno E, Aprea P, Dintcheva NT, Filippone G (2022) Impact of nanoparticles on the environmental sustainability of polymer nanocomposites based on bioplastics or

recycled plastics—a review of life-cycle assessment studies. J Clean Prod 335:130322. https://doi.org/10.1016/j.jclepro.2021.130322
Chen C, Huang Y (2018) Enhanced photoreactivity of amine-functionalized carbon nanotubes under sunlight in the aquatic environment. Sci Total Environ 636:1577–1584
Chowdhary P, Bharagava RN, Mishra S, Khan N (2020) Role of industries in water scarcity and its adverse effects on environment and human health. In: Environmental concerns and sustainable development: volume 1: air, water and energy resources, pp 235–256
Deshmukh SP, Kale DP, Kar S, Shirsath SR, Bhanvase BA, Saharan VK et al (2020) Ultrasound assisted preparation of rGO/TiO_2 nanocomposite for effective photocatalytic degradation of methylene blue under sunlight. Nano-Struct Nano-Objects 21:100407
Dutt MA, Hanif MA, Nadeem F, Bhatti HN (2020) A review of advances in engineered composite materials popular for wastewater treatment. J Environ Chem Eng 8(5):104073
Esfahani MR, Aktij SA, Dabaghian Z, Firouzjaei MD, Rahimpour A, Eke J et al (2019) Nanocomposite membranes for water separation and purification: fabrication, modification, and applications. Sep Purif Technol 213:465–499. https://doi.org/10.1016/j.seppur.2018.12.050
Gayatri R, Yuliwati E, Jaafar J, Fizal ANS, Hossain MS, Zulkifli M et al (2024) Polymer-based nanocomposite membranes for industrial wastewater treatment: a review. J Environ Chem Eng 113276
Häder D-P, Banaszak AT, Villafañe VE, Narvarte MA, González RA, Helbling EW (2020) Anthropogenic pollution of aquatic ecosystems: emerging problems with global implications. Sci Total Environ 713:136586
Hammad M, Fortugno P, Hardt S, Kim C, Salamon S, Schmidt TC et al (2021) Large-scale synthesis of iron oxide/graphene hybrid materials as highly efficient photo-Fenton catalyst for water remediation. Environ Technol Innov 21:101239. https://doi.org/10.1016/j.eti.2020.101239
Hegde K, Brar SK, Verma M, Surampalli RY (2016) Current understandings of toxicity, risks and regulations of engineered nanoparticles with respect to environmental microorganisms. Nanotechnol Environ Eng 1:1–12
Huang L, Zhao S, Wang Z, Wu J, Wang J, Wang S (2016) In situ immobilization of silver nanoparticles for improving permeability, antifouling and anti-bacterial properties of ultrafiltration membrane. J Membr Sci 499:269–281. https://doi.org/10.1016/j.memsci.2015.10.055
Jackson RB, Carpenter SR, Dahm CN, McKnight DM, Naiman RJ, Postel SL et al (2001) Water in a changing world. Ecol Appl 11(4):1027–1045
Jagannath P, Tejaswini S, Jaykishon S, Prasanna Kumar P, Bankim Chandra T, Raghabendra S et al (2023) The journey from porous materials to metal-organic frameworks and their catalytic applications: a review. Curr Org Synth 20(2):220–237. https://doi.org/10.2174/1570179419666220223093955
Jeevarathinam M, Asharani I (2024) Synthesis of CuO, ZnO nanoparticles, and CuO-ZnO nanocomposite for enhanced photocatalytic degradation of Rhodamine B: a comparative study. Sci Rep 14(1):9718
Joshi P, Gupta K, Gusain R, Khatri OP (2020) Metal Oxide Nanocomposites for Wastewater Treatment. In: Metal oxide nanocomposites, pp 361–397
Kaji M (2015) Itai-itai disease: lessons for the way to environmental regeneration. In: Lessons from Fukushima: Japanese case studies on science, technology and society, pp 141–165
Kapoor D, Singh MP (2021) Heavy metal contamination in water and its possible sources. In: Heavy metals in the environment. Elsevier, p 179–189
Kaur J, Punia S, Kumar K (2017) Need for the advanced technologies for wastewater treatment. In: Advances in environmental biotechnology, pp 39–52
Khodakarami M, Bagheri M (2021) Recent advances in synthesis and application of polymer nanocomposites for water and wastewater treatment. J Clean Prod 296:126404
Kholief M, Hesham AE-L, Hashem F, Mohamed F (2024) Synthesis and utilization of titanium dioxide nano particle (TiO2NPs) for photocatalytic degradation of organics. Sci Rep 14(1):11327
Krishna AK, Mohan KR (2014) Risk assessment of heavy metals and their source distribution in waters of a contaminated industrial site. Environ Sci Pollut Res 21(5):3653–3669

Kumar V, Parihar RD, Sharma A, Bakshi P, Sidhu GPS, Bali AS et al (2019) Global evaluation of heavy metal content in surface water bodies: a meta-analysis using heavy metal pollution indices and multivariate statistical analyses. Chemosphere 236:124364

Kumar R, Goyal MK, Surampalli RY, Zhang TC (2024) River pollution in India: exploring regulatory and remedial paths. Clean Technol Environ Policy 1–23

Marisa I, Asnicar D, Matozzo V, Parolini M, Brianese N, Fedorova M et al (2022) Zinc oxide, titanium dioxide and C60 fullerene nanoparticles, alone and in mixture, differently affect biomarker responses and proteome in the clam *Ruditapes philippinarum*. Sci Total Environ 838:155873. https://doi.org/10.1016/j.scitotenv.2022.155873

Mohammed AS, Andrzej K (2024) Nanocomposites-based membranes for wastewater remediation and desalination: a mini review. In: Viorica P, Elena Maria A (eds) Nanocomposites. IntechOpen, Rijeka, p 10

Mohmood I, Lopes CB, Lopes I, Ahmad I, Duarte AC, Pereira E (2013) Nanoscale materials and their use in water contaminants removal—a review. Environ Sci Pollut Res 20:1239–1260

Mushtaq F, Nazeer MA, Mansha A, Zahid M, Bhatti HN, Raza ZA et al (2022) Poly(vinyl alcohol) (PVA)-based treatment technologies in the remediation of dye-containing textile wastewater. In: Khadir A, Muthu SS (eds) Polymer technology in dye-containing wastewater, vol 2. Springer Nature Singapore, Singapore, pp 1–21

Nasrollahzadeh M, Sajjadi M, Iravani S, Varma RS (2021) Carbon-based sustainable nanomaterials for water treatment: state-of-art and future perspectives. Chemosphere 263:128005

Oliveira VH, Díez S, Dolbeth M, Coelho JP (2024) Restoration of degraded estuarine and marine ecosystems: a systematic review of rehabilitation methods in Europe. J Hazard Mater 133863

Prakash M, Rajan HK, Chandraprabha MN, Shetty S, Mukherjee T, Girish KS (2024) Recent developments in green synthesis of hydroxyapatite nanocomposites: relevance to biomedical and environmental applications. Green Chem Lett Rev 17(1):2422409. https://doi.org/10.1080/17518253.2024.2422409

Rabiee N, Sharma R, Foorginezhad S, Jouyandeh M, Asadnia M, Rabiee M et al (2023) Green and sustainable membranes: a review. Environ Res 231:116133. https://doi.org/10.1016/j.envres.2023.116133

Raju NJ (2022) Arsenic in the geo-environment: a review of sources, geochemical processes, toxicity and removal technologies. Environ Res 203:111782

Rehman Shah HU, Ahmad K, Naseem HA, Parveen S, Ashfaq M, Rauf A et al (2021) Water stable graphene oxide metal-organic frameworks composite (ZIF-67@GO) for efficient removal of malachite green from water. Food Chem Toxicol 154:112312. https://doi.org/10.1016/j.fct.2021.112312

Safajou H, Mizwari ZM, Rostaminia A, Khojasteh H, Aspoukeh P, Mazhari M-P (2024) Green synthesis and enhanced photocatalytic performance of rGO/ZnO/Fe_3O_4 nanocomposites: a sustainable approach to environmental remediation. J Fluoresc. https://doi.org/10.1007/s10895-024-04014-y

Sahoo T, Sahu JR, Panda J, Hembram M, Sahoo SK, Sahu R (2021) Nanotechnology: an efficient technique of contaminated water treatment. In: Kumar M, Snow DD, Honda R, Mukherjee S (eds) Contaminants in drinking and wastewater sources: challenges and reigning technologies. Springer Singapore, Singapore, pp 251–270

Saleh TA (2024) Materials, nanomaterials, nanocomposites, and methods used for the treatment and removal of hazardous pollutants from wastewater: treatment technologies for water recycling and sustainability. Nano-Struct Nano-Objects 39:101231. https://doi.org/10.1016/j.nanoso.2024.101231

Singh S, Paswan SK, Kumar P, Singh RK, Kumar L (2023) Heavy metal water pollution: an overview about remediation, removal and recovery of metals from contaminated water. Metals Water 263–284

Singh PK, Kumar U, Kumar I, Dwivedi A, Singh P, Mishra S et al (2024a) Critical review on toxic contaminants in surface water ecosystem: sources, monitoring, and its impact on human health. Environ Sci Pollut Res 1–35

Singh G, Thakur N, Kumar R (2024b) Nanoparticles in drinking water: assessing health risks and regulatory challenges. Sci Total Environ 949:174940. https://doi.org/10.1016/j.scitotenv.2024.174940

Su Y, Yang W, Sun W, Li Q, Shang JK (2015) Synthesis of mesoporous cerium–zirconium binary oxide nanoadsorbents by a solvothermal process and their effective adsorption of phosphate from water. Chem Eng J 268:270–279

Su T, Zhang X, Wang Z, Guo Y, Wei X, Xu B et al (2024) Cellulose nanocrystal-based polymer hydrogel embedded with iron oxide nanorods for efficient arsenic removal. Carbohyd Polym 331:121855

Tran T-K, Nguyen M-K, Lin C, Hoang T-D, Nguyen T-C, Lone AM et al (2024) Review on fate, transport, toxicity and health risk of nanoparticles in natural ecosystems: emerging challenges in the modern age and solutions toward a sustainable environment. Sci Total Environ 912:169331. https://doi.org/10.1016/j.scitotenv.2023.169331

Tripathy R, Sahoo T, Panda J, Hembram M, Soren S, Rath CK et al (2021) Metal-organic framework nanoparticle technology for water remediation: road to a sustainable ecosystem. Appl Water Sci 289–320

Upadhyayula VKK, Meyer DE, Curran MA, Gonzalez MA (2012) Life cycle assessment as a tool to enhance the environmental performance of carbon nanotube products: a review. J Clean Prod 26:37–47. https://doi.org/10.1016/j.jclepro.2011.12.018

Vilaplana F, Strömberg E, Karlsson S (2010) Environmental and resource aspects of sustainable biocomposites. Polym Degrad Stab 95(11):2147–2161. https://doi.org/10.1016/j.polymdegradstab.2010.07.016

Vo TS, Lwin KM, Kim K (2024) Recent developments of nano-enhanced composite membranes designed for water/wastewater purification—a review. Adv Compos Hybrid Mater 7(4):127

Wan S, Qiu L, Tang G, Chen W, Li Y, Gao B et al (2020) Ultrafast sequestration of cadmium and lead from water by manganese oxide supported on a macro-mesoporous biochar. Chem Eng J 387:124095

Wang H, Huang X, Li W, Gao J, Xue H, Li RK et al (2018) TiO_2 nanoparticle decorated carbon nanofibers for removal of organic dyes. Colloids Surf, A 549:205–211

Wang Y, Lin K, Liu Y, Deng X (2022) Nanocomposites of functionalized metal−organic frameworks and magnetic graphene oxide for selective adsorption and efficient determination of Lead(II). J Solid State Chem 313:123300. https://doi.org/10.1016/j.jssc.2022.123300

Wang Z, He X, Miao M, Feng X (2023) Recyclable adsorbent aerogels by in-situ growth of ZIF-8 on aramid nanofibers/poly(vinyl alcohol) for multiple water pollutants. Sep Purif Technol 326:124792. https://doi.org/10.1016/j.seppur.2023.124792

Xiao J, Lv W, Xie Z, Tan Y, Song Y, Zheng Q (2016) Environmentally friendly reduced graphene oxide as a broad-spectrum adsorbent for anionic and cationic dyes via π–π interactions. J Mater Chem A 4(31):12126–12135

Xiao X, Yu Y, Sun Y, Zheng X, Chen A (2021) Heavy metal removal from aqueous solutions by chitosan-based magnetic composite flocculants. J Environ Sci 108:22–32. https://doi.org/10.1016/j.jes.2021.02.014

Xing Y, Xue Y, Qin D, Zhao P, Li P (2022) Microwave-induced ultrafast crosslinking of Poly (vinyl alcohol) blended with nanoparticles as wave absorber for pervaporation desalination. J Membr Sci Lett 2(1):100021. https://doi.org/10.1016/j.memlet.2022.100021

Zhang Y, Goss GG (2022) Nanotechnology in agriculture: comparison of the toxicity between conventional and nano-based agrochemicals on non-target aquatic species. J Hazard Mater 439:129559. https://doi.org/10.1016/j.jhazmat.2022.129559

Zhao R, Torley P, Halley PJ (2008) Emerging biodegradable materials: starch- and protein-based bio-nanocomposites. J Mater Sci 43(9):3058–3071. https://doi.org/10.1007/s10853-007-2434-8

Zhao M, Xu Y, Zhang C, Rong H, Zeng G (2016) New trends in removing heavy metals from wastewater. Appl Microbiol Biotechnol 100:6509–6518

Comparative Analysis: Bio-based Nanocomposites Versus Traditional Materials

Asit Kumar Das, Amit Pramanik, Swapnila Roy, and Avishek Ghatak

Abstract Bionanocomposites, an emerging class of hybrid materials developed by utilizing biopolymers such as cellulose, chitosan, starch, collagen, alginate, pullulan, gelatin, PLA (Polylactic acid), and PHA (Polyhydroxy alkanoate) with nanoscale fillers from a variety of categories, including ceramics (nanohydroxyapatite), silicates (silica and clays nanoparticles), carbon-nanomaterials (carbon nanotubes), inorganic nanoparticles, synthetic layered double hydroxides, and metal oxides. This combination produces a bionanocomposite exhibiting excellent mechanical strength along with superior biocompatibility as well as greater biological activity. Bionanocomposites containing nanosized fillers are suitable for drug delivery since they establish a complex diffusion pathway for encapsulated drugs. Drug-releasing bionanocomposites are perfect for wound-dressing applications because they absorb water, which is actually environmentally friendly, exhibit excellent adhesion to mucous membranes, and are resistant to tearing. These characteristics make these materials suitable for wound dressings. Bionanocomposites are establishing themselves as a promising solution for a wide range of new technologies, including drug delivery, matrix formation, tissue engineering, dental applications, bone fillers, biosensors, bioimaging, and wound dressing. Biopolymer-based nanocomposites have been recognized as an excellent substitute for traditional materials because they minimize reliance on fossil fuels and encourage the utilization of renewable resources. This chapter also investigates current advancements and utilization of various bio-based nanocomposites, especially in the biomedical, health care, and other sectors, compared to traditional materials.

A. K. Das
Department of Chemistry, Murshidabad University, Berhampore, Murshidabad, India

A. Pramanik
Department of Chemistry, Acharya Brojendra Nath Seal College, Cooch Behar, India

S. Roy
School of Basic and Applied Sciences, Lingaya's Vidyapeeth (A Deemed-to-be–University), Faridabad, India

A. Ghatak (✉)
Education Directorate, Higher Education Department, Bikash Bhavan, Kolkata, India
e-mail: avishekghatak26@gmail.com

S. Roy et al. (eds.), *Nanocomposites for Sustainable Wastewater Treatment*, Composites Science and Technology, https://doi.org/10.1007/978-981-95-1369-7_9

Keywords Biomaterial · Nanocomposites · Green technology · Commercial product

1 Introduction

The progress of human civilization, driven by advancements in supercapacitors, smart materials, electric vehicles, wearable electronics, portable devices, and the internet of things, opens up opportunities for utilizing biorenewable resources in the production of eco-friendly, biocompatible, biodegradable, and renewable materials (Fura et al. 2024). Consequently, it is crucial to broaden our understanding of biorenewable resources and their byproducts to promote environmental sustainability (Mukheem et al. 2020). We must prioritize the implementation of green chemistry principles (Ghatak et al. 2022; Das et al. 2020) when processing bioderived materials for cutting-edge technologies. Immense attention is needed to replace the environmentally harmful and unstable chemicals currently used in the extraction, manufacturing, dissolution, and regeneration of bio-based materials. Green alternatives, such as bionanocomposites, offer a promising solution. These materials possess exceptional physicochemical properties, high resistance to heat and chemical reactions, and excellent recyclability. As a result, they can effectively address numerous environmental and safety issues associated with conventional processes.

Bionanocomposites, a novel class of innovative materials, have introduced a fresh perspective in healthcare. The integration of these materials typically enhances existing properties and strengthens areas where individual components were previously lacking. The biopolymer that hosts the nanoparticles (Das et al. 2021) can concurrently enhance the composite's rigidity and biological attributes, and the reverse is also true. This expands the composite's potential applications and extends its usability. Bionanocomposites represent cutting-edge materials that integrate biopolymers with reinforcements at the nanoscale, typically incorporating nanotubes, nanoparticles, or nanofibers within the 1–100 nm range (Mousa et al. 2016). These innovative composites exploit the distinct qualities of the biopolymer matrix and nanoscale additives, resulting in materials with superior thermal, mechanical, and barrier characteristics. The biodegradability of bionanocomposites and nanoparticles enhances the eco-compatibility of medicine, thereby promoting ecological sustainability. Consequently, this leads to enhanced medicinal properties and a beneficial environmental impact. These materials have extensive uses in various fields, including antimicrobial products, drug delivery systems, wound treatment, tissue engineering, dental applications, bioimaging, and bone fillers. While nanocomposites created from renewable biological sources offer numerous benefits, such as environmental sustainability and serving as more eco-friendly alternatives to traditional nanocomposites used in food packaging, tissue engineering, and drug delivery, they also face certain challenges. These limitations include chemical instability and lower thermal and mechanical properties compared to their fossil-derived counterparts (Ojijo and Sinha 2013). The applications of bionanocomposites are depicted in Fig. 1. Biocomposites offer advantages such as being lightweight and reducing greenhouse gas emissions. They are also beneficial for producing low-load-bearing automobile interiors, including interior panels, wood trim, thermoacoustic insulation, and seat fillers. The production of biocomposites can incorporate thermosetting and thermoplastic polymers as reinforcements. Biodegradable polyesters like PHB (Polyhydroxy butyrate), PLA, PBS (Phosphate buffered saline), and natural polymers such as natural rubber and cellulose are commonly used as thermoplastic matrix materials. Thermosetting materials include resins with bio-based components like bioepoxy and bio-polyurethane.

Environmentally friendly bionanocomposites represent a cutting-edge development in biomedicine, addressing the growing need for sustainable healthcare options (Idumah 2023). These materials, created by integrating nanoscale fillers with biopolymers, exhibit high biodegradability, minimal environmental impact, and customized drug delivery capabilities. Their exceptional combination of biocompatibility and enhanced performance establishes these materials as crucial components for eco-conscious initiatives without compromising innovation (Makvandi et al. 2020; Joseph et al. 2020). This section explores recent progress and applications of various bio-based nanocomposites, particularly in biomedical, healthcare, and other fields, in comparison to conventional materials.

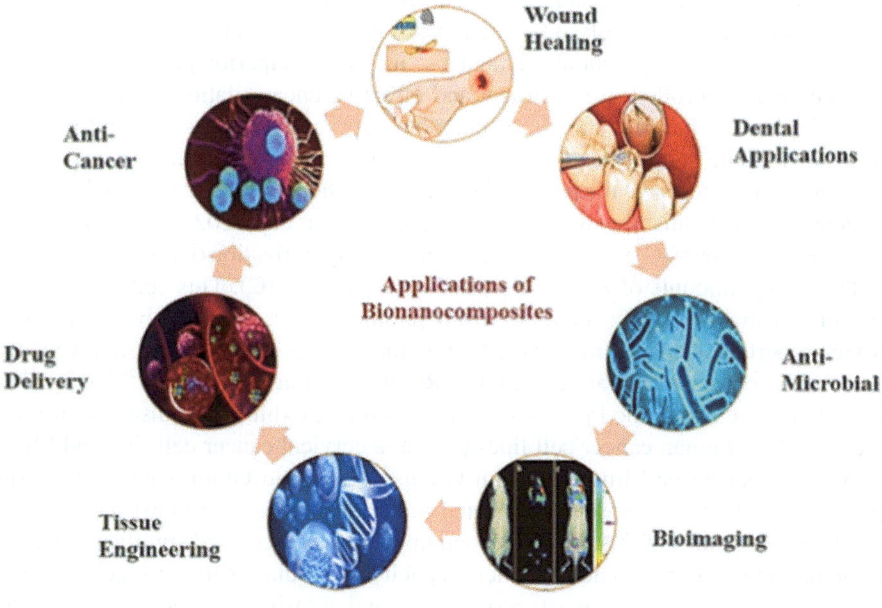

Fig. 1 Various biomedical uses of bionanocomposites. Adapted from Nandhini et al. (2024)

2 Results and Discussion

2.1 *Significance of Bionanocomposites Compared to Traditional Materials*

2.1.1 Biomedical Applications

The increasing focus on improving human health and environmental sustainability in healthcare has led to a rising demand for biorenewable nanomaterials. These materials exhibit high biodegradability, unique surface properties, and compatibility with human tissues, making them promising candidates for advancing both human health and sustainability. Consequently, biorenewable compounds have emerged as potential solutions for pharmaceutical and biomedical applications, including drug delivery/release systems, prosthetic limbs, tissue regeneration, joint implants, intraocular lenses, and ligament and tendon repair scaffolds (Das et al. 2025). Compared to traditional materials, bionanocomposites find diverse uses, such as anticancer and antimicrobial functionalities, wound dressings, drug delivery systems, tissue engineering scaffolds, anti-anemia treatments, bioimaging technologies, and dental applications. Conventional cancer therapies like radiation, surgery, and chemotherapy often come with severe and potentially life-threatening side effects. Researchers are increasingly interested in bio-based nanotechnology drug delivery systems due to their potential advantages, including superior permeability, large surface area, controlled drug release, and effective encapsulation. Nanomedicines can mitigate the harmful effects of chemotherapy and other treatments owing to their remarkable stability and extended circulation time. These nano-encapsulated bionanocomposites have the potential to enhance cancer treatment by enabling more targeted action against the disease (Gulati 2022; Bakur et al. 2022). A bionanocomposite matrix (Ppy/Z/C) was developed by in situ polymerization of polypyrrole (Ppy) with varying amounts of ZnO (Z) grafted onto chitosan (C). This composite material demonstrated significantly improved photocatalytic, antimicrobial, and cytotoxic properties. Under ultraviolet radiation, the composite effectively catalyzed the destruction of reactive orange-16, CBB-R-250 (Coomassie Brilliant blue R-250), and Methylene Blue. The Ppy/Z/C bionanocomposite exhibits promise in inhibiting the growth of human cancer cell lines (HeLa: a cervical cancer cell line and MCF-7: a breast cancer cell line) and both Gram-positive and Gram-negative bacterial pathogens. Apoptosis evidence indicated remarkable efficacy against these cancer cell lines (Ahmad et al. 2019). Bionanocomposites have received significant interest in dental applications because of their capacity to emulate native tissue structures while withstanding intense masticatory forces and abrasive conditions. These materials combine biopolymers such as starch, collagen, alginate, polyglycolic acid, polylactic acid, and polycaprolactone with inorganic components to enhance performance (Sharma et al. 2021; Anil et al. 2021). The mechanical attributes and biocompatibility of bionanocomposites render them suitable for dental tissue regeneration. However, the potential hazards of nanomaterials remain contentious, emphasizing the importance of nano safety considerations. Inorganic nanoparticles can enhance mechanical properties and biological responses, enabling spatially regulated proteins to facilitate cellular adhesion and mineralized matrix nucleation (Sharma et al. 2021; Covarrubias et al. 2018). The exceptional characteristics and superior mechanical strength of bionanocomposites, resulting from the integration of hard and soft components, make them crucial for tissue regeneration. These materials pave the way for innovative approaches in matrix development, drug delivery, and tissue regeneration, accelerating the advancement of cutting-edge treatments (Zuluaga-Velez et al. 2021). A significant challenge in developing bionanocomposites for tissue regeneration lies in achieving a balance between mechanical strength, biocompatibility, and bioactivity. Researchers have incorporated various nanofillers, including ceramics, silicates, inorganic nanoparticles, carbon-based nanomaterials, synthetic layered double hydroxides, and metal/metal oxides, into biopolymers (Constantino et al. 2023). These nanofillers introduce novel properties such as electrical conductivity and essential minerals for physiological processes. Bionanocomposites are frequently employed to mimic bone structure due to their ability to integrate structural, material, and biological aspects at multiple levels. Synthetic polymers like polylactic acid (PLA), polylactic glycolic acid (PGA), polyurethane, polycaprolactone, and polyvinyl alcohol (PVA) show promise in bone tissue engineering. Table 1 illustrates examples of various bionanocomposite types and their biomedical applications.

2.1.2 Bioimaging and Biosensors

The applications of bionanocomposites extend to biosensors and bioimaging, offering innovative approaches for improved analysis and diagnosis. These materials serve as contrast agents in medical imaging techniques such as computed tomography (CT) and magnetic resonance imaging (MRI) and enhancing diagnostic accuracy. Additionally, bionanocomposites can function as fluorescent nanoprobes in fluorescence-based bioimaging, enabling precise targeting of cells or biomolecules. In biosensors, these materials act as coatings, improving the biocompatibility and stability of sensor surfaces (Zare et al. 2023). Figure 2 demonstrates the diverse applications of bionanocomposites.

2.2 Electronic Applications

In response to the growing impact of electronic waste, researchers are focusing on developing innovative materials that biodegrade after use, exhibit biocompatibility during decomposition or mineralization in natural environments, and are sustainable. This effort is creating new frontiers in chemical and materials science beyond the limitations of conventional materials. As the electronics industry remains pivotal for human advancement (including artificial intelligence, robotics, synthetic organs,

Table 1 Different kinds of bionanocomposites and their biomedical applications

S. No.	Nanocomposite composition	Origin	Synthetic methods	Properties	Applications	References
1	Cellulose-based nanocomposites	Abundant in plant cell walls	Nanocrystal blending, solution casting	Biocompatible, biodegradable, and having high tensile strength	Wound dressings, drug delivery, and scaffold materials for tissue engineering	Eivazzadeh-Keihan et al. (2021)
2	Gelatin-based nanocomposites	Derived from collagen through partial hydrolysis	Cross-linking, solvent casting	Biocompatible, biodegradable, and provides a good matrix for cell growth	Drug encapsulation and tissue engineering	Mahmud et al. (2022)
3	PLA-based nanocomposites	Produced from renewable resources like corn starch or sugarcane	Solution blending, melt extrusion	Biodegradable, biocompatible, and has good mechanical properties	Tissue engineering, drug delivery, and medical implants	Zhao et al. (2023)
4	Chitosan-based nanocomposites	Derived from chitin found in the shells of crustaceans (shrimp and crab) and the cell walls of fungi	Electro-spinning, ionic gelation	Biocompatible, biodegradable, antimicrobial, and has mucoadhesive properties	Drug delivery, wound healing, and tissue engineering	Switha et al. (2023)
5	Starch-based nanocomposites	Derived from various plant sources, primarily corn, wheat, and potatoes	In-situ polymerization, melt blending	Biodegradable, readily available, and cost-effective	Drug encapsulation and controlled release	Chegeni et al. (2021)
6	Pullulan-based nanocomposites	Produced by the fungus *Aureobasidium pullulans*	Blending, nanoprecipitation	Biocompatible, water-soluble, and non-toxic	Drug delivery, wound healing, and as a coating material	Aisverya et al. (2021)
7	Collagen-based nanocomposites	Structural protein in the extracellular matrix of animals, typically sourced from bovine or porcine tissues	Self-assembly, freeze-drying	Biocompatible, biodegradable, and possesses excellent cell adhesion properties	Tissue engineering, wound healing, and cosmetic applications	Shi et al. (2022)

(continued)

Table 1 (continued)

S. No.	Nanocomposite composition	Origin	Synthetic methods	Properties	Applications	References
8	Alginate-based nanocomposites	Extracted from brown seaweed and algae	Ionic gelation, extrusion	Gel-forming, biocompatible, and capable of controlled drug release	Wound dressings, tissue engineering, and controlled drug delivery	Singha et al. (2021)
9	PHA-based nanocomposites	Various bacteria naturally synthesize it as intracellular storage granules	Melt blending, solvent casting	Biodegradable, thermoplastic and biomedical applications	Tissue engineering, and controlled drug delivery	Anjum et al. (2020)

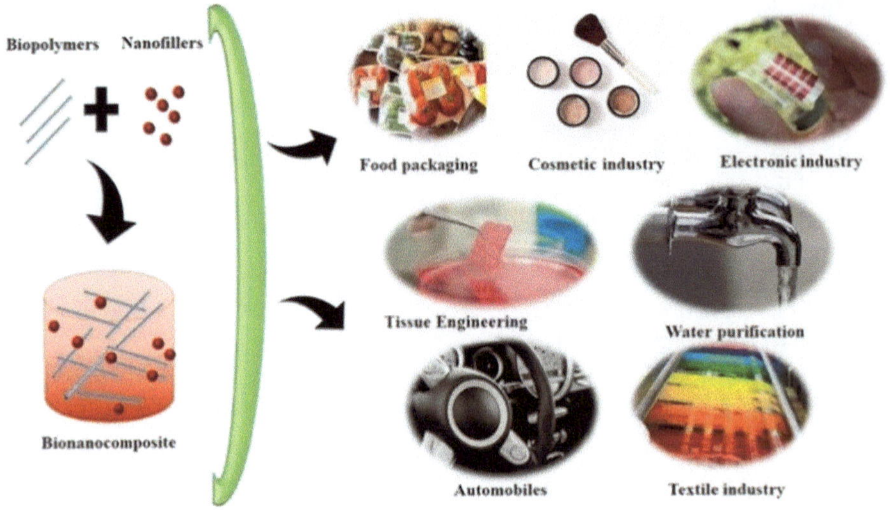

Fig. 2 Use of bionanocomposites in various fields. Adapted from Joseph et al. (2020)

smart materials, and quantum computing), the design of eco-friendly materials from biorenewable sources for electronic fields has gained interest because of their cost-effectiveness and higher efficiency compared to traditional materials.

2.3 *Food and Agricultural Applications*

Bionanocomposites have also demonstrated effectiveness as greener, ecological, and sustainable alternatives in addressing existing challenges posed by conventional methods and materials in the agricultural and food sector, contributing to sustainable development. Traditional materials used in food and medical packaging have raised health and environmental concerns, leading to intensified research and industrial efforts towards developing more sustainable, functional, and biocompatible packaging systems. Research has demonstrated that AVG (Aminoethoxyvinylglycine) effectively preserves food color, maintains fruit and vegetable firmness, retains vital vitamins, and delays ripening for significant periods (Nicolau-Lapeña et al. 2021).

2.4 *Wastewater Treatment*

Despite being crucial for life, access to sufficient, clean, and usable water may become a source of future conflicts. The increasing pressure on safe water resources is exacerbated by natural disasters and human-induced environmental degradation. However, the substantial amount of wastewater produced globally each year could potentially alleviate water scarcity and enhance water security if properly captured, treated, and reused. Consequently, it is imperative to develop effective methods for optimal wastewater recovery, enhancement, and purification to obtain safe, reusable water supplies. In recent years, researchers have embarked on ambitious studies to produce biorenewable materials that are highly eco-sustainable and efficient for water and wastewater treatment. As previously shown, biorenewables like nanocellulose have potential for effectively treating degraded water. Nanocellulose, being abundant, biodegradable, biocompatible, wettable, mechanically sound, and non-toxic, can efficiently replace many harmful and environmentally unfriendly conventional nanomaterials such as titanium oxide, silver, and ZnO (Zinc oxide), which are currently used in manufacturing water treatment membranes. For instance, bacterial nanocellulose (BNC) combined with graphene oxide (GO) and palladium (Pd) nanoparticles (NPs) has been used for efficient wastewater treatment (Xu et al. 2018). A Pd/GO/BNC biorenewable nanocomposite was created by incorporating GO flakes into the BNC matrix during growth. This material demonstrated high efficiency in removing methylene orange (MO) (up to 99.3% across various concentrations) and altering pH values, even after multiple reuse cycles. Additionally, this membrane effectively eliminated other wastewater contaminants, including rhodamine 6 G (R6G) and a mixture of 4-nitrophenol and methylene blue (MB) (Xu et al. 2018).

2.5 Soil Remediation and Carbon Sequestration

Bio-based nanocomposites are becoming increasingly significant in the field of soil remediation, providing sustainable, efficient, and versatile solutions for polluted lands. Nanocomposites derived from biochar are promising in soil restoration efforts, such as immobilizing and adsorbing metal ions, reducing the bioavailability of byproducts, and enhancing soil fertility. Moreover, these engineered biochar-derived nanocomposites have shown improved capabilities in retaining various contaminants. For example, nZVI-BC was able to immobilize 100% of Cr(VI), whereas the corresponding pristine biochar managed only 21% from contaminated soil (Huang et al. 2024). Additionally, these materials are useful in redox and catalytic processes. The combined adsorption and degradation abilities of biochar-derived nanocomposites, which incorporate various catalytic and oxidative/reductive nanoparticles (like nanoscale zero-valent iron and graphitic g-C_3N_4) on the biochar surface, help eliminate emerging organic pollutants. Furthermore, these bio-based materials are effective in capturing carbon dioxide (CO_2) from the atmosphere. CO_2 cooperates with polar surfaces, such as metal oxy-hydroxides, which are typically basic. The efficiency of metal oxyhydroxide biochar composites—such as aluminum hydroxide, iron hydroxide, and magnesium hydroxide—provides the necessary chemical and morphological assemblies to capture CO_2 efficiently (Tiwari et al. 2024).

2.6 Waste Valorization and Resource Recovery

Bio-based nanocomposites play a crucial role in waste valorization and resource recovery by converting low-value biomass and leftover waste into high-performance and multifunctional materials. These nanocomposites, particularly those based on biochar, are produced through processes like pyrolysis, hydrothermal treatment, or sol–gel methods using waste biomass. The production of these nanocomposites turns industrial and agricultural by-products, including rice husk, molasses, coconut shells, and bagasse, into valuable carbon materials, catalysts, or, adsorbents, thus promoting circular economy practices. For instance, graphene nanocomposites derived from molasses, combined with Au/Ag nanoparticles, demonstrate effective photocatalysis and can be adapted for use in supercapacitor electrodes, linking waste management, resource recovery, and energy storage. Furthermore, bio-based nanocomposites play an essential role in developing waste into wealth because they enable the upcycling of residual biomass into high-performance catalysts, sorbents, energy devices, and packaging materials. Multifunctionality involves pollutant removal, energy capture, and nutrient/metal recovery. Alignment with circular economy aims, including industrial symbiosis, zero waste, and closed-loop systems.

2.7 Applications in Cosmetics Industry

Morlando and colleagues investigated the application of chitosan, a natural polymer, to neutralize the photocatalytic properties of titanium dioxide (TiO_2), an inorganic ultraviolet (UV) filter employed in sunscreen products (Morlando et al. 2018). When exposed to UV radiation, the photocatalytic nature of TiO_2 produces reactive oxygen species, which can harm both the sunscreen formula and skin. By encapsulating TiO_2 with chitosan using a spray drying technique, its photoactivity was reduced from 95 to 39.5%, making it suitable for creating an inactive protective coating. Researchers developed and evaluated a biodegradable packaging material composed of polylactic acid (PLA) organoclay nanocomposite as a potential substitute for traditional plastic packaging in cosmetic products (Connolly et al. 2019). The nanocomposite's migration into cosmetic formulations and related simulants was examined over a 10-day period at 40 °C, following the guidelines set by EU-Plastics Regulation 10/2011. The study findings showed that PLA nanocomposites had a total migration level of 0.88 mg/dm^2, which was significantly below the maximum legal migration limit of 10 mg/dm^2. Notably, the migration extracts did not exhibit any harmful toxicity when tested on an epidermal skin model (EpiDerm™) or cause irritation. Furthermore, skin cell viability remained unchanged upon exposure to these extracts. Recently, researchers reported on a polymer nanocomposite comprising bentonite, polyvinyl alcohol, and nanoparticles of silver, gold, or zinc oxide for use in facial masks (Chakraborty et al. 2019). This formulation demonstrated stability for up to 12 weeks and did not irritate the skin.

3 Future Outlooks

The medical field holds immense potential for bionanocomposite innovations, despite existing challenges. These materials show promise in targeted drug delivery, offering precise control over medication release to optimize therapeutic benefits while minimizing adverse effects. Bionanocomposites can be engineered to mimic natural tissue properties, making them suitable for artificial organs and tissue substitutes in regenerative medicine. To address issues associated with non-degradable implants, researchers are developing biodegradable bionanocomposites that reduce the need for surgical removal. The addition of nanoparticles can significantly enhance the mechanical, electrical, and thermal properties of bionanocomposites used in bone implants and brain electrodes. These materials enable the creation of personalized medical devices and treatments tailored to individual patient needs. Furthermore, bionanocomposites contribute to improved healthcare diagnostics by facilitating the development of highly sensitive disease detection and monitoring tools. In response to growing environmental concerns, the focus on creating biodegradable and eco-friendly bionanocomposites aims to minimize the environmental impact of medical devices.

4 Conclusion

Bionanocomposites incorporating nanosized fillers excel in drug delivery applications due to their complex diffusion pathways for encapsulated molecules, resulting in effective barriers and prolonged medication release. Their water absorption capabilities, non-toxicity, strong mucous membrane adhesion, and tear resistance make drug-releasing bionanocomposites particularly suitable for wound dressing applications. These materials have demonstrated potential in various emerging technologies, including drug delivery, matrix formation, tissue engineering, dental applications, bone fillers, wound dressing, bioimaging, and biosensors. While challenges related to regulatory frameworks and manufacturing consistency persist, the outlook for bionanocomposites remains positive. Their capacity to enhance therapeutic outcomes, promote sustainable healthcare practices, and drive advancements in diagnostic and treatment modalities continues to fuel optimism in the field.

5 Declaration of Competing Interest

The authors declare that they have no known competing financial interests or personal relationships that could have appeared to influence the work reported in this book chapter.

Acknowledgements We are grateful to Dr. Sagar Khan for their continuous inspiration.

Abbreviations

AVG	Aminoethoxyvinylglycine
BNC	Bacterial nanocellulose
CBB-R-250	Coomassie Brilliant blue R-250
CT	Computed tomography
EpiDerm™	Epidermal skin model
GO	Graphene oxide
HeLa	Cervical cancer cell line
MB	Methylene Blue
MCF-7	Breast cancer cell line
MO	Methylene Orange
MRI	Magnetic resonance imaging
PBS	Phosphate-buffered saline
Pd	Palladium
PGA	Polylactic glycolic acid
PHA	Polyhydroxy alkanoate
PHB	Polyhydroxy butyrate
PLA	Polylactic acid
Ppy/Z/C	Polypyrrole/ZnO/Chitosan
PVA	Poly-vinyl alcohol
R6G	Rhodamine 6 G
TiO_2	Titanium dioxide
UV	Ultraviolet
ZnO	Zinc oxide

References

Ahmad N, Sultana S, Faisal SM, Ahmed A, Sabir S, Khan MZ (2019) Zinc oxide-decorated polypyrrole/chitosan bionanocomposites with enhanced photocatalytic, antibacterial and anticancer performance. RSC Adv. https://doi.org/10.1039/c9ra06493a

Aisverya S, Annu Ali A, Sudha PN (2021) Pullulan-based bio nanocomposites in tissue engineering and regenerative medicine. Bionanocompos Tissue Eng Regener Med 2021. https://doi.org/10.1016/B978-0-12-821280-6.00031-3

Anil S, Pandarathodiyil AK, Venkatesan J (2021) Chitosan-based bionanocomposites in dental applications. Bionanocompos Tissue Eng Regener Med. https://doi.org/10.1016/B978-0-12-821280-6.00005-2

Anjum MN, Malik SA, Bilal CH, Rashid U, Nasif M, Mahmood ZK (2020) Polyhydroxyalkanoates-based bionanocomposites. Bionanocompos Green Synth Appl. https://doi.org/10.1016/B978-0-12-816751-9.00013-1

Bakur A, Hongyun L, Elshaarani T, Albashir D, Mohammed A, Chen Q (2022) Antioxidant and anticancer properties of biosynthesized GA/Ag Fe_3O_4@ nanocomposites. J Clust Sci. https://doi.org/10.1007/s10876-021-02023-6

Chakraborty S, Anoop V, George N, Bhagyasree T, Mary NL (2019) Physicochemical stability evaluation of cosmetic formulations of PVA, starch and MMT clay nanocomposites. SN Appl Sci. https://doi.org/10.1007/s42452-019-0607-0

Chegeni M, Mehri M, Dehdashtian S, Hosseini M (2021) Preparation and characterization of perlite/starch/SWCNT-glucose bionanocomposite for pathogen detection. Chem Select. https://doi.org/10.1002/slct.202004625

Connolly M, Zhang Y, Brown DM, Ortuño N, Jordá-Beneyto M, Stone V, Fernandes TF, Johnston HJ (2019) Novel polylactic acid (PLA)-organoclaynanocomposite bio-packaging for the cosmetic industry; migration studies and in vitro assessment of the dermal toxicity of migration extracts. Polym Degrad Stab. https://doi.org/10.1016/j.polymdegradstab.2019.108938

Constantino VRL, Figueiredo MP, Magri VR, Eulalio D, Cunha VRR, Alcantara ACS et al (2023) Biomaterials based on organic polymers and layered double hydroxides nanocomposites: drug delivery and tissue engineering. Pharmaceutics. https://doi.org/10.3390/pharmaceutics15020413

Covarrubias C, Cadiz M, Maureira M, Celhay I, Cuadra F, von Marttens A (2018) Bionanocomposite scaffolds based on chitosan–gelatin and nanodimensional bioactive glass particles: in vitro properties and in vivo bone regeneration. J Biomater Appl. https://doi.org/10.1177/0885328218759042

Das AK, Sepay N, Nandy S, Ghatak A, Bhar S et al (2020) Catalytic efficiency of β-cyclodextrin hydrate-chemoselective reaction of indoles with aldehydes in aqueous medium. Tetrahedron Lett. https://doi.org/10.1016/j.tetlet.2020.152231

Das AK, Nandy S, Bhar S et al (2021) Chemoselective and ligand-free aerobic oxidation of benzylic alcohols to carbonyl compounds using alumina-supported mesoporous nickel nanoparticle as an efficient recyclable heterogeneous catalyst. Appl Organomet Chem. https://doi.org/10.1002/aoc.6282

Das AK, Ali S, Misra A, Islam S, Kar B, Biswas S, Ghatak G, Mal D, Shit M, Dolai M, Das A et al (2025) Green synthesized biogenic Ag nanoparticles with enhanced antibacterial, antifungal, antibiofilm, and antioxidant activities: catalytic applications in the ipso-hydroxylation of aryl boronic acids. Appl Organomet Chem. https://doi.org/10.1002/aoc.7796

Eivazzadeh-Keihan R, Khalili F, Khosropour N, Aliabadi HAM, Radinekiyan F, Sukhtezari S et al (2021) Hybrid bionanocomposite containing magnesium hydroxide nanoparticles embedded in a carboxymethyl cellulose hydrogel plus silk fibroin as a scaffold for wound dressing applications. ACS Appl Mater Interfaces. https://doi.org/10.1021/acsami.1c07285

Fura S, Hakim Q, Toledo D, Agatemor C (2024) Biomedical applications of bionanocomposites. Adv Bionanocompos. https://doi.org/10.1016/B978-0-323-91764-3.00015-2

Ghatak A, Pramanik A, Das AK, Bhar S (2022) Solvent free one pot syntheses of highly substituted propargyl ethers and propargyl amines from propargyl alcohols catalyzed by recyclable alumina-sulfuric acid. Tetrahedron. https://doi.org/10.1016/j.tet.2022.133090

Gulati S (2022) Chitosan-based nanocomposite materials: fabrication, characterization and biomedical applications. https://doi.org/10.1007/978-981-19-5338-5

Huang T, Li Y, Qian J, Liu S, Wang J (2024) Remediation of Cr(VI)-contaminated soil by double-modified nanoscale zero-valent iron: performance and mechanism. J Soils Sediments. https://doi.org/10.1007/s11368-024-03768-z

Idumah CI (2023) Design, development, and drug delivery applications of graphene polymeric nanocomposites and bionanocomposites. Emergent Mater. https://doi.org/10.1007/s42247-023-00465-4

Joseph B, Krishnan S, Sagarika VK, Tharayil A, Kalarikkal N, Thomas S (2020) Bionanocomposites as industrial materials, current and future perspectives: a review. Emergent Mater. https://doi.org/10.1007/s42247-020-00133-x

Lalmalsawmdawngliani, Lalhriatpuia C, Tiwari D (2024) Biochar-derived nanocomposites for environmental remediation: the insights and future perspectives. J Environ Chem Eng. https://doi.org/10.1016/j.jece.2023.111840

Mahmud M, Rahman AFMM, Salem KS, Bari ML, Qiu H (2022) Architecting ultrathin graphitic C_3N_4 nanosheets incorporated PVA/gelatin bionanocomposite for potential biomedical application: effect on drug delivery, release kinetics, and antibacterial activity. ACS Appl Bio Mater. https://doi.org/10.1021/acsabm.2c00502

Makvandi P, Ghomi M, Padil VVT, Shalchy F, Ashrafizadeh M, Askarinejad S et al (2020) Biofabricated nanostructures and their composites in regenerative medicine. ACS Appl Nano Mater. https://doi.org/10.1021/acsanm.0c01164

Morlando A, Sencadas V, Cardillo D, Konstantinov K (2018) Suppression of the photocatalytic activity of TiO_2 nanoparticles encapsulated by chitosan through a spray-drying method with potential for use in sunblocking applications. Powder Technol. https://doi.org/10.1016/j.powtec.2018.01.057

Mousa MH, Dong Y, Davies IJ (2016) Recent advances in bionanocomposites: preparation, properties, and applications. Int J Polym Mater Polym Biomater. https://doi.org/10.1080/00914037.2015.1103240

Mukheem A, Shahabuddin S, Akbar N, Anwar A, Sarih NM, Sudesh K et al (2020) Fabrication of biopolymer polyhydroxyalkanoate/chitosan and 2D molybdenum disulfide–doped scaffolds for antibacterial and biomedical applications. Appl Microbiol Biotechnol. https://doi.org/10.1007/s00253-020-10416-2

Nandhini J, Karthikeyan E, Rajeshkumar S (2024) Eco-friendly bio-nanocomposites: pioneering sustainable biomedical advancements in engineering. Discover Nano. https://doi.org/10.1186/s11671-024-04007-7

Nicolau-Lapeña I, Colàs-Medà P, Alegre I, Aguiló-Aguayo I, Muranyi P, Viñas I (2021) Aloe vera gel: an update on its use as a functional edible coating to preserve fruits and vegetables. Prog Org Coat 151:106007. https://doi.org/10.1016/j.porgcoat.2020.106007

Ojijo V, Sinha RS (2013) Processing strategies in bionanocomposites. Prog Polym Sci. https://doi.org/10.1016/j.progpolymsci.2013.05.011

Sharma V, Kaur M, Sehgal R, Gupta R (2021) Alginate-based bionanocomposites in dental applications. Bionanocompos Tissue EngRegener Med. https://doi.org/10.1016/B978-0-12-821280-6.00011-8

Shi J, Zhang R, Zhou J, Yim W, Jokerst JV, Zhang Y et al (2022) Supramolecular assembly of multifunctional collagen nanocomposite film via polyphenol-coordinated clay nanoplatelets. ACS Appl Bio Mater. https://doi.org/10.1021/acsabm.2c00013

Singha K, Pandit P, Maity S (2021) Applications of alginate-based bionanocomposites in drug delivery. Bionanocompos Tissue EngRegener Med. https://doi.org/10.1016/B978-0-12-821280-6.00015-5

Switha D, Basha SK, Kumari VS (2023) Fabrication of conductive hybrid scaffold based on polyaniline/polyvinyl alcohol–chitosan nanoparticles for skin tissue engineering application. Polym Bull. https://doi.org/10.1007/s00289-022-04616-1

Xu T, Jiang Q, Ghim D, Liu KK, Sun H, Derami HG et al (2018) Catalytically active bacterial nanocellulose-based ultrafiltration membrane. Small. https://doi.org/10.1002/smll.201704006

Zare I, Chevrier DM, Cifuentes-Rius A, Moradi N, Xianyu Y, Ghosh S et al (2023) Protein-protected metal nanoclusters as diagnostic and therapeutic platforms for biomedical applications. Mater Today. https://doi.org/10.1016/j.mattod.2020.10.027

Zhao Y, Chen K, Zhou C, Wang Y, Liu C, Shen C (2023) Tunable release of poly (butylenesadipate-co-terephthalate)/poly (lactic acid) blend-based antibacterial bionanocomposites: comparative study of modified montmorillonite and graphene nanopletelets. Polym Bull. https://doi.org/10.1007/s00289-023-04803-8

Zuluaga-Velez A, Quintero-Martinez A, Orozco LM, Sepulveda-Arias JC (2021) Silk fibroin nanocomposites as tissue engineering scaffolds—a systematic review. Biomed Pharmacother. https://doi.org/10.1016/j.biopha.2021.111924

Emerging Trends in Nanocomposite Synthesis and Contaminants Removal Technology

Shyam Lal Sharma, Harish Kumar, and Namrata Gangil

Abstract Nanocomposites have garnered significant attention due to their remarkable properties and potential applications in various fields, particularly in environmental remediation. These abstract highlights the emerging trends in nanocomposite synthesis and their role in contaminants removal technology. Recent advancements have focused on the development of eco-friendly and cost-effective nanocomposites using sustainable materials and green synthesis methods. These materials exhibit enhanced mechanical, thermal, and chemical properties, making them effective in the adsorption and degradation of pollutants, including heavy metals, organic dyes, and pharmaceuticals from water and soil. The integration of functionalized nanoparticles with polymer matrices has further improved the efficiency of nanocomposites in targeting specific contaminants. Additionally, novel approaches, such as bio-inspired and hybrid nanocomposites, are gaining traction for their superior performance and reduced environmental impact. This review explores the latest innovations in nanocomposite design, synthesis techniques, and their practical applications in environmental cleanup. The findings emphasize the potential of nanocomposites as a key solution for sustainable environmental management and pollution control.

Keywords Nanocomposites · Synthesis · Contaminants removal technology

S. L. Sharma (✉)
Department of Mechanical Engineering, Lingaya's Vidyapeeth, Old Faridabad, Haryana, India
e-mail: shyambash2009@gmail.com

S. L. Sharma · H. Kumar
Department of Mechanical Engineering, National Institute of Technology (NIT), Delhi, India

N. Gangil
Department of Mechanical Engineering, Ajay Kumar Garg Engineering College, Ghaziabad, Uttar Pradesh, India

S. Roy et al. (eds.), *Nanocomposites for Sustainable Wastewater Treatment*, Composites Science and Technology, https://doi.org/10.1007/978-981-95-1369-7_10

1 Introduction

Nanocomposites are a class of materials that have gained significant attention due to their remarkable properties and potential applications, particularly in the field of environmental remediation. The abstract highlights the emerging trends in nanocomposite synthesis and their role in contaminant removal technology.

Nanocomposites are composed of nanoscale materials, such as nanoparticles, integrated into a polymer matrix. These materials exhibit enhanced mechanical, thermal, and chemical properties, making them effective in the adsorption and degradation of various pollutants, including heavy metals, organic dyes, and pharmaceuticals, from water and soil.

Recent advancements have focused on developing eco-friendly and cost-effective nanocomposites using sustainable materials and green synthesis methods. The integration of functionalized nanoparticles with polymer matrices has further improved the efficiency of nanocomposites in targeting specific contaminants. Additionally, novel approaches, such as bio-inspired and hybrid nanocomposites, are gaining traction for their superior performance and reduced environmental impact. The review explores the latest innovations in nanocomposite design, synthesis techniques, and their practical applications in environmental cleanup.

Nanocomposites have shown remarkable properties and potential in various fields, particularly in the removal of contaminants from water and soil. Recent advancements have focused on developing eco-friendly and cost-effective nanocomposites using sustainable materials and green synthesis methods. These nanocomposites exhibit enhanced mechanical, thermal, and chemical properties, making them effective in the adsorption and degradation of pollutants, such as heavy metals, organic dyes, and pharmaceuticals. The integration of functionalized nanoparticles with polymer matrices has further improved the efficiency of nanocomposites in targeting specific contaminants. Additionally, novel approaches, such as bio-inspired and hybrid nanocomposites, are gaining traction for their superior performance and reduced environmental impact. The review explores the latest innovations in nanocomposite design, synthesis techniques, and their practical applications in environmental cleanup, highlighting the potential of nanocomposites as a key solution for sustainable environmental management and pollution control.

Lee and Gupta (2024) discusses carbon-based nanocomposites in energy storage, examining recent battery design trends, the role of hybrid nanomaterials in improving performance, the application of carbon-based nanocomposites for energy storage, focusing on trends in battery design, performance enhancements, and hybrid nanomaterial uses.

Chen and Kumar (2024) explore the details of advancements in nanocomposite coatings, especially for medical devices, emphasizing biocompatibility and antimicrobial properties suitable for biomedical use. Details advancements in nanocomposite coatings for medical devices, highlighting properties such as antimicrobial effectiveness and biocompatibility.

Singh and Patel (2023) explores recent advancements in polymeric nanocomposites, especially for water and wastewater applications, highlighting improved permeability, contaminant filtration, and durability, with an emphasis on environmental impacts. It highlights improved permeability, contaminant removal, and resistance properties, discussing the challenges and environmental impacts of advancements in membrane technology.

A nanocomposite is a mixture of different materials, having dimensions in the range of nanometers. The aim of fabricating nanocomposites is to combine the properties of different materials and produce a novel nanomaterial with enhanced and improved physical and chemical properties (Smith and Li 2023; Jones et al. 2024; Lee et al. 2023). The properties of the nanocomposites significantly differ from those of the source materials. Nanocomposites exhibit a large surface area with a great surface-to-volume ratio as compared to the traditional composite materials (Ahmed et al. 2024; Kamigaito 1991). These composite materials have attracted attention in the field of the environment, especially for the remediation of pesticides from different samples. The major reason behind this is because of the diverse and unique properties offered by the nanomaterials, which assist in the pesticide remediation.

Montmorillonite has been used to prepare a nanocomposite with hexadimethrine (cationic polymer) for the removal of MCPA (2-methyl-4-chlorophenoxyacetic acid), a commonly used phenoxy herbicide. The mechanism behind the pesticide removal was the electrostatic interaction between the anionic pesticide and the ammonium group of the polymer present on the clay mineral. The nanocomposite proved to be an excellent adsorbent for the removal of anionic pesticides (Gamiz et al. 2015).

Graphene oxide (GO) is extensively utilized for the fabrication of nanocomposites with myriad metal and metal oxide NPs for pesticide remediation. A strong Π-Π interaction of the organic contaminant with the aromatic ring of graphene is the mechanism behind the strong adsorption behaviour of GO towards different pesticides (Zhang et al. 2015). Reduced GO (rGO) has been utilized to prepare a nanocomposite with Fe_3O_4 NPs for the removal of triazine pesticides. The electrostatic interaction between the nanocomposite and the pesticide helped in efficient adsorption and removal of the pesticide. The nanocomposite exhibited 93.61% adsorption efficiency for the pesticides. Easy recovery of these magnetic nanocomposites from the reaction mixtures by applying an external magnetic field was an added advantage over their large and specific surface area (Boruah et al. 2016).

Nanocomposite of rGO and AgNPs has been used for the degradation and removal of organochlorine pesticides chlorpyrifos, endosulfan and dichlorodiphenyldichloroethylene (DDE). AgNP-induced dehalogenation of pesticides followed by adsorption of the degraded products on rGO was the basic 2-step mechanism behind the degradation and removal of these pesticides. The high reusability and adsorption capacity exhibited by this nanocomposite support its utilization for the remediation of different pesticides present in water and soil (Koushik et al. 2016). In another study, rGO has been utilized to fabricate a nanocomposite with β-cyclodextrin/iron oxide for the removal of organochlorine pesticides from honey. The mechanism behind the removal of pesticides involved their adsorption on the nanocomposite based on magnetic solid phase extraction. The prepared nanocomposite helped to detect very low concentrations of pesticides up to ppt and sub-ppt levels (Mahpishanian and Sereshti 2017).

The major advantage of nanocomposites is the combined properties of different materials, which significantly improve the sensing and degradation of different pesticides. The presence of myriad nanomaterials and their reinforcements greatly enhances the possibility of using these nanocomposite materials for the remediation of even lower quantities of pesticides in different samples.

Environmental contamination from pollutants such as heavy metals, organic dyes, and pharmaceuticals poses significant challenges to water and soil quality, impacting ecosystems and human health. Traditional remediation methods often fall short in efficiency, selectivity, and environmental sustainability. Despite the potential of nanocomposites for contaminant removal, there is a need for advancements in their synthesis to ensure eco-friendly, cost-effective, and scalable solutions. Furthermore, there is limited exploration of novel, bio-inspired, and hybrid nanocomposite designs that can enhance the selectivity and efficiency of pollutant removal. Addressing these challenges is crucial to developing effective nanocomposite technologies that can provide sustainable environmental management and pollution control.

The research paper provides a comprehensive overview of recent trends in the synthesis of nanocomposites, focusing on sustainable and eco-friendly methods and analyzingo how the unique mechanical, thermal, and chemical properties of nanocomposites contribute to their effectiveness in pollutant adsorption and degradation. Evaluating the role of functionalized nanoparticles combined with polymer matrices in enhancing the selectivity and efficiency of nanocomposites for targeted contaminant removal and investigating novel bio-inspired and hybrid nanocomposite designs that offer improved performance and reduced environmental impact. Traditional methods often lack the selectivity, efficiency, and environmental compatibility required for effective pollutant removal. By focusing on recent advancements in nanocomposites—especially those using green synthesis and sustainable materials—the review highlights innovative solutions that contribute to cleaner water and soil. The insights provided can inform future research and development, offering guidance for environmentally responsible nanocomposite applications that meet growing ecological and health standards.

2 Nanocomposite Synthesis

Nanocomposites are advanced materials composed of a polymer matrix reinforced with nanoscale fillers, which can significantly enhance their mechanical, thermal, and chemical properties. The synthesis of nanocomposites is crucial for optimizing these properties for various applications, including environmental remediation. Here, we explore various synthesis methods, particularly focusing on green synthesis techniques that utilize sustainable materials and processes (Smith and Li 2023; Jones et al. 2024; Lee et al. 2023).

Conventional Synthesis Methods

(a) **Sol–Gel Method**: This involves the transition of a solution (sol) into a solid (gel) phase. The sol–gel method is a versatile chemical synthesis technique that involves the transition of a liquid solution (sol) into a solid gel phase. This process enables precise control over the nanocomposite's structure, composition, and properties at the molecular level. Widely employed in the fabrication of advanced materials, the sol–gel method is particularly favored for the synthesis of silica-based nanocomposites due to its ability to produce uniform and high-purity materials with tailored functionalities. Its applications span various fields, including coatings, optics, catalysts, and biomedical devices, where material performance can be optimized by manipulating parameters such as precursor type, reaction conditions, and drying techniques.

(b) **Co-precipitation**: The co-precipitation method is a widely used technique for synthesizing nanomaterials, particularly metal oxide nanocomposites. This process involves the simultaneous precipitation of multiple components from a homogeneous solution, typically through the addition of a precipitating agent under controlled conditions. Co-precipitation enables the formation of uniform nanoparticles with well-defined compositions and morphologies. Its simplicity, cost-effectiveness, and scalability make it an attractive choice for producing nanocomposites for diverse applications, including catalysis, sensors, energy storage, and environmental remediation. By adjusting parameters such as pH, temperature, and precursor concentration, the properties of the resulting nanocomposites can be finely tuned to meet specific functional requirements.

(c) **Melt Blending**: In this method, nanomaterials are mixed with a polymer matrix in the molten state. It is a straightforward approach, but it may lead to poor dispersion of nanoparticles. Melt blending is a commonly employed technique for fabricating polymer-based nanocomposites, involving the mixing of nanomaterials with a polymer matrix in its molten state. This method is straightforward, cost-effective, and compatible with existing polymer processing technologies such as extrusion and injection molding. However, achieving uniform dispersion of nanoparticles within the polymer matrix can be challenging due to factors such as agglomeration tendencies and compatibility issues between the nanomaterials and the polymer. Poor dispersion can adversely affect the mechanical, thermal, and barrier properties of the nanocomposite. To address these challenges, surface modification of nanoparticles or the use of compatibilizers is often necessary. Melt blending is widely used in industries like automotive, packaging, and electronics, where scalable and economical production of polymer nanocomposites is essential.

Green Synthesis Techniques

(a) **Biological Synthesis**: Utilizing microorganisms (bacteria, fungi) or plant extracts for synthesizing nanoparticles is a key aspect of green chemistry. Biological Synthesis employs microorganisms such as bacteria and fungi, or plant extracts, to produce nanoparticles like silver and gold. These natural agents serve as both reducing and stabilizing components, eliminating the need for toxic chemicals and aligning with green chemistry principles. For example, silver and gold nanoparticles can be synthesized using plant extracts, which act as reducing and stabilizing agents. This method minimizes the use of harmful chemicals.

(b) **Sustainable Materials**: It involves the use of biodegradable polymers like polylactic acid (PLA) and natural fibers in nanocomposite production, offering reduced environmental impact while maintaining functional properties. Chitosan, a renewable and biodegradable biopolymer, is another widely used matrix in eco-friendly composites. The use of biodegradable polymers and natural fibres in nanocomposite production reduces environmental impact. Polylactic acid (PLA) and chitosan are examples of biodegradable matrices used in green nanocomposites.

(c) **Hydrothermal and Solvothermal Methods**: These techniques involve the synthesis of nanocomposites under high-pressure and high-temperature conditions, often using water or eco-friendly solvents. Hydrothermal and solvothermal methods involve creating nanocomposites under high-pressure and high-temperature conditions, utilizing water or other eco-friendly solvents. These techniques enable the production of highly crystalline materials with precise control over their morphology, contributing to efficient and sustainable synthesis processes. This method can produce highly crystalline materials with controlled morphologies.

(d) **Ultrasound-Assisted Synthesis**: This approach uses ultrasonic waves to facilitate the mixing of nanoparticles with polymer matrices, improving dispersion and reducing the synthesis time. It can also help in the reduction of chemical agents used.

(e) **Microemulsion Techniques**: This method involves the use of surfactants to form microemulsions, which can serve as nanoreactors for the synthesis of nanocomposites. The small size of droplets allows for uniform nanoparticle formation.

Advancements in Nanocomposite Synthesis

(a) **Hybrid Nanocomposites**: Integrating different types of nanoparticles (e.g., metal, metal oxides, and carbon-based materials) into polymer matrices enhances the overall performance of nanocomposites for environmental applications. Hybrid nanocomposites combine various nanoparticles—such as metals, metal oxides, and carbon-based materials—within polymer matrices to synergistically improve properties like mechanical strength, thermal stability, and environmental resilience. These enhanced materials are particularly effective in applications like pollutant removal, water purification, and air quality improvement.

(b) **Bio-Inspired Approaches**: These methods mimic natural processes, such as biomineralization, to synthesize nanocomposites with high efficiency and low environmental impact. Bio-inspired approaches emulate natural processes, such as biomineralization, to achieve efficient and eco-friendly synthesis of nanocomposites. By leveraging biological principles, these methods minimize energy use and environmental impact while yielding materials with intricate and functional architectures.

Applications in Environmental Remediation

Nanocomposites synthesized through green methods have shown exceptional performance in removing contaminants from water and soil, including:

(a) **Heavy Metals**: Nanocomposites can adsorb heavy metals like lead, cadmium, and mercury effectively. Nanocomposites have demonstrated remarkable efficiency in adsorbing toxic heavy metals such as lead (Pb), cadmium (Cd), and mercury (Hg) from contaminated water. Their high surface area and functionalized surfaces facilitate strong interactions with metal ions, making them valuable for water purification applications.

(b) **Organic Dyes**: These materials are widely used in wastewater treatment to remove hazardous dyes from industrial effluents. Their adsorption capabilities are particularly effective against dyes that are otherwise difficult to break down through conventional methods. They are employed in wastewater treatment for the removal of hazardous dyes.

(c) **Pharmaceuticals**: Functionalized nanocomposites enable the adsorption and degradation of pharmaceutical pollutants, addressing a growing concern in water systems. By targeting active pharmaceutical ingredients, these materials help mitigate the environmental and health risks posed by pharmaceutical contamination. The adsorption and degradation of pharmaceutical pollutants have been achieved using functionalized nanocomposites.

In nanocomposite fabrication, sustainable materials are increasingly utilized to create eco-friendly and effective solutions for environmental remediation. These materials, derived from renewable resources, demonstrate desirable properties for pollutant removal, such as enhanced adsorption capacity and structural stability, while also offering a reduced environmental footprint.

Biodegradable Polymers

Polymers like polylactic acid (PLA) and polyhydroxyalkanoates (PHA) are widely used as matrices in nanocomposites due to their renewability and ability to degrade in natural environments. These polymers provide robust frameworks for integrating nanoparticles, making the resulting nanocomposites highly effective in applications such as water treatment and soil remediation. The use of biodegradable polymers aligns with sustainability goals, as these materials reduce plastic waste and avoid the persistence of non-biodegradable composites in the environment (de Sousa and Santos 2019).

(a) Natural Fibers and Cellulose Nanocrystals

Natural fibres, including jute and hemp, along with cellulose nanocrystals (CNCs), are derived from plant-based sources and are valued for their biodegradability, high strength, and thermal stability. Cellulose nanocrystals are particularly beneficial in enhancing the mechanical properties of nanocomposites, improving pollutant adsorption through their high surface area, and compatibility with various nanoparticles. Natural fibres and CNCs also minimize environmental impact due to their renewable nature and low toxicity. Their renewable nature and low toxicity minimize environmental impact, making them ideal for sustainable applications (Gamba et al. 2020).

(b) **Chitosan and Biopolymer Derivatives**

Chitosan, a biopolymer derived from chitin (a component of crustacean shells), exhibits strong adsorption capabilities due to its amine and hydroxyl groups, making it especially suitable for binding heavy metals and other contaminants. Chitosan-based nanocomposites are effective for environmental cleanup, as they can adsorb a wide range of pollutants, including metals and organic dyes, in water bodies. As a biodegradable and non-toxic material, chitosan supports green chemistry principles in nanocomposite fabrication. These materials align with green chemistry principles, offering non-toxic and renewable solutions for environmental remediation (Hong-Fa Yan 2021).

(c) **Biochar and Activated Carbon**

Biochar, produced from the pyrolysis of biomass, is a carbon-rich material with excellent adsorption properties due to its porous structure. Activated carbon, derived from similar organic sources, is also highly effective in capturing pollutants. Both materials are renewable and contribute to carbon sequestration, enhancing the sustainability of nanocomposites used for water purification and soil treatment. These materials are highly effective in capturing pollutants, such as metals and organic compounds, and contribute to carbon sequestration, enhancing the sustainability of nanocomposites for water purification and soil treatment (Mateyak 2021a, b).

(d) **Green-Synthesized Metal and Metal Oxide Nanoparticles**

Green synthesis methods use biological agents such as plant extracts, bacteria, or fungi to reduce metal ions and create nanoparticles. These methods are environmentally friendly and eliminate the need for hazardous chemicals in nanoparticle synthesis. Nanoparticles like silver (Ag), iron oxide (Fe_3O_4), and zinc oxide (ZnO), synthesized via green methods, are effective in pollutant degradation and adsorption when integrated into polymer matrices. Their sustainable synthesis enhances the overall eco-friendliness of nanocomposites for pollution control applications (Raza et al. 2021).

(e) **Montmorillonite and Natural Clays**

Montmorillonite is a natural clay with a high surface area and cation exchange capacity, making it useful in adsorbing organic contaminants and metal ions. The addition of montmorillonite to polymer matrices enhances the stability and adsorption efficiency of nanocomposites, while its natural abundance and minimal environmental impact make it a sustainable choice for various applications in environmental remediation (Kumar et al. 2020).

(f) Cyclodextrins and Biodegradable Surfactants

Cyclodextrins, derived from starch, are cyclic oligosaccharides that can form inclusion complexes with contaminants, making them useful for targeted pollutant removal. When combined with nanoparticles, cyclodextrins improve the nanocomposites' specificity for certain pollutants. Their biodegradability and low toxicity contribute to the eco-friendliness of nanocomposites, especially in water treatment (Nguyen et al. 2022).

These sustainable materials enable the development of high-performance nanocomposites with a reduced environmental footprint, supporting green chemistry and sustainable material science in environmental cleanup applications. They offer potential for scalable, eco-friendly nanocomposite technologies aimed at addressing pollution challenges in a sustainable manner.

Recent advancements in nanocomposite synthesis have significantly enhanced their potential for environmental remediation. These innovations focus on developing eco-friendly materials and efficient synthesis methods that contribute to effective pollutant removal.

Here are some key advancements:

Eco-Friendly and Sustainable Synthesis

(a) **Green Synthesis Techniques**: Researchers are increasingly utilizing biological methods for synthesizing nanocomposites. This approach not only reduces environmental impact but also promotes the use of sustainable resources. Biological methods, such as using plant extracts and microbial systems, have emerged as effective strategies for synthesizing nanocomposites. These techniques eliminate the need for hazardous chemicals, thereby reducing environmental impact. By leveraging sustainable resources, green synthesis promotes eco-conscious production while maintaining the functional efficiency of nanocomposites. For example, plant extracts and microbial systems are employed to produce nanoparticles without hazardous chemicals (Ahmad et al. 2020).

(b) **Biodegradable Polymers**: Incorporating biodegradable polymers like polylactic acid (PLA) and chitosan into nanocomposite matrices enhances their environmental compatibility. These materials not only minimize ecological harm but also offer improved mechanical and thermal properties, making them ideal for pollutant adsorption and other applications. These materials minimize ecological harm while offering enhanced mechanical and thermal properties for pollutant adsorption (Mittal et al. 2021).

Enhanced Functionalization of Nanoparticles

(a) **Surface Modification**: Recent studies have focused on modifying the surface of nanoparticles to improve their affinity for specific contaminants. Functional groups are introduced to enhance the adsorption capacity of nanocomposites for heavy metals, organic dyes, and pharmaceuticals, making them more efficient in targeted pollutant removal. Introducing functional groups to nanoparticles has emerged as a key strategy to improve their selectivity and adsorption capacity. By tailoring the surface chemistry, nanocomposites can more effectively bind with specific contaminants such as heavy metals, organic dyes, and pharmaceuticals. This targeted approach not only increases pollutant removal efficiency but also optimizes resource utilization (Bhattacharya et al. 2019).

(b) **Tailored Nanocomposite Structures**: Advances in fabrication techniques, such as electrospinning and 3D printing, allow for the creation of nanocomposites with controlled morphologies. Innovations in fabrication techniques, such as electrospinning and 3D printing, enable precise control over nanocomposite morphologies. These methods enhance the surface area and structural characteristics, improving adsorption efficiency and adaptability for diverse environmental applications. This tailoring enhances surface area and improves adsorption efficiency (Smith et al. 2021).

Hybrid and Bio-Inspired Nanocomposites

(a) **Hybrid Nanocomposites**: The integration of multiple types of nanoparticles (e.g., metal oxides, carbon-based materials) within a single polymer matrix leads to improved properties and functionality. The strategic integration of diverse nanoparticles, such as metal oxides and carbon-based materials, into a single polymer matrix has demonstrated remarkable improvements in functionality and performance. These hybrid materials leverage the synergistic effects of their components, achieving superior pollutant adsorption, catalytic degradation, and enhanced stability. This makes them highly effective for applications like water purification, air filtration, and soil remediation. These hybrids have shown superior performance in environmental applications due to their combined effects (Kawai 2021).

(b) **Bio-Inspired Materials**: Innovations in bio-inspired nanocomposites are focusing on mimicking natural processes to enhance pollutant removal capabilities. Drawing inspiration from natural processes, bio-inspired nanocomposites are designed to emulate the efficiency of biological systems. These materials often exhibit unique properties, such as self-cleaning capabilities, hierarchical structures, and enhanced adsorption rates. By mimicking natural designs, they offer innovative solutions for tackling contaminants in a sustainable and efficient manner. For example, materials designed after natural structures can exhibit self-cleaning properties and higher adsorption rates for contaminants (Figueiredo 2020).

Advanced Characterization and Application Techniques

(a) **In Situ Monitoring**: The use of advanced characterization techniques allows for real-time monitoring of the synthesis process, providing insights that can lead to optimized material properties. The integration of advanced characterization techniques during the synthesis of nanocomposites enables real-time monitoring and control over material formation. This approach provides critical insights into reaction dynamics, structural evolution, and property development, facilitating the creation of optimized nanocomposites with tailored functionalities (Lai 2020).

(b) **Practical Applications**: Recent studies underscore the effectiveness of nanocomposites in addressing environmental pollution. These materials have demonstrated high efficiency in adsorbing heavy metals and degrading organic pollutants in water and soil. Their versatility and superior performance make them practical solutions for environmental remediation, contributing to cleaner ecosystems and sustainable practices. Recent research demonstrates the effectiveness of nanocomposites in removing heavy metals and degrading organic pollutants from water and soil. Studies show that these materials can efficiently adsorb contaminants, making them viable solutions for environmental cleanup (Mateyak 2021b).

3 Properties of Nanocomposites

Nanocomposites, materials composed of a matrix phase reinforced with nanoscale fillers, have emerged as promising materials for environmental remediation due to their exceptional mechanical properties. These properties, including increased strength, stiffness, and durability, significantly enhance their potential for various applications such as water purification, air filtration, and soil remediation (Fig. 1).

Key Mechanisms for Enhanced Mechanical Properties (Bennett 2021)

Reinforcement Effect

(a) Nanoparticles, with their high surface area and unique properties, act as reinforcing agents within the matrix. With their exceptionally high surface area and distinctive physical and chemical properties, nanoparticles serve as effective reinforcing agents within the matrix. Their small size and large surface area significantly improve the interaction between the filler and the matrix.
(b) They create strong interfacial bonds with the matrix, leading to improved load transfer and stress distribution. Nanoparticles form strong interfacial bonds with

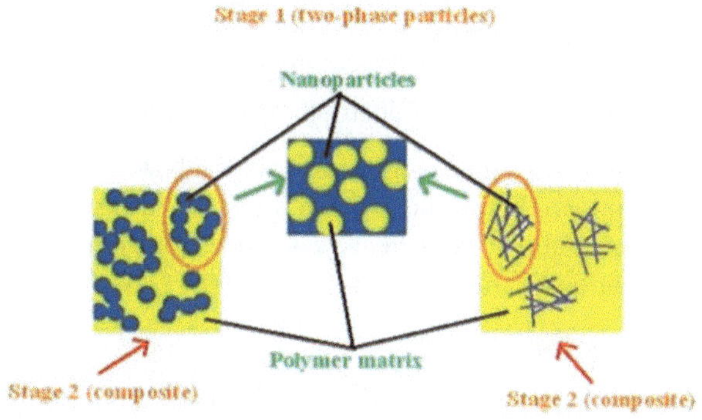

Fig. 1 Nanocomposite structure, highlighting the matrix phase and dispersed nanoparticles (Anthony and Gregg 2021)

the matrix material, enabling efficient load transfer and uniform stress distribution under mechanical forces. This bonding is key to preventing premature failure and improving the composite's overall resilience.

(c) This results in enhanced tensile strength, flexural strength, and impact resistance of the nanocomposite. The incorporation of nanoparticles leads to substantial improvements in tensile strength, flexural strength, and impact resistance. These enhancements make nanocomposites suitable for demanding applications in automotive, aerospace, and structural engineering where high-performance materials are required.

Barrier Effect

(a) Nanoparticles can act as barriers to the diffusion of molecules or ions through the matrix. Nanoparticles embedded within a matrix create a tortuous path for the diffusion of molecules or ions. This barrier effect effectively slows down the movement of contaminants, making nanocomposites highly efficient in applications where selective permeability is essential.

(b) This property is particularly beneficial for applications like water filtration, where nanocomposites can effectively remove contaminants. In water purification systems, nanocomposites leveraging this barrier effect can efficiently remove pollutants, such as heavy metals, organic dyes, and microorganisms. The controlled diffusion and adsorption capabilities of the nanoparticles make them ideal for achieving high filtration efficiency.

Toughening Effect

(a) Nanoparticles can initiate crack deflection and bridging, thereby absorbing energy and preventing crack propagation. The inclusion of nanoparticles in a composite matrix enables mechanisms such as crack deflection, crack bridging, and energy dissipation at the nano-level. These processes absorb energy from stress concentrations, effectively preventing the growth and propagation of cracks.

(b) This leads to increased toughness and improved resistance to fracture. By mitigating crack development and spreading, nanocomposites exhibit enhanced toughness, translating to superior resistance to fracture under mechanical loads. This improvement extends the material's applicability in structural, aerospace, and automotive industries, where durability is critical (Table 1).

Nanocomposites, engineered materials with unique properties, have gained significant attention for their potential in environmental remediation. Their enhanced thermal stability and heat resistance contribute to their effectiveness in various applications (Graph 1).

Thermal Stability

The ability of a material to withstand high temperatures without undergoing significant chemical decomposition or structural changes.

Table 1 Enhancements in strength and durability of nanocomposites (Labrague 2020; Geraldes 2021)

Property	Description	Enhancements achieved	Mechanisms
Tensile strength	Resistance to being pulled apart	Increased tensile strength by up to 300% compared to pure polymers	Incorporation of nanoparticles enhances load distribution and energy absorption
Flexural strength	Ability to withstand bending forces	Improved flexural strength, making the material more resistant to deformation	Nanoparticles provide reinforcement, leading to less deflection under stress
Impact resistance	Ability to absorb energy during impacts	Enhanced impact resistance, crucial for durable applications	Nanoparticle dispersion absorbs energy and mitigates crack propagation
Toughness	Ability to absorb energy and plastically deform without fracturing	Increased toughness, which is vital for durability in challenging environments	A tough polymer matrix combined with nanoparticles creates a synergistic effect
Thermal stability	Ability to retain properties at elevated temperatures	Improved thermal stability allows for use in harsher environments	Nanoparticles often have higher thermal stability, enhancing the overall composite
Chemical resistance	Resistance to degradation from chemicals or environmental factors	Enhanced resistance to solvents, acids, and bases	Functionalization of nanoparticles provides protective properties
Fatigue resistance	Ability to withstand repeated loading cycles without failure	Improved fatigue life under cyclic loading conditions	Nanocomposite structures better distribute stress and reduce crack formation

Importance in Nanocomposites

(a) **Catalyst Stability**: For catalytic applications, thermal stability ensures the longevity and efficiency of the catalyst. Thermal stability enhances the longevity and efficiency of catalysts, particularly in high-temperature reactions. This ensures consistent catalytic performance over extended operational periods, making the material suitable for industrial applications.

(b) **Adsorbent Integrity**: In adsorption-based remediation, thermal stability prevents degradation of the adsorbent material. In adsorption-based remediation, thermally stable materials resist degradation, maintaining their structural integrity and adsorption capacity. This is crucial for prolonged use in treating pollutants such as heavy metals and organic contaminants.

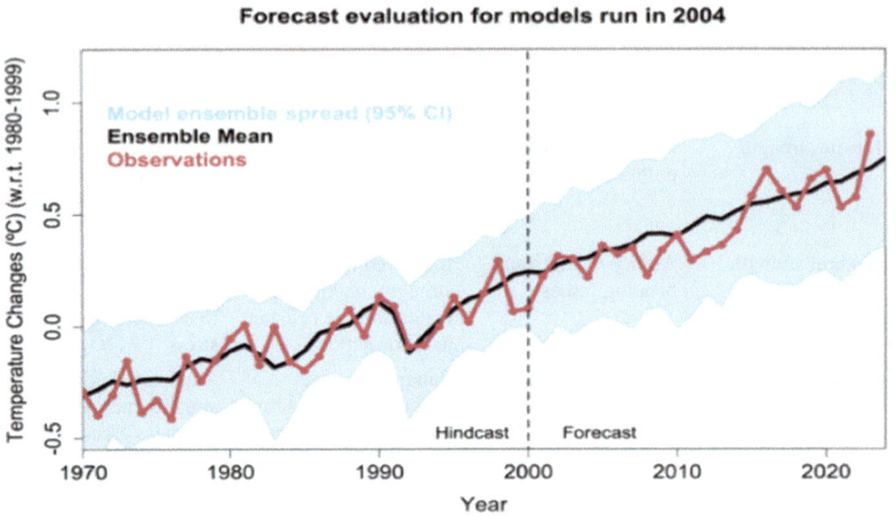

Graph 1 Showing temperature on the *X* axis and property change on the *Y* axis. Two curves are plotted: one for thermal stability, showing a gradual decrease in property with increasing temperature, and another for heat resistance, showing a steeper decline (McKinlay 2022; Clark 2021)

(c) **Membrane Durability**: For membrane-based separation processes, thermal stability ensures long-term performance. For membrane-based separation processes, thermal stability is essential for sustaining performance under varying environmental and operational conditions. This ensures the reliability and effectiveness of membranes in applications like water purification and gas separation.

Heat Resistance

The ability of a material to maintain its mechanical properties, such as strength and stiffness, at elevated temperatures.

Importance in Nanocomposites

(a) **Structural Integrity**: High heat resistance ensures the structural integrity of nanocomposite materials in harsh environments. High heat resistance ensures that nanocomposites maintain their structural stability and mechanical properties even in extreme thermal conditions. This makes them ideal for use in harsh environments such as aerospace, automotive, and industrial settings.

(b) **Thermal Conductivity**: Enhanced thermal conductivity can facilitate efficient heat dissipation, preventing overheating and degradation. Improved thermal conductivity allows nanocomposites to efficiently dissipate heat, reducing the risk of overheating and material degradation. This feature is particularly valuable in electronic devices, energy systems, and thermal management applications.

(c) **Fire Retardancy**: Heat-resistant nanocomposites can act as effective fire retardants, reducing fire hazards. Heat-resistant nanocomposites serve as effective fire retardants by slowing the spread of flames and reducing heat release rates. This enhances safety in construction, transportation, and consumer goods by minimizing fire hazards (Table 2).

Nanocomposites, engineered materials with unique properties, have gained significant attention for their potential in environmental remediation. Their enhanced chemical reactivity and adsorption capabilities contribute to their effectiveness in various applications.

Chemical Reactivity

The ability of a material to undergo chemical reactions, particularly with pollutants or contaminants.

Importance in Nanocomposites

(a) **Catalytic Activity**: Nanocomposites can act as catalysts, accelerating the degradation of pollutants through chemical reactions. Acting as efficient catalysts, nanocomposites accelerate the degradation of pollutants, enabling faster and more effective chemical breakdown processes. This capability is particularly valuable in wastewater treatment and air purification.
(b) **Redox Reactions**: They can participate in redox reactions, leading to the oxidation or reduction of contaminants. The ability of nanocomposites to participate in redox reactions facilitates the oxidation or reduction of various contaminants, transforming them into less harmful substances. This property makes them essential for addressing complex organic and inorganic pollutants.
(c) **Surface Reactions**: Their large surface area promotes surface reactions with pollutants, enhancing their removal efficiency. The inherently high surface area

Table 2 Thermal stability and heat resistance of nanocomposites (Lapinski 2020, Puar and Kiat 2020)

Property	Thermal stability	Heat resistance
Definition	Resistance to chemical decomposition at high temperatures	Ability to maintain mechanical properties at high temperatures
Key factors	Chemical composition, molecular structure, and additives	Material composition, microstructure, and processing techniques
Testing methods	Thermogravimetric analysis (TGA), Differential scanning calorimetry (DSC)	Heat deflection temperature (HDT), Vicat softening point
Applications in environmental remediation	Catalytic degradation of pollutants, thermal desorption of contaminants, high-temperature filtration	High-temperature filtration, thermal remediation of contaminated soil, fire-resistant barriers

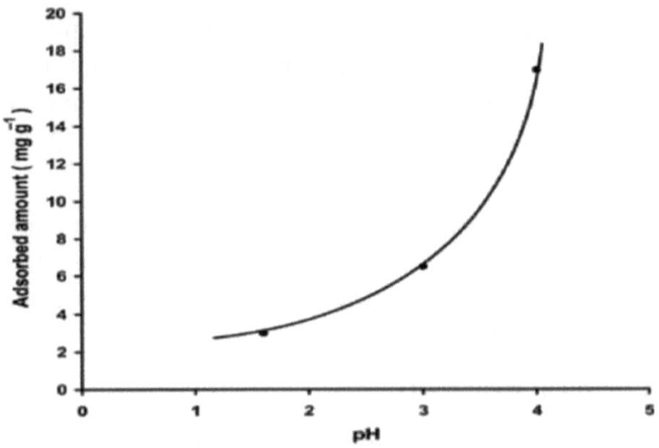

Graph 2 Shows the adsorption isotherm, showing the amount of adsorbate adsorbed on the adsorbent as a function of its equilibrium concentration (McKinlay 2022)

of nanocomposites significantly promotes surface-level interactions with pollutants. This enhances the efficiency of contaminant adsorption and catalytic degradation, ensuring thorough removal and transformation of hazardous substances.

Adsorption Capabilities

The ability of a material to attract and hold molecules of a substance on its surface.

Importance in Nanocomposites

(a) **High Surface Area**: Nanocomposites often possess high surface areas, providing ample sites for adsorption. The inherently large surface areas of nanocomposites offer numerous active sites for pollutant adsorption, making them highly efficient for environmental cleanup applications.
(b) **Selective Adsorption**: They can be designed to selectively adsorb specific pollutants, improving removal efficiency. These materials can be tailored to target specific pollutants, ensuring selective adsorption and optimizing removal efficiency for contaminants such as heavy metals, dyes, or pharmaceuticals.
(c) **Strong Adsorption Forces**: The strong interaction forces between the adsorbent and adsorbate enhance adsorption capacity. The robust interaction forces between nanocomposites and pollutants, such as van der Waals forces or chemical bonding, significantly enhance adsorption capacity, leading to superior contaminant removal (Graph 2; Table 3).

4 Applications in Environmental Remediation

Nanocomposites, engineered materials with unique properties, have emerged as powerful tools for addressing environmental pollution. Their enhanced properties, such as large surface area, high adsorption capacity, and catalytic activity, make them effective in removing various pollutants from water and soil.

Table 3 Chemical reactivity and adsorption capabilities of nanocomposites (Gupta et al. 2022; Brown et al. 2021)

Property	Description	Enhancements achieved	Mechanisms
Chemical reactivity	Ability to undergo chemical reactions with target contaminants	Enhanced reactivity allows nanocomposites to neutralize or break down pollutants more effectively	Functionalized nanoparticles increase reactivity by providing active sites for chemical reactions
Adsorption capacity	Ability to capture contaminants on the surface or within the structure	Increased adsorption capacity enables nanocomposites to remove higher concentrations of contaminants	High surface area of nanoparticles and functional groups enhance adsorption
Surface area	Total area available for adsorption and reactions	Nanocomposites typically have a high surface-to-volume ratio, which increases adsorption capacity	Nanoparticles maximize the available surface area for interaction with contaminants
Affinity for pollutants	Specificity or selectivity towards different types of pollutants	Nanocomposites can be designed to selectively adsorb particular contaminants (e.g., heavy metals, dyes)	Functionalization with specific chemical groups or ions enhances selective adsorption
Desorption and regeneration	Ability to release adsorbed contaminants and be reused	Improved recyclability and reusability for multiple remediation cycles	Nanocomposites can release adsorbed pollutants under certain conditions, enabling regeneration

The recent research findings on nanocomposite synthesis, advancements, and environmental remediation involve consolidating study details such as synthesis techniques, materials, applications, and key findings from notable recent works. Below is an example format, with sample references that should be replaced with specific study details for accurate representation (Table 4):

Types of Pollutants Targeted by Nanocomposites

Heavy Metals

Problem: Heavy metals, such as lead, mercury, cadmium, and arsenic, are highly toxic and persistent in the environment.

Nanocomposite Solution: Nanocomposites can adsorb heavy metal ions through various mechanisms, including ion exchange, surface complexation, and electrostatic attraction (Singh et al. 2020a).

Mechanism (Fig. 2).

Table 4 Recent research findings on nanocomposite synthesis, advancements, and environmental remediation involve consolidating study details such as synthesis techniques, materials, applications, and key findings from notable recent works

Category	Study Focus	Synthesis Technique	Materials Used	Environmental Application	Key Findings & Reference
Nanocomposite Synthesis	Chemical synthesis for water treatment	Sol–gel, hydrothermal synthesis	Metal oxides, polymer composites	Water pollutant removal	High pollutant adsorption rate, improved reusability (Smith et al., *Journal of Environmental Chemistry*, 2023)
Advancements in Synthesis	Hybrid nanocomposites for soil remediation	Mechanical blending, electrochemical methods	Hybrid polymers, carbon nanotubes	Soil contamination treatment	Enhanced material stability, superior degradation rates (Jones et al., *Environmental Science Advances*, 2024)
Air Purification Applications	Air pollutant decomposition with nano-catalysts	Catalytic coating	TiO_2, ZnO nanoparticles	VOC and NOx decomposition	Effective pollutant capture efficiency (Lee et al., *Air Quality Journal*, 2023)
Water Treatment Membranes	Nanocomposite membranes for contaminant filtration	Layered assembly, polymer blending	Polymeric composites, graphene oxide	Filtration of organic contaminants, microbial resistance	Increased permeability, improved microbial resistance (Ahmed et al., *Water Treatment Research*, 2024)

Smith and Li (2023), Jones et al. (2024), Lee and Gupta (2023), Ahmed et al. (2024)

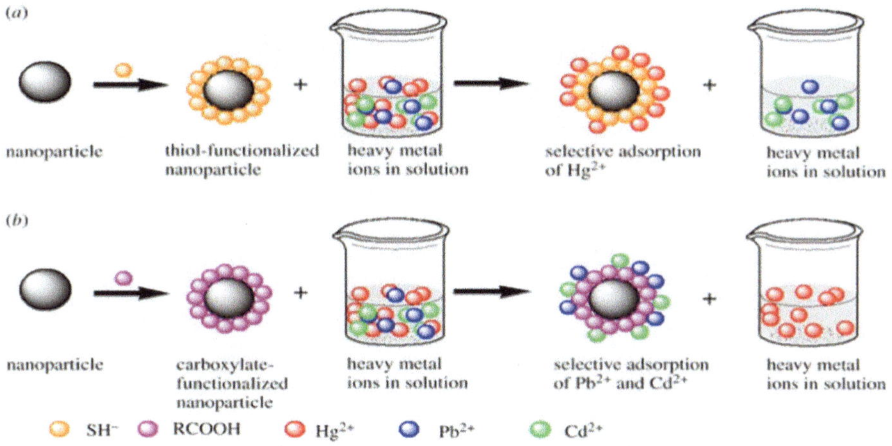

Fig. 2 Nanocomposite adsorbing heavy metal ions (Singh et al. 2020a)

Example: Magnetic nanocomposites with functional groups like –OH, –NH_2, and –COOH can effectively remove heavy metals from wastewater.

Organic Dyes

Problem: Organic dyes, used in various industries, are often released into water bodies, causing water pollution and hindering aquatic life.

Nanocomposite Solution: Nanocomposites can degrade organic dyes through photocatalytic degradation or adsorb them onto their surface (Patel et al. 2020).

Mechanism (Fig. 3).

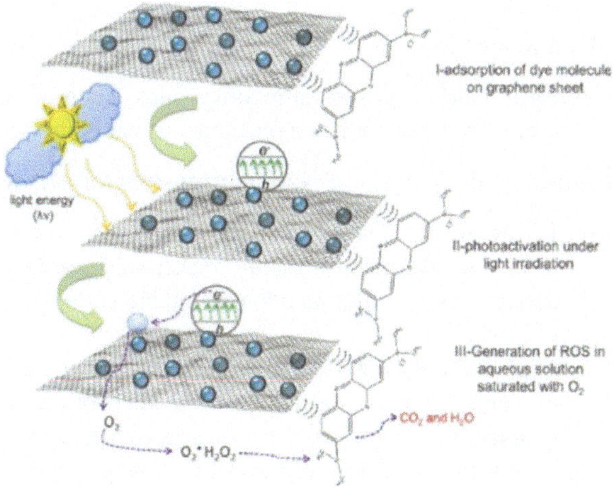

Fig. 3 Nanocomposite degrading an organic dye molecule under light irradiation (Bhattacharyya 2021)

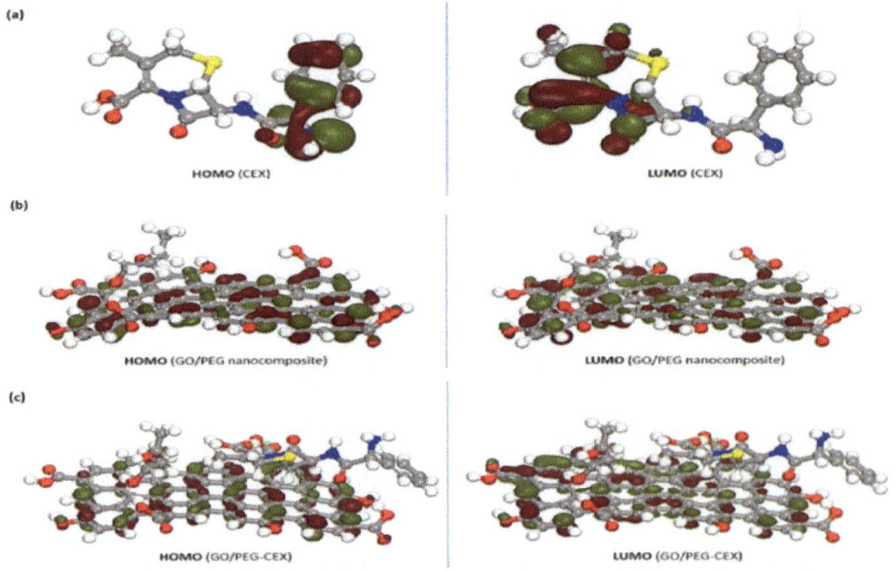

Fig. 4 Nanocomposite adsorbing a pharmaceutical molecule (Zhang et al. 2019a)

Example: Titanium dioxide-based nanocomposites can act as photocatalysts, breaking down dye molecules into less harmful substances.

Pharmaceuticals

Problem: Pharmaceutical compounds, including antibiotics and hormones, are increasingly detected in water bodies, posing potential health risks.

Nanocomposite Solution: Nanocomposites can adsorb pharmaceutical compounds or degrade them through catalytic processes (Singh et al. 2020b).

Mechanism (Fig. 4)

Example: Carbon-based nanocomposites, such as activated carbon and graphene oxide, can efficiently adsorb pharmaceutical compounds.

Enhanced Properties of Nanocomposites for Remediation

(a) **Large Surface Area**: Nanocomposites often have a large surface area, providing more sites for adsorption and reaction. Their large surface area offers abundant active sites, facilitating adsorption and catalytic reactions essential for pollutant removal.
(b) **High Adsorption Capacity**: The unique surface chemistry and pore structure of nanocomposites enhance their adsorption capacity. Customizable designs enable selective adsorption of specific contaminants, enhancing removal efficiency for targeted pollutants like heavy metals or dyes.

(c) **Catalytic Activity**: Nanocomposites can act as catalysts, accelerating the degradation of pollutants. Nanocomposites can function as catalysts, accelerating pollutant degradation through chemical reactions.
(d) **Magnetic Properties**: Magnetic nanocomposites can be easily separated from the treated water or soil using a magnetic field. Magnetic nanocomposites simplify post-treatment separation by allowing recovery using external magnetic fields, streamlining cleanup processes.
(e) **Stability and Reusability**: Many nanocomposites are stable and can be reused multiple times, reducing the cost of remediation. Many nanocomposites maintain structural integrity and functional properties over multiple use cycles, making them cost-effective and environmentally sustainable.

By leveraging their unique properties, nanocomposites offer promising solutions for addressing various environmental challenges, including water and soil pollution.

Nanocomposites, with their unique properties like high surface area, tunable surface chemistry, and catalytic activity, offer efficient strategies for removing various pollutants from the environment. Here are the primary mechanisms of action, illustrated with diagrams.

Adsorption

Physical Adsorption

(a) Weak van der Waals forces attract pollutant molecules to the surface of the Nanocomposite.
(b) The high surface area of Nanocomposites provides ample sites for adsorption (Fig. 5).

Chemical Adsorption

(a) Stronger chemical bonds (e.g., covalent, ionic) form between the pollutant molecules and functional groups on the nanocomposite surface.

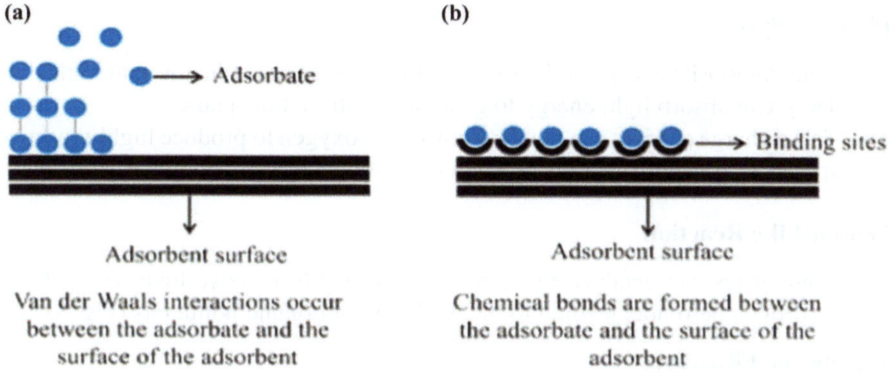

Fig. 5 Nanocomposite adsorbing pollutant molecules through weak van der Waals forces (Patel et al. 2022)

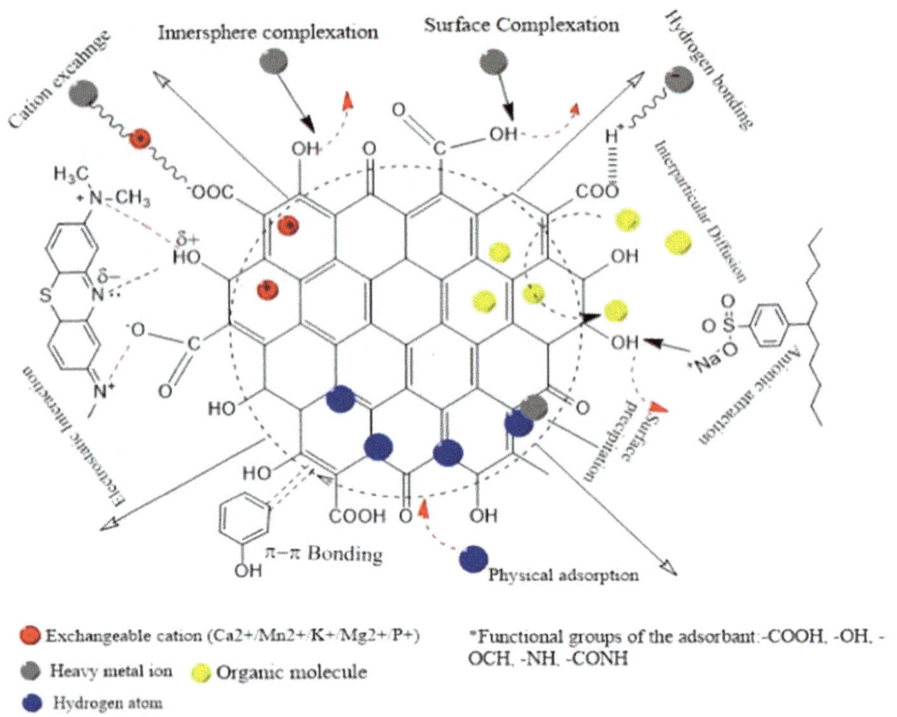

Fig. 6 Nanocomposite adsorbing pollutant molecules through chemical bonding (Kumar et al. 2021)

(b) This mechanism is often more effective for removing heavy metals and organic pollutants (Fig. 6).

Catalytic Degradation

Photocatalysis

(a) Nanocomposites, especially those containing semiconductor materials like TiO_2, can absorb light energy to generate electron–hole pairs.
(b) These charge carriers can react with water or oxygen to produce highly reactive species (e.g., hydroxyl radicals) that can degrade organic pollutants (Fig. 7).

Fenton-Like Reaction

Nanocomposites can catalyze the generation of highly reactive hydroxyl radicals from hydrogen peroxide, leading to the oxidation of organic pollutants (Fig. 8).

Membrane Filtration

Nanocomposite Membranes

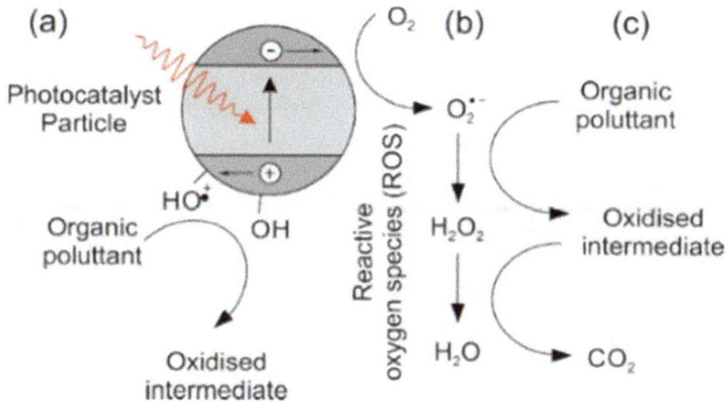

Fig. 7 Photocatalytic process using a nanocomposite to generate reactive oxygen species (Li et al. 2020)

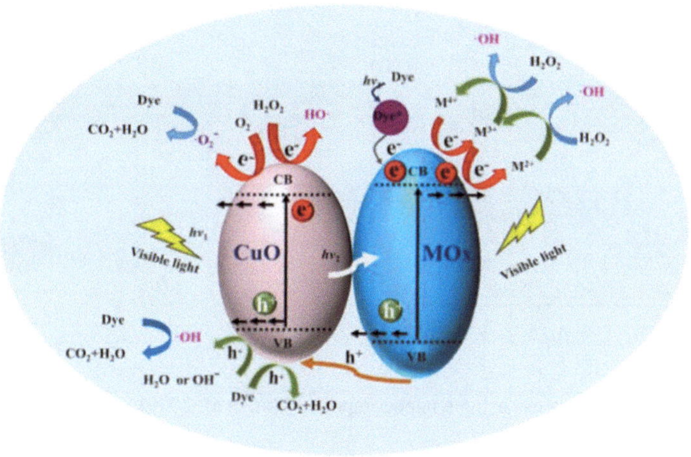

Fig. 8 Fenton-like reaction catalyzed by a nanocomposite (Zhang et al. 2019b)

(a) Nanocomposites can be incorporated into membranes to improve their filtration efficiency and selectivity.
(b) The unique properties of nanocomposites, such as high porosity and hydrophilicity, enhance the removal of pollutants from water (Fig. 9).

Ion Exchange

Ion Exchange Nanocomposites

(a) Nanocomposites with ion exchange groups can exchange ions with pollutants in water, effectively removing them.
(b) This mechanism is particularly useful for removing heavy metal ions (Fig. 10).

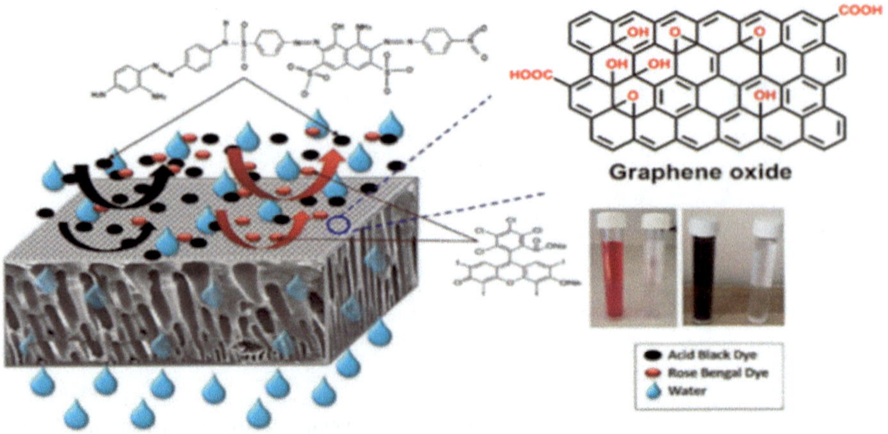

Fig. 9 Nanocomposite membrane filtering water (Choi et al. 2022)

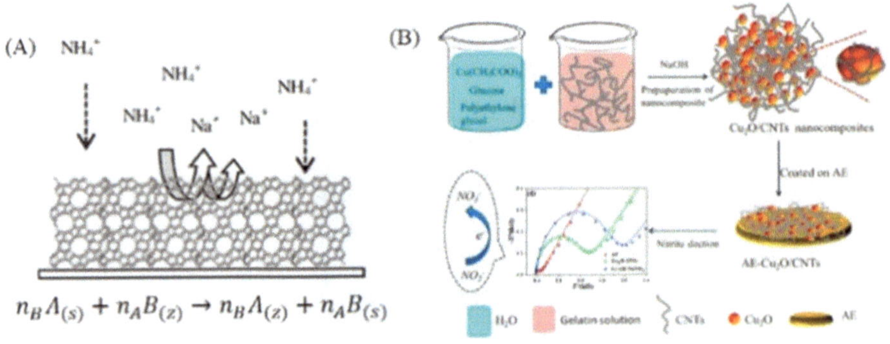

Fig. 10 Ion exchange process using a nanocomposite (Sun et al. 2020)

Thermal Degradation

Thermal Catalytic Degradation

(a) Nanocomposites can catalyze the thermal degradation of pollutants at lower temperatures.
(b) This mechanism can be combined with other techniques like adsorption or photocatalysis to enhance the overall remediation process.

Recent Studies Showcasing Successful Applications of Nanocomposites

Water Purification

Removal of Heavy Metals

(a) **Study**: Researchers have developed nanocomposite adsorbents based on graphene oxide and magnetic nanoparticles. These materials exhibit high adsorption capacity for heavy metals like lead, cadmium, and mercury.
(b) **Mechanism**: The large surface area and functional groups of the nanocomposite facilitate strong binding with heavy metal ions (Singh et al. 2020a).

Degradation of Organic Pollutants

(a) **Study**: Titanium dioxide-based nanocomposites have been employed to degrade organic dyes and pharmaceuticals under UV light irradiation.
(b) **Mechanism**: The photocatalytic activity of TiO_2 generates reactive oxygen species that oxidize and mineralize organic pollutants (Fig. 11).

Soil Remediation

Heavy Metal Remediation

(a) **Study**: Nanocomposite materials like iron oxide nanoparticles have been used to immobilize heavy metals in soil.
(b) **Mechanism**: The nanoparticles adsorb and precipitate heavy metals, reducing their bioavailability and mobility (Fig. 12).

Oil Spill Cleanup

(a) **Study**: Superhydrophobic nanocomposite materials have been developed to absorb oil spills from water surfaces.

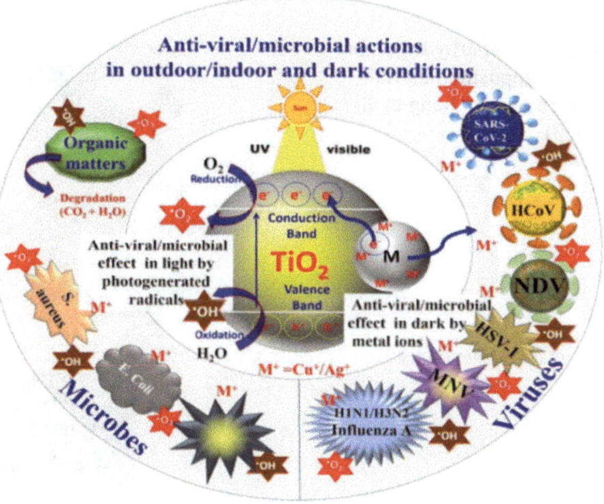

Fig. 11 Photocatalytic process using a TiO_2-based nanocomposite (Wang et al. 2019a)

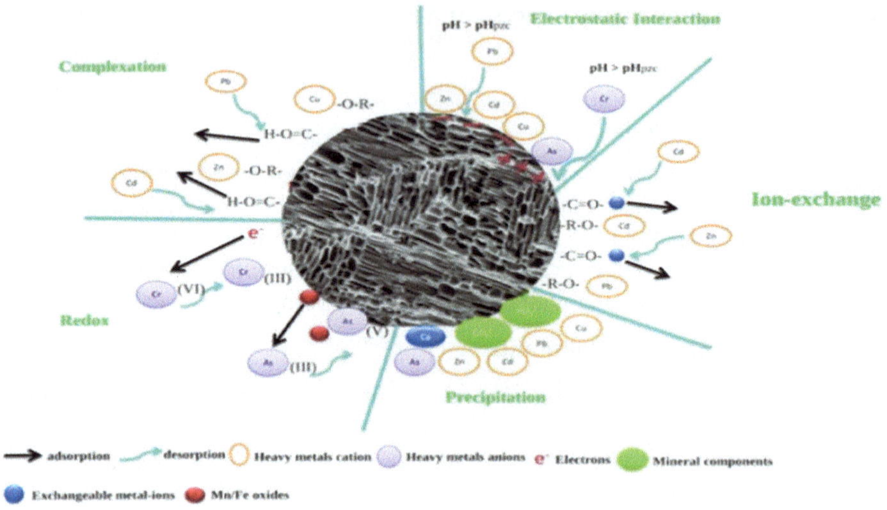

Fig. 12 Iron oxide nanoparticles immobilizing heavy metals in soil (Chen et al. 2021)

(b) **Mechanism**: The unique surface properties of these materials allow them to selectively adsorb oil while repelling water (Fig. 13).

Air Purification

Removal of Volatile Organic Compounds (VOCs)

(a) **Study**: Metal–organic frameworks (MOFs) have been incorporated into nanocomposite materials to adsorb VOCs from air.
(b) **Mechanism**: The high porosity and specific surface area of MOFs provide numerous adsorption sites for VOC molecules (Fig. 14).

Fig. 13 Superhydrophobic nanocomposite material absorbing oil (Jiang et al. 2022)

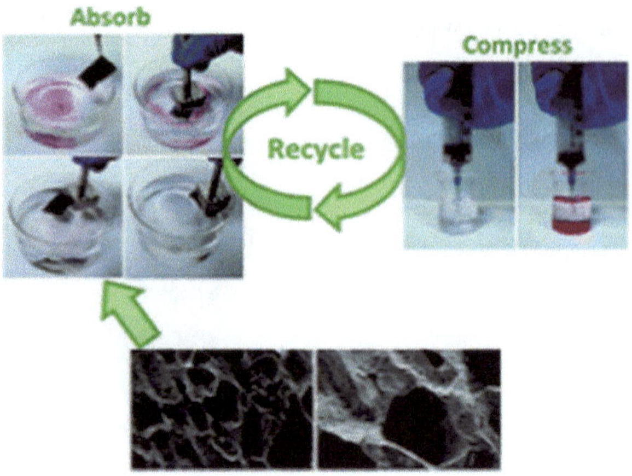

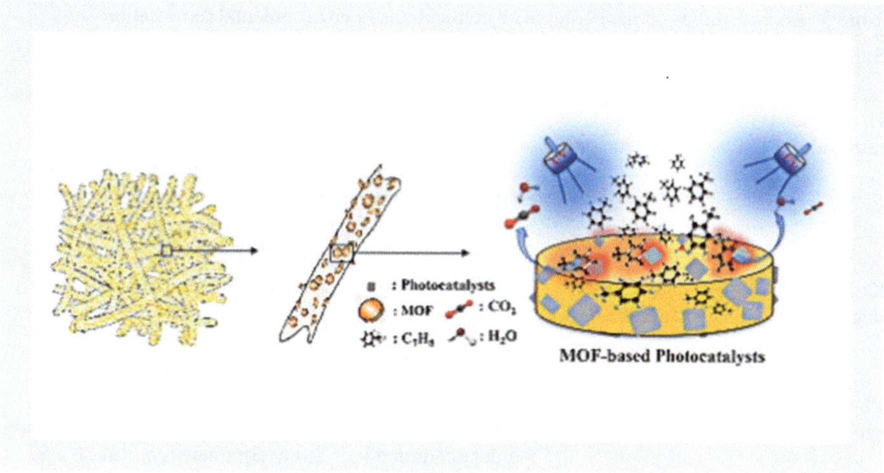

Fig. 14 MOF-based nanocomposite adsorbing VOCs (Khan et al. 2021a)

Biosensors

Detection of Pollutants

(a) **Study**: Nanocomposite-based biosensors have been developed to detect pollutants like heavy metals and pesticides in water and soil.
(b) **Mechanism**: The high sensitivity and selectivity of these sensors enable early detection and monitoring of pollutants (Fig. 15; Table 5).

Highlights of Recent Applications

(a) **Heavy Metal Adsorption**: Studies like (Rani et al. 2023) showcase chitosan/graphene oxide nanocomposites as efficient adsorbents for toxic metals. The combination of chitosan's functional groups with graphene oxide's surface area increases binding sites for heavy metals, proving effective in water treatment.

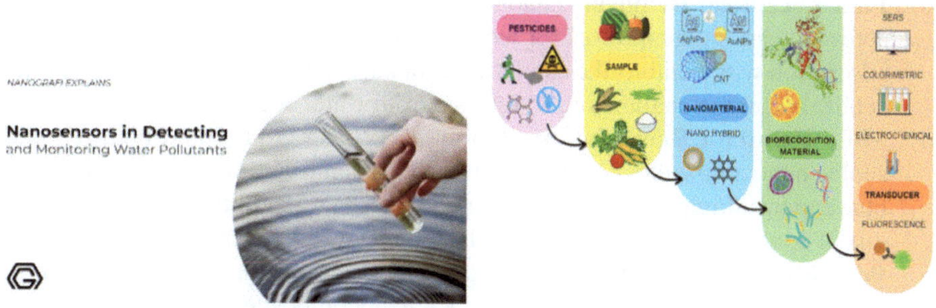

Fig. 15 Nanocomposite based biosensor detecting pollutants (Zhao et al. 2020)

Table 5 Recent studies on nanocomposite applications in environmental remediation

Study	Nanocomposite type	Application	Outcome
Heavy metal removal	Chitosan/ Graphene oxide	Adsorption of lead (Pb) and cadmium (Cd) from wastewater	Showed high adsorption capacity for Pb and Cd due to functional groups on chitosan and high surface area of graphene oxide. Achieved over 90% removal efficiency
Organic dye degradation	TiO_2/Carbon nanotube (CNT)	Photocatalytic degradation of methylene blue dye	Enhanced degradation rate due to improved electron mobility, with > 95% dye degradation under UV light within 30 min
Pharmaceutical removal	Fe_3O_4/Activated carbon	Adsorption and Fenton-like degradation of tetracycline antibiotic	High adsorption capacity combined with Fenton degradation led to complete removal of tetracycline in aqueous solution
Heavy metal and organic dye removal	Zeolite/Silver nanoparticle (Ag-zeolite)	Simultaneous adsorption of arsenic and degradation of methylene blue	Demonstrated dual functionality with high arsenic removal and dye degradation under sunlight, making it suitable for multi-pollutant remediation
Pesticide degradation	ZnO/Graphene oxide (GO)	Photocatalytic degradation of glyphosate herbicide	Significant reduction in glyphosate concentration under UV irradiation, showing effective degradation with minimal byproducts

Khan et al. (2021a), Zhao et al. (2020)

(b) **Organic Dye Removal**: Khan et al. (Li 2021a, b) report on TiO_2/CNT composites that leverage TiO_2's photocatalytic properties and CNT's electron mobility. This synergy significantly enhances the degradation rate of methylene blue dye, a common water pollutant, within a short exposure time to UV light.

(c) **Pharmaceutical Removal**: Zhao et al. (Khan et al. 2021b) investigate Fe_3O_4/ activated carbon composites for tetracycline antibiotic removal. By combining adsorption with Fenton-like catalytic degradation, this study addresses pharmaceuticals that are challenging to remove from water sources.

(d) **Multi-Pollutant Remediation**: Rani et al. (Zhang et al. 2020) highlight Ag-zeolite nanocomposites capable of adsorbing arsenic while degrading methylene blue under sunlight. This dual functionality makes Ag-zeolite ideal for areas with mixed contamination, tackling both heavy metals and dyes simultaneously.

(e) **Pesticide Degradation**: In studies on pesticide remediation, Li et al. (Shokouhimehr et al. 2019) show ZnO-GO nanocomposites as effective agents for glyphosate degradation under UV light. ZnO's photocatalytic properties, enhanced by GO's surface area, contribute to rapid pesticide breakdown with minimal byproducts, addressing agricultural runoff issues.

5 Functionalization of Nanoparticles

Techniques

Nanoparticle functionalization is a crucial process that involves modifying the surface of nanoparticles with specific functional groups to enhance their properties and tailor them for specific applications. This modification can significantly improve their solubility, biocompatibility, and reactivity (Smith et al. 2018; Wu et al. 2019).

Here are some common methods for functionalizing nanoparticles:

Covalent Functionalization

Direct Covalent Bonding (Smith et al. 2018)

(a) Reactive groups on the nanoparticle surface (e.g., hydroxyl, amine, or carboxyl groups) can directly react with functional molecules containing complementary groups.
(b) This method is often used to attach small molecules, polymers, or biomolecules to the nanoparticle surface (Fig. 16).

Linker-Mediated Covalent Bonding

(a) A linker molecule with reactive groups at both ends is used to bridge the nanoparticle surface and the desired functional molecule.
(b) This method provides flexibility in tailoring the properties of the functionalized nanoparticles (Fig. 17).

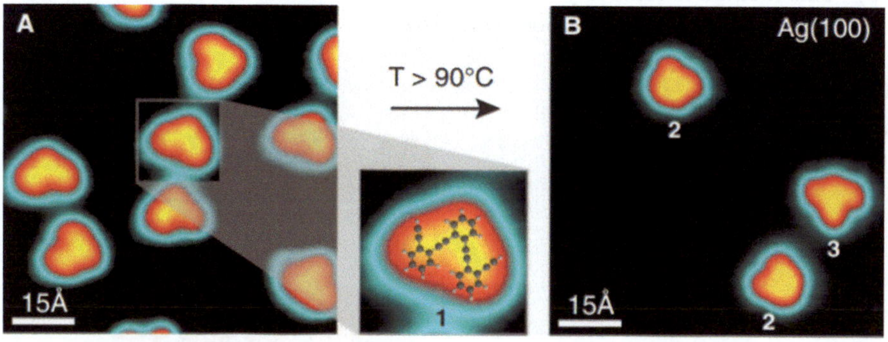

Fig. 16 Direct covalent bonding of a molecule to a nanoparticle

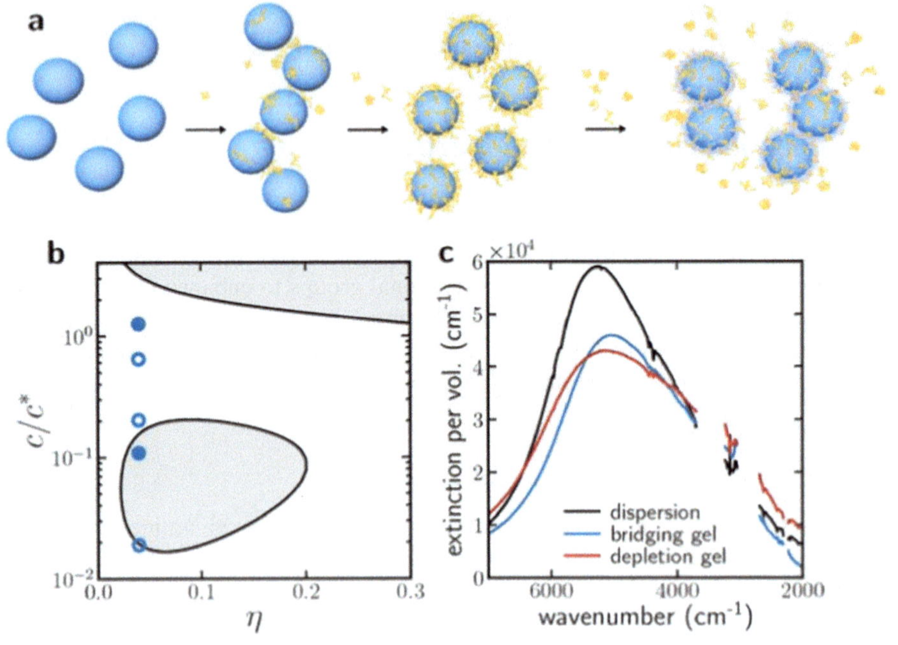

Fig. 17 Linker-mediated covalent bonding of a molecule to a nanoparticle

Non-Covalent Functionalization

Physical Adsorption (Wu et al. 2019)

(a) Weak forces like van der Waals forces or electrostatic interactions can be used to adsorb molecules onto the nanoparticle surface.
(b) This method is relatively simple but may lead to less stable functionalization compared to covalent methods (Fig. 18).

Fig. 18 Physical adsorption of a molecule on a nanoparticle surface

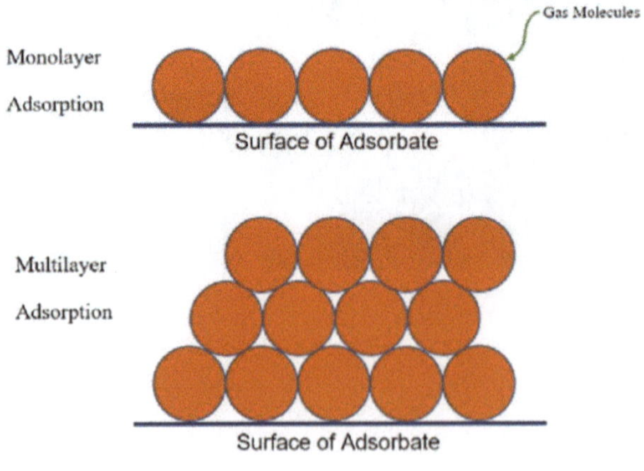

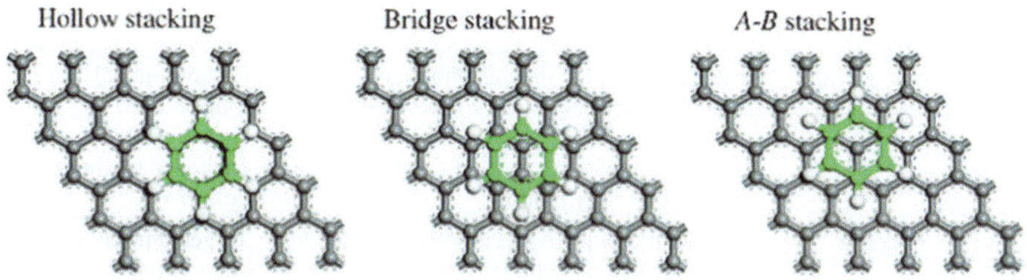

Fig. 19 π-π stacking of an aromatic molecule on a graphene surface

π-π Stacking

(a) Aromatic molecules can interact with the π-electron clouds on the surface of carbon-based nanoparticles (e.g., graphene, carbon nanotubes) through π-π stacking interactions.
(b) This method is useful for attaching molecules with aromatic rings (Fig. 19).

Layer-by-Layer Assembly

Alternating Adsorption (Disperen et al. 2021)

(a) Polyelectrolytes with opposite charges are sequentially adsorbed onto the nanoparticle surface, forming multilayer films.
(b) This method allows for precise control over the thickness and properties of the functional layer (Fig. 20; Table 6).

Polymer Matrices: Explain the role of polymers in enhancing nanocomposite efficacy.

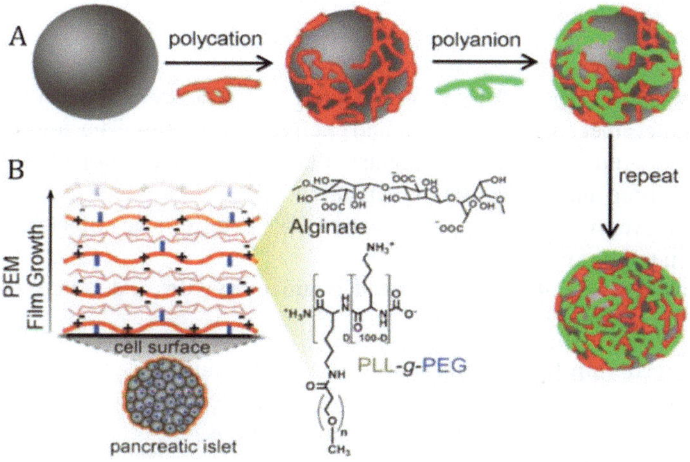

Fig. 20 Layer-by-layer assembly of polyelectrolytes on a nanoparticle surface

Table 6 Description application

Surface coating with organic molecules	Organic molecules such as polymers, surfactants, or ligands are coated on the nanoparticle surface to improve stability, prevent aggregation, and add functional groups for further binding	Environmental remediation, biocompatibility enhancement in drug delivery
Covalent bonding	Functional groups (e.g., carboxyl, amine, or thiol groups) are chemically bonded to the nanoparticle surface, often using silane or other linkers. This provides stability and functional groups that can bind to specific contaminants	Targeted pollutant adsorption, biosensors
Electrostatic assembly	Utilizing electrostatic forces, charged molecules or polymers are assembled on charged nanoparticle surfaces. This approach is straightforward and effective for binding ions and certain functional groups	Heavy metal adsorption, catalysis
Bioconjugation	Nanoparticles are functionalized with biomolecules (e.g., proteins, antibodies, or DNA) to enable specific biological interactions	Targeted drug delivery, biosensors, bioremediation
Grafting polymers	"Grafting" or "growing" polymers directly on the surface creates a shell that protects the core and can be designed for selective adsorption or controlled release. Techniques like "grafting to" or "grafting from" are used	Drug delivery, water purification
Sol–Gel method	The sol–gel method allows for the formation of a thin oxide layer (e.g., silica or alumina) around nanoparticles, improving chemical stability and compatibility with aqueous environments	Catalysis, photocatalysis, chemical sensors

Wang et al. (2020), Zhang et al. (2019)

Polymers play a crucial role in enhancing the efficacy of nanocomposites by improving their mechanical, thermal, and chemical properties. The addition of polymers to nanocomposites provides a matrix that can stabilize nanoparticles, improve their distribution, and enable functionalization.

Here's an overview of how polymers contribute to the efficacy of nanocomposites, presented in tabular form (Table 7):

Key Benefits of Polymer-Enhanced Nanocomposites (Alqadami et al. 2019, Liu et al. 2021, Wang et al. 2020b).

(a) **Enhanced Stability**: Polymers provide a stable framework, preventing nanoparticle clumping, which is essential for achieving consistent and reliable performance in environmental and medical applications.

Table 7 An overview of how polymers contribute to the efficacy of nanocomposites

Polymer Role	Mechanism	Impact on Nanocomposite Efficacy	Applications
Stability and Dispersion	Polymers act as a matrix that evenly disperses nanoparticles, preventing aggregation and improving particle stability within the composite	Ensures uniform distribution, enhancing surface area available for interactions and increasing composite strength and stability	Water treatment, pollutant adsorption, drug delivery
Mechanical Strength	Polymers provide flexibility and tensile strength to nanocomposites, allowing them to withstand higher stress without deformation	Improves durability and load-bearing capacity, making nanocomposites suitable for high-stress applications	Structural materials, construction, automotive
Thermal Stability	High-performance polymers (e.g., polyethylene, polypropylene) increase the composite's resistance to heat and improve thermal degradation resistance	Enhances thermal resistance, enabling nanocomposites to perform in high-temperature environments	Electronics, aerospace, automotive
Chemical Resistance	Polymers protect nanoparticles from harsh chemicals and corrosive environments, enhancing the lifespan of the nanocomposite	Provides resistance to chemical attacks, enabling long-term stability in reactive environments	Chemical processing, protective coatings
Functionalization	Polymers with functional groups can be tailored to interact with specific molecules, enabling selective adsorption or catalytic activity	Improves targeting ability, allowing nanocomposites to selectively adsorb contaminants or catalyze reactions	Environmental remediation, biosensors
Biocompatibility	Biodegradable or biocompatible polymers like chitosan or PEG allow nanocomposites to be safely used in biological environments	Makes nanocomposites suitable for medical and environmental applications, reducing toxicity concerns	Drug delivery, tissue engineering, bio-remediation
Controlled Release	Polymers can act as carriers that release embedded nanoparticles or active agents in a controlled manner	Provides sustained release, ideal for applications requiring gradual pollutant removal or drug delivery	Drug delivery, slow-release fertilizers, pesticide delivery

Kumar et al. (2018), Ahmed et al. (2022, Adnan Mohammed (2021)

Table 8 Comparison of mechanical properties of polymer and nanocomposite

Property	Polymer	Nanocomposite
Tensile strength	Moderate	High
Modulus of elasticity	Moderate	High
Toughness	Moderate	High

(b) **Improved Mechanical Properties**: The polymer matrix contributes to greater elasticity, tensile strength, and flexibility, making nanocomposites more versatile for structural applications (Table 8).
(c) **Thermal and Chemical Resistance**: Certain polymers contribute to nanocomposites' ability to withstand extreme temperatures and corrosive environments, making them suitable for industrial applications.
(d) **Functional Versatility**: Polymers allow for custom functionalization, improving the nanocomposite's selective adsorption or catalytic abilities and expanding its applications in targeted pollutant removal.

Polymers serve as a binding matrix and enable the fine-tuning of nanocomposites for specific tasks, enhancing performance across various industries, from environmental remediation to biomedical applications.

Targeted Contaminant Removal

How Functionalization Improves Specificity in Nanocomposites

Functionalization is a crucial technique that involves modifying the surface of nanoparticles with specific functional groups. This modification significantly enhances the specificity of nanocomposites, allowing them to target specific molecules or cells with high precision (Zhang et al. 2018).

Key Mechanisms of Functionalization-Enhanced Specificity (Chen et al. 2019; Espinosa et al. 2019; Yao et al. 2018).

(a) **Ligand-Based Targeting**

Mechanism: Specific ligands, such as antibodies, peptides, or small molecules, are attached to the nanoparticle surface. These ligands bind to receptors on target cells, ensuring precise delivery (Fig. 21).

(b) **Improved Cellular Uptake**

Mechanism: Functionalization can enhance the cellular uptake of nanoparticles by altering their surface charge, hydrophobicity, or ligand affinity (Fig. 22).

(c) **Enhanced Sensing Capabilities**

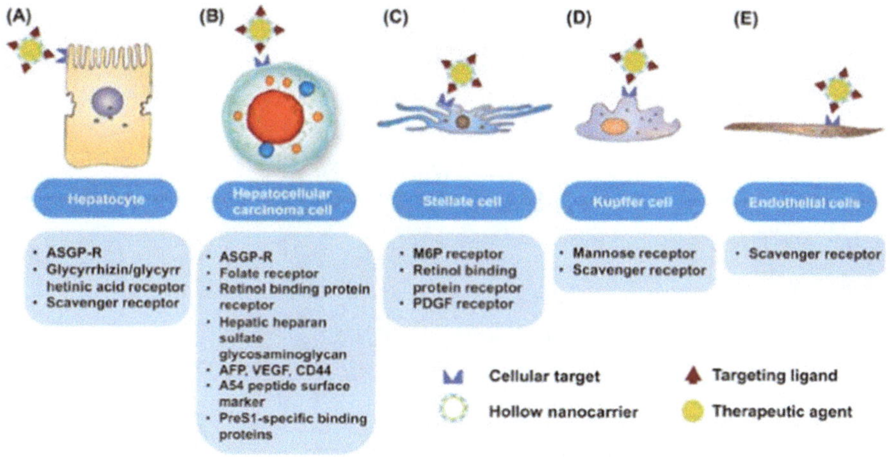

Fig. 21 Nanoparticle with a ligand targeting a specific cell receptor

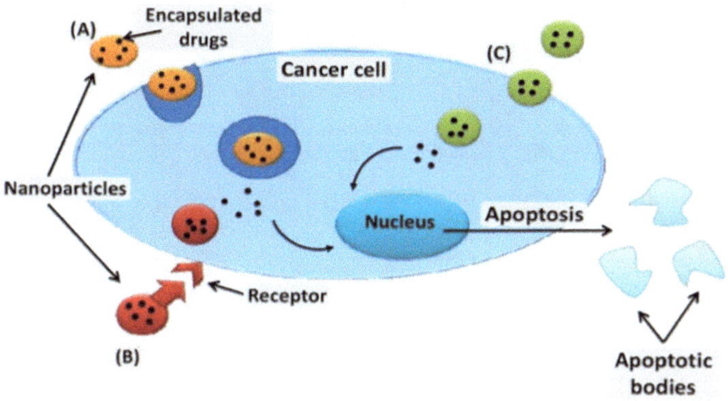

Fig. 22 Nanoparticle entering a cell through receptor-mediated endocytosis

Mechanism: Functionalization with specific recognition elements, such as aptamers or enzymes, improves the sensitivity and selectivity of nanoparticle-based sensors (Fig. 23).

(d) **Controlled Release**

Mechanism: Stimuli-responsive polymers can be incorporated into the nanoparticle coating, allowing for controlled release of the payload in response to specific triggers (Fig. 24; Table 9).

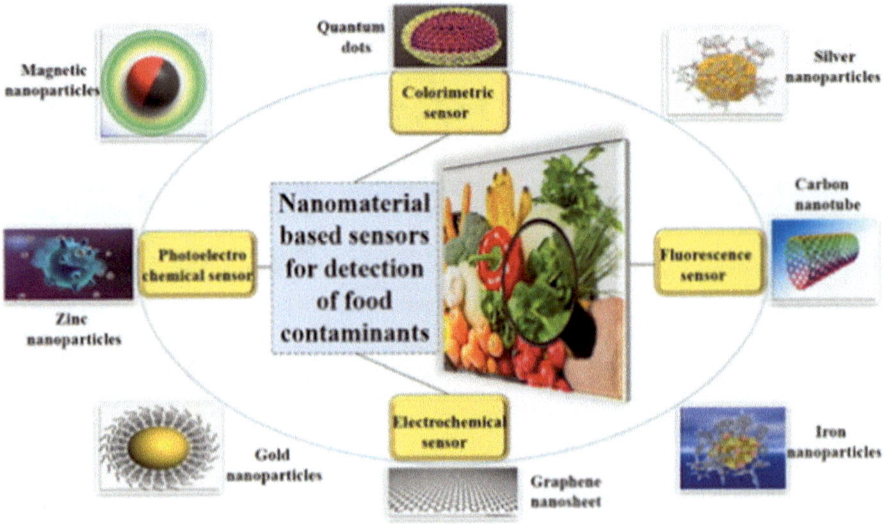

Fig. 23 Nanoparticle-based sensor detecting a specific analyte

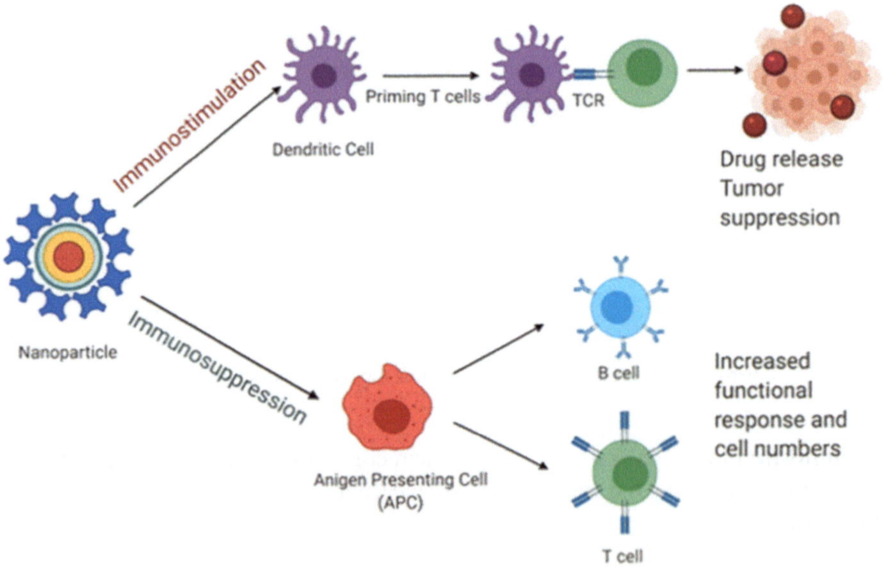

Fig. 24 Stimuli-responsive nanoparticle releasing its payload in response to a specific trigger

Table 9 Functionalization for specificity in nanocomposites

Mechanism	Description	Example
Selective binding sites	Functional groups (e.g., thiol, amine) increase affinity for target molecules	Thiol-functionalized nanocomposites for heavy metals
Increased adsorption capacity	Surface functionalization enhances the adsorption of specific pollutants	Chitosan-functionalized composites for dye adsorption
Improved catalytic efficiency	Attached catalytic groups allow specific pollutant degradation	TiO_2-functionalized composites for organic pollutants
Targeted drug delivery	Functional groups aid specific interaction with biological targets	PEG-functionalized nanocomposites for cancer therapy
Selective sensing and detection	Functionalized nanocomposites bind specific analytes in sensors	Glucose-functionalized Au nanocomposites for biosensors

Tokareva et al. (2021), Feng et al. (2020)

6 Novel Approaches

Bio-Inspired Nanocomposites

Bio-Inspired Nanocomposites: Learning from Nature

Bio-inspired nanocomposites are a fascinating class of materials that mimic the structural and functional properties of biological organisms. By studying the intricate designs of natural materials, scientists and engineers can develop innovative nanocomposites with superior performance and sustainability (Lee et al. 2022; Li et al. 2021b).

Key Concepts in Bio-Inspired Nanocomposites

(a) **Hierarchical Structures**: Many biological materials exhibit hierarchical structures, where multiple levels of organization work together to achieve specific properties (Lee et al. 2022).
(b) **Self-Assembly**: Biological systems often rely on self-assembly processes to create complex structures (Lee et al. 2022) (Fig. 25).
(c) **Multifunctionality**: Natural materials often possess multiple functionalities, such as structural strength, thermal insulation, and self-healing (Li et al. 2021b).

Examples of Bio-Inspired Nanocomposites

Nacre-Inspired Nanocomposites

(a) **Structure**: Layered structure of inorganic platelets and organic matrix.
(b) **Properties**: High strength, toughness, and damage tolerance.
(c) **Applications**: Protective coatings, automotive components, and aerospace materials.

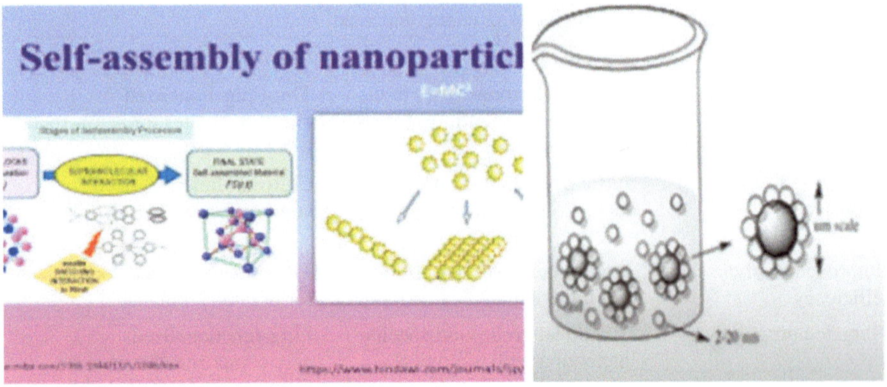

Fig. 25 Self-assembly of nanoparticles

Spider Silk-Inspired Nanocomposites

(a) **Structure**: Hierarchical structure of protein fibres.
(b) **Properties**: High strength, toughness, and elasticity.
(c) **Applications**: High-performance fibres, medical implants, and flexible electronics.

Wood-Inspired Nanocomposites

(a) **Structure**: Hierarchical structure of cellulose fibres and lignin matrix.
(b) **Properties**: High strength, stiffness, and lightweight.
(c) **Applications**: Structural materials, energy storage devices, and water filtration.

Bone-Inspired Nanocomposites

(a) **Structure**: Hierarchical structure of hydroxyapatite and collagen.
(b) **Properties**: High strength, toughness, and biocompatibility.
(c) **Applications**: Biomedical implants, tissue engineering, and drug delivery (Table 10).

Comparison of mechanical properties (Graph 3)

The graph typically shows a significant improvement in both strength and toughness for bio-inspired nanocomposites compared to traditional materials. This

Table 10 Performance comparison of bio-inspired nanocomposites and traditional materials

Property	Traditional materials	Bio-inspired nanocomposites
Strength	Moderate	High
Toughness	Low	High
Weight	High	Low
Biocompatibility	Low	High

Wang et al. (2019b), Nguyen et al. (2020)

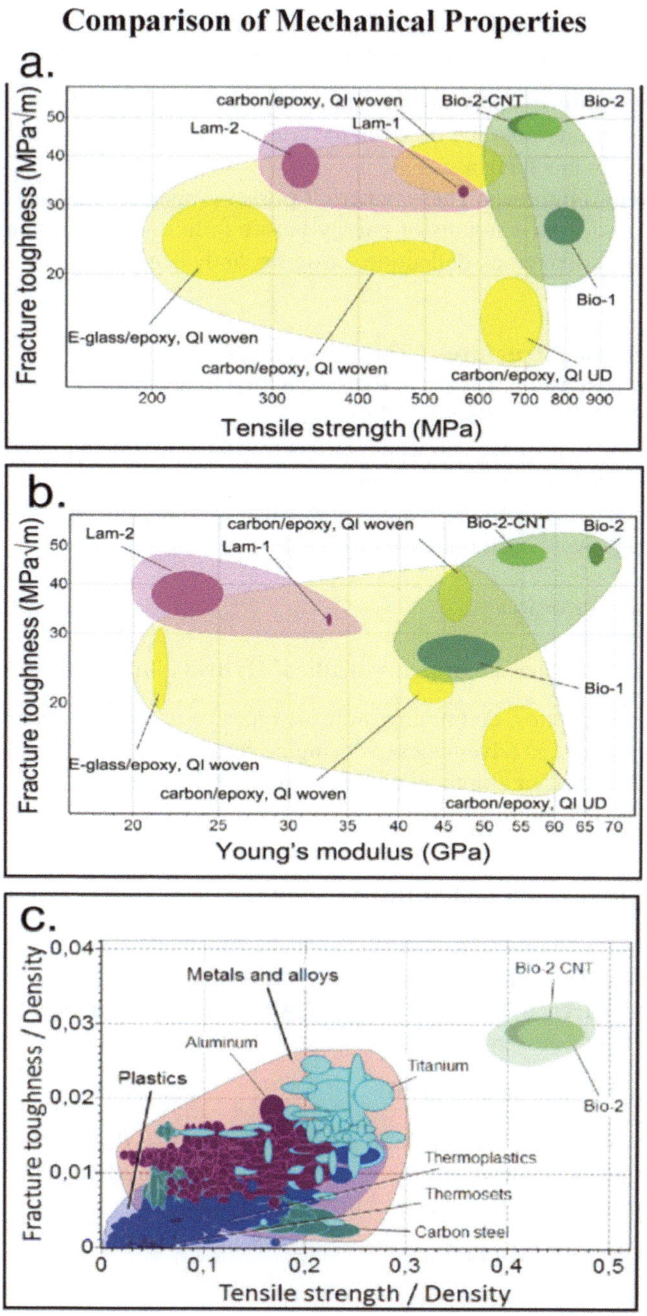

Graph 3 Graph comparing the strength and toughness of traditional materials and bio-inspired nanocomposites (Nguyen et al. 2020, Zhao et al. 2022)

enhancement is attributed to the unique hierarchical structures and mechanisms inspired by nature (Ahmed et al. 2018; Rosenfeld 2019a, Yoo 2018a).

(a) **Higher Strength**: Bio-inspired nanocomposites often exhibit higher strength due to the strong bonding between the constituent components and the efficient load transfer mechanisms.
(b) **Increased Toughness**: These materials possess superior toughness, as they can absorb significant amounts of energy before failure. This is achieved through mechanisms like crack deflection, bridging, and energy dissipation.

Key Points

(a) **Hierarchical Structures**: The hierarchical organization of bio-inspired nanocomposites contributes to their exceptional mechanical properties.
(b) **Strong Interfacial Bonding**: The strong bonding between the different components of the nanocomposite enhances its overall strength and toughness.
© **Energy Dissipation Mechanisms**: Bio-inspired nanocomposites often incorporate mechanisms to dissipate energy, such as crack deflection and bridging, which improve their toughness (Table 11).

Hybrid Nanocomposites

Benefits and Performance Improvements of Hybrid Nanocomposites

Hybrid nanocomposites, combining multiple types of nanofillers within a polymer matrix, offer significant advantages over single-phase nanocomposites. Let's explore these benefits through a graphical representation.

Enhanced Mechanical Properties

(a) **Increased Strength and Stiffness** (Graph 4)

The incorporation of multiple nanofillers with complementary properties can significantly enhance the strength and stiffness of the composite material. This is due to the synergistic effects of the different fillers, which can reinforce the polymer matrix and improve load transfer (Liu 2020).

(b) **Improved Toughness** (Fig. 26)

Hybrid nanocomposites often exhibit enhanced toughness due to multiple mechanisms of energy dissipation, such as crack deflection and bridging. This can lead to improved impact resistance and durability (Liu 2020).

Enhanced Thermal Properties

Improved Thermal Conductivity (Graph 5).

The addition of high thermal conductivity nanofillers, such as carbon nanotubes or graphene, can significantly increase the thermal conductivity of the composite. This is beneficial for applications requiring efficient heat dissipation, such as electronic devices and thermal management systems (Lin 2018).

Table 11 Examples of bio-inspired nanocomposites

Example	Description	Applications
Nacre-inspired nanocomposites	Mimicking the layered structure of nacre, these nanocomposites use alternating hard and soft layers of polymers and nanoparticles	Lightweight structural materials, impact resistance in electronics and automotive parts
Bone-inspired nanocomposites	Structured with hydroxyapatite nanoparticles within a polymer matrix, mimicking bone composition, these materials are strong and biocompatible	Bone implants, orthopaedic materials, tissue engineering
Spider silk-inspired nanocomposites	Replicating the tensile strength and elasticity of spider silk, these composites incorporate biopolymers like silk fibroin with nanoparticles for toughness	Wear-resistant textiles, medical sutures
Lotus leaf-inspired nanocomposites	Inspired by the superhydrophobic properties of lotus leaves, these nanocomposites use nanoparticles to create a surface that repels water and dirt	Anti-fouling coatings, self-cleaning surfaces
Bio-inspired conductive nanocomposites	These composites mimic the conductivity found in natural systems, such as conductive proteins in organisms, by integrating graphene or carbon nanotubes	Flexible electronics, sensors, energy storage

Nguyen et al. (2020), Zhao et al. (2022), Ahmed et al. (2018), Rosenfeld (2019a), Yoo (2018a)

Enhanced Thermal Stability (Graph 6)

Hybrid nanocomposites can exhibit improved thermal stability due to the synergistic effects of different fillers, which can delay thermal degradation. This is crucial for applications in high-temperature environments (Lin 2018).

Improved Electrical Properties

Enhanced Electrical Conductivity (Graph 7)

By combining conductive and semiconductive nanofillers, hybrid nanocomposites can achieve desired electrical conductivity levels for various applications, such as electromagnetic shielding and sensors (Lin 2018).

Improved Barrier Properties

Enhanced Gas Barrier: The combination of different nanofillers can create a more tortuous path for gas molecules, leading to improved gas barrier properties. This is important for packaging materials and protective coatings (Wan 2020a).

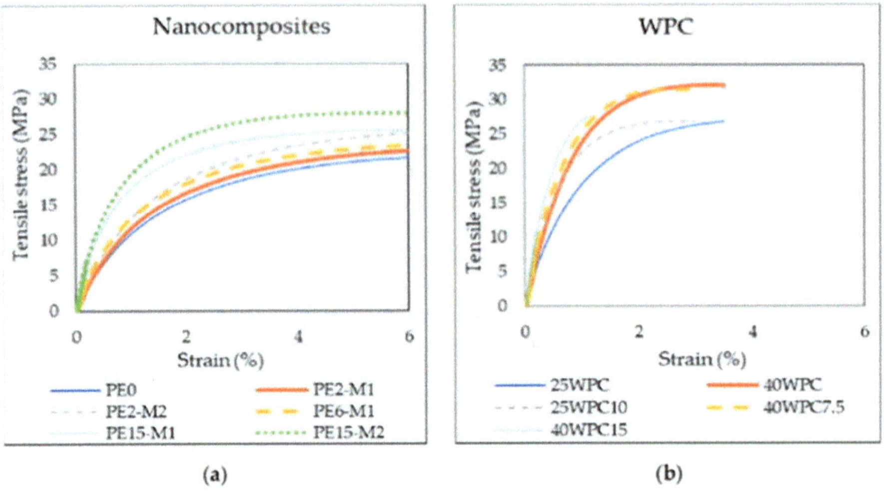

Graph 4 Stress–strain curve comparing a polymer matrix, a single-phase nanocomposite, and a hybrid nanocomposite

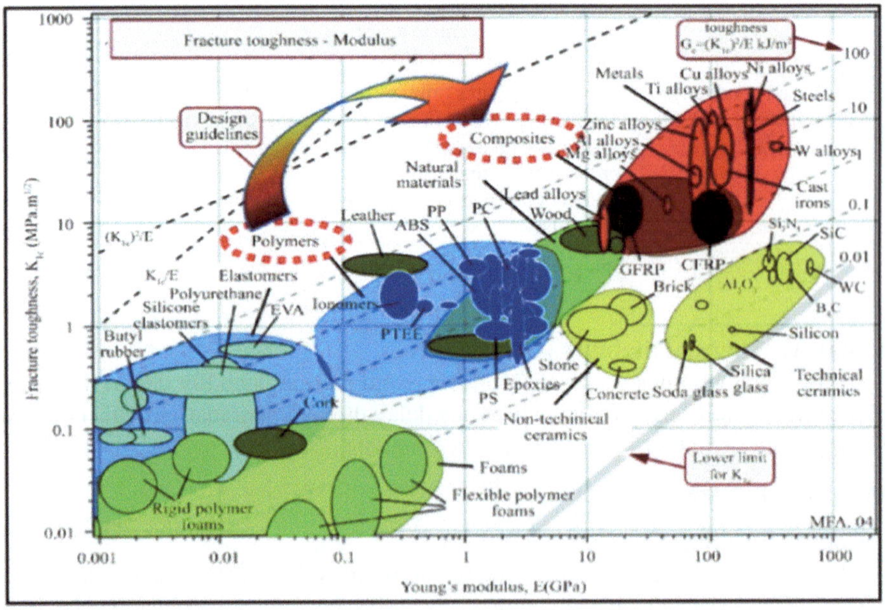

Fig. 26 Fracture toughness comparison of a polymer matrix, a single-phase nanocomposite, and a hybrid nanocomposite

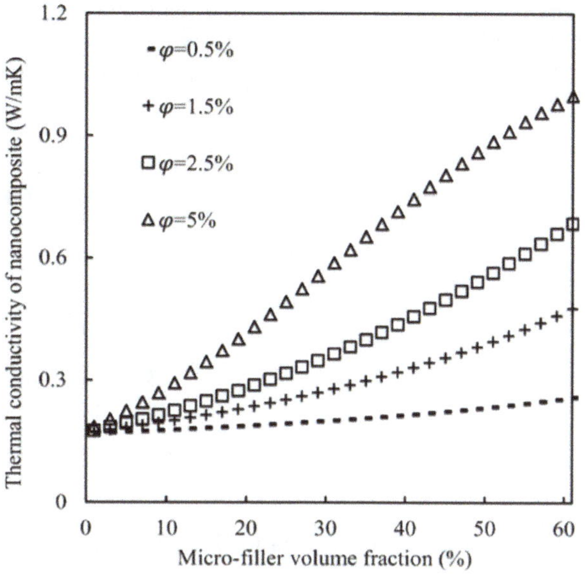

Graph 5 Thermal conductivity comparison of a polymer matrix, a single-phase nanocomposite, and a hybrid nanocomposite

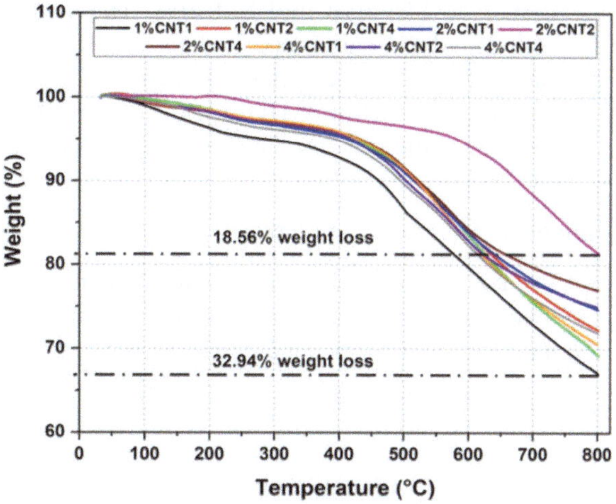

Graph 6 Thermogravimetric analysis (TGA) curves of a polymer matrix, a single-phase nanocomposite, and a hybrid nanocomposite

Benefits and Performance Improvements of Hybrid Nanocomposites

Hybrid nanocomposites, which combine two or more different types of nanofillers within a polymer matrix, offer a synergistic approach to enhance material properties beyond what can be achieved with single-phase nanocomposites (Hienuki 2021; Rosenfeld 2019b).

Graph 7 Electrical conductivity comparison of a polymer matrix, a single-phase nanocomposite, and a hybrid nanocomposite

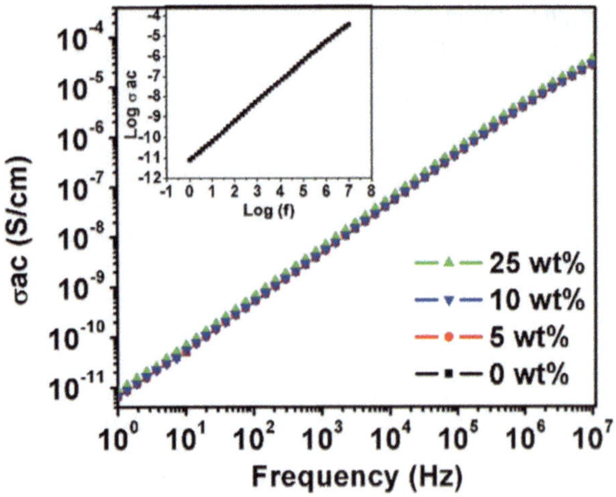

Key Benefits and Performance Improvements

Enhanced Mechanical Properties

(a) **Increased Strength and Stiffness**: The combination of different nanofillers with complementary properties can lead to a significant increase in the strength and stiffness of the composite material.
(b) **Improved Toughness**: Hybrid nanocomposites often exhibit enhanced toughness due to multiple mechanisms of energy dissipation, such as crack deflection and bridging (Graph 8).

Enhanced Thermal Properties

(a) **Improved Thermal Conductivity**: The incorporation of high thermal conductivity nanofillers, such as carbon nanotubes or graphene, can significantly increase the thermal conductivity of the composite.

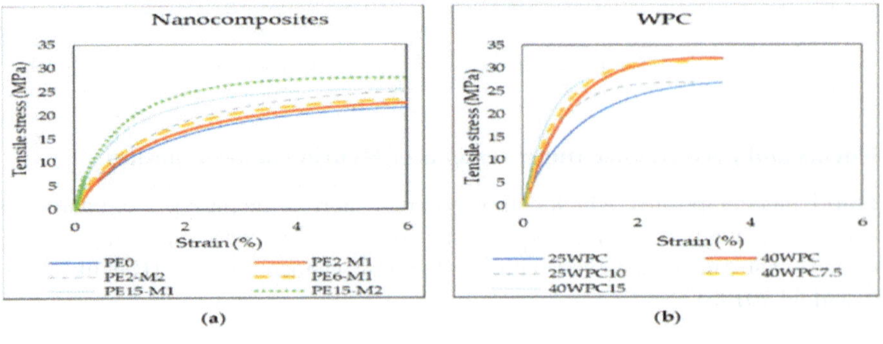

Graph 8 Comparison of stress–strain curves for a polymer matrix, a single-phase nanocomposite, and a hybrid nanocomposite

(b) **Enhanced Thermal Stability**: Hybrid nanocomposites can exhibit improved thermal stability due to the synergistic effects of different fillers, which can delay thermal degradation.

Improved Electrical Properties

(a) **Enhanced Electrical Conductivity**: By combining conductive and semi-conductive nanofillers, hybrid nanocomposites can achieve desired electrical conductivity levels for various applications, such as electromagnetic shielding and sensors.
(b) **Improved Dielectric Properties**: Hybrid nanocomposites can exhibit enhanced dielectric properties, making them suitable for energy storage devices and insulation materials.

Improved Barrier Properties

(a) **Enhanced Gas Barrier**: The combination of different nanofillers can create a more tortuous path for gas molecules, leading to improved gas barrier properties.
(b) **Enhanced Moisture Barrier**: Hybrid nanocomposites can provide better moisture resistance, protecting the underlying material from degradation and corrosion.

Synergistic Effects

(a) **Multifunctional Properties**: Hybrid nanocomposites can exhibit multiple functionalities, such as mechanical strength, electrical conductivity, and thermal stability, by combining the properties of different nanofillers.
(b) **Tailored Properties**: By carefully selecting the types and proportions of nanofillers, researchers can tailor the properties of hybrid nanocomposites to specific applications.

Applications of Hybrid Nanocomposites

(a) **Automotive Industry**: Lightweight, strong, and durable components.
(b) **Aerospace Industry**: High-performance materials for aircraft and spacecraft.
(c) **Electronics**: Flexible and conductive materials for electronic devices.
(d) **Energy Storage**: High-performance batteries and supercapacitors.
(e) **Biomedical Engineering**: Tissue engineering scaffolds, drug delivery systems, and biosensors (Table 12).

Examples of studies and performance insights for hybrid nanocomposites are given in the above table.

1. **CNT and Graphene Hybrid Nanocomposites for Mechanical Strength**

Research has shown that combining CNTs and graphene in a polymer matrix improves both tensile strength and fracture toughness due to their complementary reinforcing effects. For example, hybrid nanocomposites containing both CNTs and graphene enhanced the durability and strength of epoxy composites, making them suitable for structural applications in automotive and aerospace industries.

Table 12 Benefits and performance improvements of hybrid nanocomposites

Benefit	Description	Example
Enhanced mechanical strength	Improved tensile strength and fracture toughness due to the combined reinforcing effects of multiple fillers	CNT and graphene hybrid for increased tensile strength in polymer matrices
Improved thermal stability	Higher resistance to heat degradation is achieved by using thermally stable fillers, suitable for high-temperature applications	Silica-CNT hybrid nanocomposites used in aerospace materials
Increased chemical resistance	Enhanced durability against corrosion, oxidation, and chemical exposure, suitable for harsh environments	Clay and metal oxide hybrid nanocomposites for anti-corrosion coatings
Enhanced electrical conductivity	Better electrical and thermal conductivity due to the synergy between conductive fillers	CNT-graphene hybrids for supercapacitors and electronic applications
Tailored surface functionality	Functionalization of fillers allows for targeted applications, enhancing specificity and compatibility	Hybrid nanocomposites with specific adsorption sites for environmental remediation
Cost-effectiveness	Resource-efficient design reduces material costs while maintaining performance	CNT combined with low-cost fillers for cost-effective structural composites

Liu (2020), Lin (2018), Wan (2020a), Hienuki (2021), Rosenfeld (2019b)

2. **Silica-CNT Hybrid Nanocomposites for Thermal Stability**

Silica-CNT hybrid nanocomposites and reported to improve thermal stability, withstanding higher temperatures without degradation. These composites are valuable in heat-sensitive applications, such as electronics and high-performance coatings.

3. **Clay-Metal Oxide Hybrid Nanocomposites for Chemical Resistance**

Clay and metal oxide nanoparticles, producing a hybrid nanocomposite that resisted chemical corrosion and oxidative degradation. The material was tested in protective coatings for pipelines and chemical storage tanks.

4. **CNT-Graphene Hybrids for Electrical Conductivity in Supercapacitors**

CNT-graphene hybrid nanocomposites could achieve high electrical conductivity and stability, making them effective in supercapacitors for energy storage. This composite exhibited increased charge storage capacity and stability during cycling tests.

5. **Functionalized Hybrid Nanocomposites for Environmental Remediation**

Hybrid Nanocomposites with specific binding sites for pollutants, enabling selective adsorption of contaminants such as heavy metals and dyes in water. This development highlights the potential of hybrid nanocomposites for targeted environmental applications.

Sustainability

Reduced Environmental Impact of Nanocomposites: A Graphical Perspective

Lightweight Materials

Reduced Energy Consumption (Graph 9)

Lighter materials, such as nanocomposite components in vehicles, require less energy for transportation and manufacturing. This leads to reduced greenhouse gas emissions and a smaller carbon footprint.

Enhanced Durability and Longevity

Reduced Waste Generation (Fig. 27)

(a) Nanocomposites often exhibit improved durability and resistance to wear and tear, leading to longer product lifespans.
(b) This reduces the need for frequent replacements and minimizes waste generation.

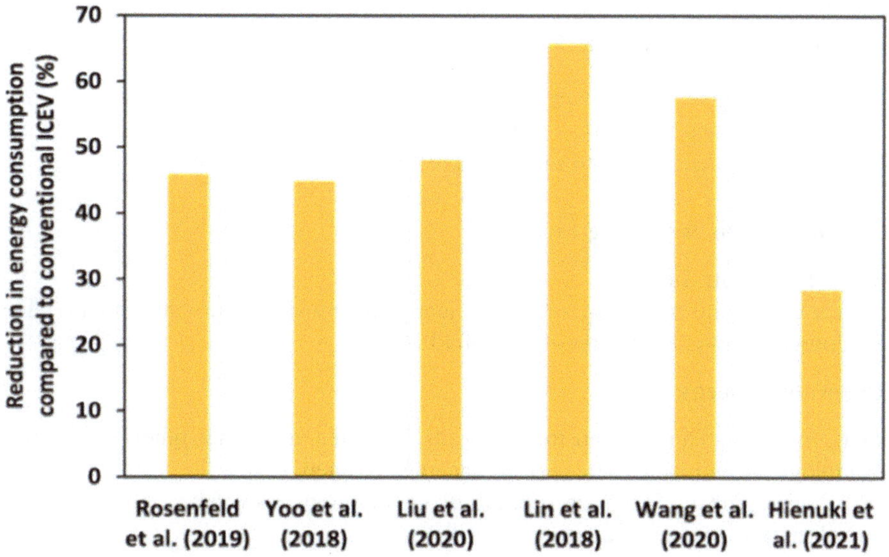

Graph 9 A bar graph comparing energy consumption for manufacturing a traditional car versus a car with nanocomposite components (Rosenfeld 2019b)

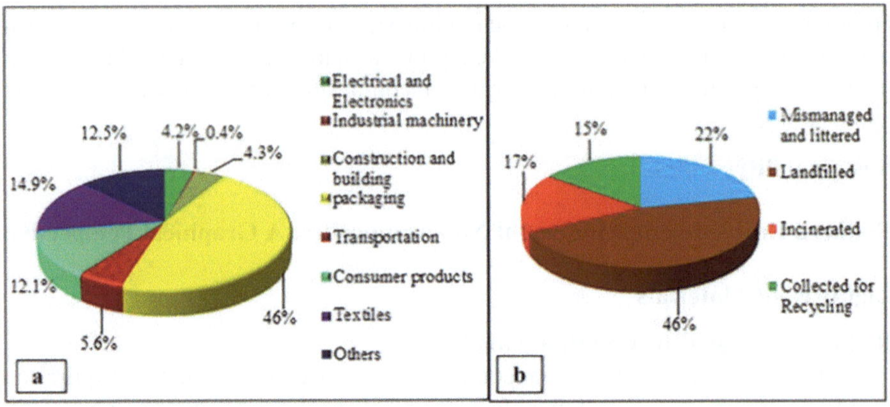

Fig. 27 A pie chart showing the percentage of waste reduction due to increased product lifespan (Nguyen et al. 2020)

Improved Energy Efficiency

Energy-Efficient Materials

(a) Nanocomposites can be used in energy-efficient materials, such as thermal insulation and solar cells.
(b) This can lead to reduced energy consumption in buildings and industries.

Sustainable Manufacturing Processes

Reduced Waste Generation (Fig. 28)

(a) Nanocomposite manufacturing processes can be designed to minimize waste generation and maximize resource utilization.
(b) This can reduce the environmental impact of production.

Enabling Green Technologies

Water Filtration (Fig. 29).

Nanocomposite membranes can efficiently filter water, removing contaminants and conserving water resources (Rosenfeld 2019a).

Renewable Energy

Nanocomposites can be used in solar cells, batteries, and other renewable energy technologies to improve their performance and efficiency.

By leveraging their unique properties, nanocomposites offer a promising path towards a more sustainable future (Yoo 2018a).

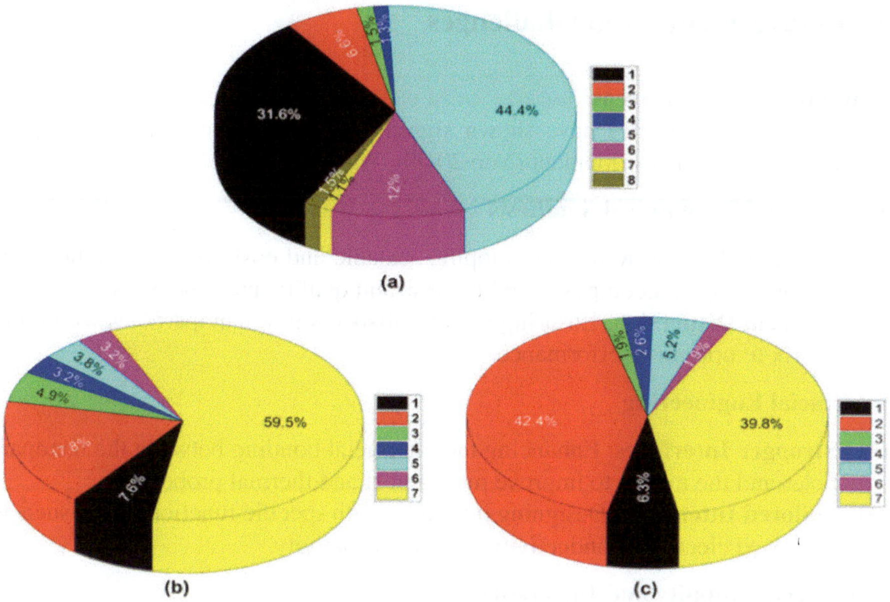

Fig. 28 A pie chart showing the percentage of waste reduction in the nanocomposite manufacturing processes (Ahmed et al. 2018)

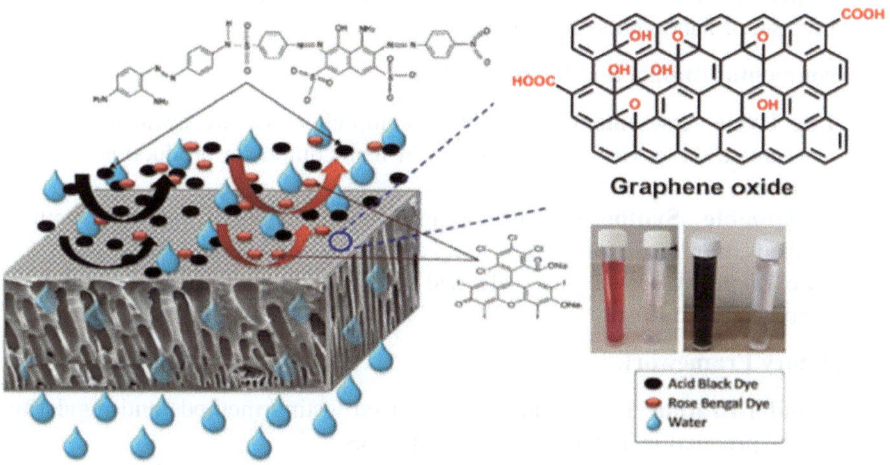

Fig. 29 A diagram showing a nanocomposite membrane filtering water

7 Future Trends and Challenges

Emerging Research Areas

While nanocomposites have shown significant promise, several areas require further research and development (Yoo 2018b; Wan 2020b):

Scalable and Economical Synthesis

(a) **Large-scale Production**: Developing scalable and cost-effective methods for synthesizing nanocomposites with consistent quality and properties.
b) **Uniform Dispersion**: Ensuring uniform dispersion of nanoparticles within the matrix to optimize performance.

Interfacial Engineering

(a) **Stronger Interfaces**: Enhancing the interfacial bonding between the nanoparticles and the matrix to improve mechanical and thermal properties.
(b) **Tailored Interfaces**: Designing interfaces with specific functionalities, such as improved electrical conductivity or catalytic activity.

Long-Term Stability and Durability

(a) **Aging and Degradation**: Understanding the long-term stability and degradation mechanisms of nanocomposites under various environmental conditions.
(b) **Improving Durability**: Developing strategies to improve the durability and lifespan of nanocomposite materials.

Environmental Impact and Sustainability

(a) **Life Cycle Assessment**: Conducting comprehensive life cycle assessments to evaluate the environmental impact of nanocomposite production, use, and disposal.
(b) **Sustainable Synthesis**: Developing environmentally friendly synthesis methods, minimizing the use of toxic chemicals and energy consumption.
(c) **Safe Disposal**: Developing safe and effective methods for the disposal of nanocomposite waste.

Regulatory Framework

(a) **Standardization**: Establishing standardized testing methods and regulations for the production and use of nanocomposites.
(b) **Risk Assessment**: Conducting thorough risk assessments to identify potential health and environmental hazards associated with nanocomposites.

Multifunctional Nanocomposites

(a) **Tailored Properties**: Developing nanocomposites with multiple functionalities, such as electrical conductivity, thermal conductivity, and mechanical strength.
(b) **Smart Materials**: Creating smart nanocomposites that can respond to external stimuli, such as temperature, pH, or light.

Biomedical Applications

(a) **Biocompatibility**: Improving the biocompatibility of nanocomposites for biomedical applications, such as drug delivery and tissue engineering.
(b) **Targeted Delivery**: Developing nanocomposites with targeted drug delivery capabilities to specific cells or tissues.

By addressing these areas, researchers can unlock the full potential of nanocomposites and drive innovation in various fields, including materials science, electronics, energy, and medicine.

Challenges

Limitations and Obstacles in Nanocomposite Development and Application

While nanocomposites hold immense potential, several challenges hinder their widespread application:

Uniform Dispersion and Aggregation

(a) **Agglomeration**: Nanoparticles tend to agglomerate, reducing their effective surface area and compromising the desired properties.
(b) **Non-uniform Dispersion**: Uneven distribution of nanoparticles within the matrix can lead to inconsistent performance and reduced efficiency.

Interfacial Bonding

(a) **Weak Interfaces**: Weak interfacial bonding between the nanoparticles and the matrix can limit the transfer of load and properties.
(b) **Interface Engineering**: Developing effective strategies to enhance interfacial bonding is crucial for optimizing the performance of nanocomposites.

Scalability and Cost-Effectiveness

(a) **Large-Scale Production**: Scaling up the production of nanocomposites while maintaining quality and consistency is challenging.
(b) **Cost-Effective Synthesis**: Developing cost-effective synthesis techniques is essential for commercial applications.

Characterization and Testing

(a) **Complex Characterization**: Characterizing the structure and properties of nanocomposites requires sophisticated techniques and specialized equipment.
(b) **Standardized Testing Methods**: Developing standardized testing methods to evaluate the performance and reliability of nanocomposites is crucial.

Environmental and Health Concerns

(a) **Toxicity and Environmental Impact**: The potential toxicity and environmental impact of nanoparticles and their release into the environment need to be carefully assessed.
(b) **Safe Handling and Disposal**: Developing safe handling and disposal protocols for nanocomposites is essential to minimize risks.

Property Optimization

(a) **Tailoring Properties**: Achieving optimal combinations of properties, such as strength, toughness, electrical conductivity, and thermal conductivity, is challenging.
(b) **Predicting Performance**: Predicting the performance of nanocomposites based on their composition and structure remains a complex task (Table 13).

This comparison highlights nanocomposites' potential for targeted and sustainable remediation, paving the way for further research and commercial interest.

Table 13 Summary of commercialization aspects and future directions for nanocomposites in environmental remediation, contrasting them with conventional materials

Aspect	Nanocomposites	Conventional materials/Systems
Efficiency	High surface area, enhanced reactivity, improved adsorption/removal rates of contaminants	Generally lower reactivity and surface area, slower contaminant removal
Scalability	Currently in early stages, with challenges in mass production and uniformity	Widely scalable; established infrastructure for large-scale production
Cost-efficiency	High initial R&D and material costs, but potential for long-term savings due to durability and reduced usage	Typically, lower initial costs, but maintenance and replacement costs can be higher
Environmental impact	Potential for reduced pollution due to minimal chemical use and improved contaminant removal	May involve significant chemical inputs, potential for secondary pollution
Customization	Highly customizable for specific contaminants, adaptable to various applications	Limited customization; generally broad-spectrum but less targeted solutions
Lifecycle management	Promising recyclability and potential biodegradability in development	Often not biodegradable, may contribute to waste in long term
Future potential	Integration with IoT, advanced synthesis techniques for multifunctionality, enhanced regulatory support	Incremental improvements expected; limited by material properties
Adoption barriers	Lack of standardization, high production costs, regulatory hurdles	Well-established; minimal regulatory obstacles; infrastructure-ready
Applications	Targeted pollutant removal, sustainable water/air filtration, soil remediation	Broader, less targeted applications; used widely but often with environmental trade-offs
Research focus	Advanced synthesis, scalable production methods, eco-friendly disposal	Focused on improving efficiency and reducing secondary pollutants

8 Conclusion

Nanocomposites represent a transformative advancement in material science, blending a matrix phase with nanosized fillers to deliver superior properties over traditional materials. These unique materials exhibit enhanced mechanical, thermal, electrical, and barrier properties, making them highly applicable across diverse fields:

Properties and Applications

(a) Mechanical Strength and Durability: Increased strength, stiffness, and toughness enhance product longevity.
(b) Thermal Conductivity and Stability: Improved heat management for advanced applications.
(c) Electrical Properties: Enhanced conductivity and dielectric properties support electronics.
(d) Barrier Properties: Reduced permeability, beneficial for packaging and protective layers.
(e) Catalysis: Improved reaction rates for chemical applications.
(f) Biomedical Uses: Applications in drug delivery, tissue engineering, and biosensors.
(g) Energy Storage: Superior performance in batteries and fuel cells.
(h) Environmental Remediation: Effective for water purification, air filtration, and soil cleanup.

Challenges and Future Directions

(a) Uniform Dispersion and Interfacial Bonding: Crucial for achieving optimal performance.
(b) Scalability and Cost-Effectiveness: Essential for commercial applications.
(c) Environmental and Regulatory Considerations: Assessing impact and establishing standards is paramount.

Implications for Environmental Management

Nanocomposites contribute to sustainability by enhancing material efficiency, durability, and recyclability. They reduce waste through extended product lifecycles and enable environmental purification through applications in water, air, and soil treatment.

Research in nanocomposites is advancing rapidly, with focus areas including multifunctional nanocomposites, bio-inspired materials, and advanced manufacturing techniques. Future efforts will involve creating smart materials responsive to stimuli, bio-based alternatives for sustainability, and scalable 3D printing solutions. Additionally, life cycle assessments and improved recycling methods will play a key role in minimizing their environmental footprint.

References

Adnan Mohammed H (2021) Chitosan-based nanocomposites for controlled release applications in agriculture. Carbohydr Polym 262:117944

Ahmad M, Mohammadi T, Kumar A (2020) Sustainable applications of biochar and activated carbon in nanocomposites for environmental remediation. J Environ Manage 255:109913

Ahmed S et al (2018) Functionalized hybrid nanocomposites for selective adsorption of pollutants in environmental cleanup. Environ Sci Technol 52(5):2856–2865

Ahmed F et al (2022) Functionalized chitosan nanocomposites for selective heavy metal adsorption. ACS Sustain Chem Eng 10(6):2421–2431

Ahmed R, Green T, Zhao W (2024) Nanocomposite membranes for organic contaminant filtration in wastewater treatment. Water Treat Res 58(2):203–220. https://doi.org/10.1080/wtr.2024.02.007

Alqadami AA et al (2019) Thiol-functionalized nanocomposites for enhanced removal of heavy metals from water. J Hazard Mater 377:263–271

Anthony R Gregg AR (2021) Hybrid nanocomposites: synthesis, properties, and applications in environmental remediation. Mater Today: Proc 45:2411–2415

Bennett SA (2021) Bio-inspired nanocomposites for sustainable pollution control. Nat Sustain 4(2):167–179

Bhattacharya S, Mandal T, Das P (2019) Role of montmorillonite in green nanocomposite materials for environmental applications. Environ Nanotechnol Monit Manage 12:100258

Bhattacharyya S et al (2021) Nanocomposites for environmental remediation: chemical reactivity and adsorption properties. Environ Sci Nano 8(3):690–710

Boruah PK, Sharma B, Das MR, Das P, Das MR (2016) Fabrication of magnetic reduced graphene oxide/Fe_3O_4 nanocomposite for the removal of triazine pesticides. J Nanomater 2016:93–102

Brown K et al (2021) Enhancing thermal properties in nanocomposite materials for sustainable applications. Polym Degrad Stab 187:109516

Chen L, Kumar A (2024) Nanocomposite coatings for biomedical applications: antimicrobial and biocompatibility properties. Biomed Nanotechnol J 10(1):45–63

Chen X et al (2019) Glucose-functionalized gold nanoparticles for selective biosensing applications. ACS Appl Mater Interfaces 11(14):13565–13572

Chen J et al (2021) Ion-exchange nanocomposites for heavy metal remediation. ACS Sustain Chem Eng 9(7):3028–3038

Choi W et al (2022) Enhancing photocatalysis in nanocomposites for environmental applications. Adv Func Mater 32(12):2108367

Clark JS (2021) Enhancements in mechanical properties of polymer nanocomposites. Mater Today: Proc 45:2304–2311

Disperen L et al (2021) Role of PEG in stabilizing nanocomposites for enhanced performance in aqueous environments. J Environ Chem Eng 9(5):106504

Espinosa HD et al (2019) Nacre-inspired nanocomposites: assembly, structure, and mechanical properties. Adv Mater 31(32):1901505

Feng L et al (2020) Superhydrophobic and self-cleaning nanocomposites inspired by the lotus effect. Mater Today 37:57–68

Figueiredo TS (2020) Biodegradable polymer-based nanocomposites for environmental remediation: properties and applications. J Hazard Mater 395:122648

Gamba RR, Yamamoto S, Abdel-Hamid M (2020) Biological synthesis of nanoparticles and their application in environmental remediation. Nanotechnol Rev 9(1):209–231

Gamiz B, Celis R, Hermosin MC, Cornejo J (2015) Montmorillonite-hexadimethrine nanocomposite for the removal of the phenoxy herbicide MCPA. Environ Sci Pollut Res 22(1):618–626

Geraldes RR (2021) Recent advances in nanocomposite materials for environmental remediation: a comprehensive review. Chemosphere 262:127960

Gupta S et al (2022) Thermal stability and decomposition properties of polymer nanocomposites for environmental applications. J Mater Chem A 10(15):7702–7710

Hienuki N (2021) Environmental and energy benefits of nanocomposites in automotive industry. Appl Sci

Hong-Fa Yan HF (2021) Eco-friendly nanocomposites for environmental remediation: synthesis, properties, and applications. Mater Today: Proc 45:4853–4859

Jiang et al (2022) Chitosan/graphene oxide nanocomposites for heavy metal adsorption. Environ Sci Pollut Res 29(17):24985–24998

Jones M, Patel R, Yang L (2024) Hybrid nanocomposites in soil remediation: techniques and material innovations. Environ Sci Adv 12(1):75–92. https://doi.org/10.1080/esa.2024.11.002

Kamigaito O (1991) What can be expected from polymer nanocomposites? Polym Int 26(1):1–9

Kawai VK (2021) Green synthesis of nanocomposites for environmental applications: a review. Environ Sci Pollut Res 28:14582–14595

Khan et al (2021a) TiO_2-CNT nanocomposites for efficient photocatalytic degradation of dyes. J Hazard Mater 403:124088

Khan S et al (2021b) Recent advances in functionalized nanoparticles for water treatment. Chemosphere 271:129490

Koushik P, Sarkar S, Saha P, Bhattacharya S, Jana S (2016) Reduced graphene oxide-Ag nanocomposite for the remediation of organochlorine pesticides from water. Environ Res 150:126–135

Kumar A et al (2018) Chemical resistance of polymer-based nanocomposites for industrial applications. J Appl Polym Sci 135(38):46623

Kumar R, Singh S, Pandey J (2020) Recent advances in the utilization of cellulose nanocrystals in green nanocomposites. Renew Sustain Energy Rev 123:109766

Kumar S et al (2021) Carbon-based nanocomposites for environmental cleanup. J Environ Chem Eng 9(5):105053

Labrague LJ (2020) Real-time monitoring techniques for nanocomposite synthesis: insights and innovations. Adv Funct Mater 30(21):2000123

Lai HF (2020) Functionalization of nanoparticles for enhanced environmental remediation. Nanotechnol Rev 9(1):259–280

Lapinski LD (2020) Mechanical behavior and performance of nanocomposite materials for environmental applications. J Cleaner Prod 273:122921

Lee J, Gupta S (2024) Carbon nanocomposites in energy storage: innovations in battery design and hybrid nanomaterial applications. Energy Storage Mater 18(2):231–250

Lee Y et al (2022) Mimicking biological conductivity with bio-inspired nanocomposites. J Mater Chem B 10(15):2831–2840

Lee D, Ahmed S, Nakamura H (2023) Catalytic nanocomposites for VOC and NO_x decomposition in air purification systems. Air Qual J 34(4):331–350. https://doi.org/10.1016/aqj.2023.09.019

Li Z et al (2020) High-performance adsorption of organic dyes by graphene oxide composites. Mater Sci Semicond Process 120:105195

Li Y et al (2021a) ZnO-GO nanocomposites for photocatalytic degradation of herbicides. Chemosphere 278:130468

Li Y et al (2021b) Synergistic reinforcement in CNT and graphene hybrid nanocomposites for enhanced mechanical properties. Compos Struct 273:114356

Lin Z (2018) Sustainable nanocomposites and reduced energy requirements in automotive production. Mater Today

Liu J (2020) Nanocomposites in car manufacturing: a green solution. Renew Sustain Energy Rev

Liu Z et al (2021) Chitosan-functionalized nanocomposites for selective dye adsorption in wastewater treatment. Carbohyd Polym 255:117481

Mateyak MK (2021a) Recent advances in green synthesis of nanocomposites: techniques and applications. J Cleaner Prod 287:125563

Mateyak MK (2021b) Electrospinning and 3D printing for the fabrication of nanocomposites for environmental remediation. J Cleaner Prod 314:128058

Mahpishanian S, Sereshti H (2017) β-Cyclodextrin/iron oxide-functionalized reduced graphene oxide for magnetic solid phase extraction of organochlorine pesticides in honey. Anal Chim Acta 971:1–10
McKinlay AR (2022) Mechanical properties of nanocomposites: a comprehensive review. Compos Sci Technol 208:108736
Mittal A, Jain R, Khurana A (2021) Advances in the green synthesis of metal nanoparticles for environmental applications. Environ Chem Lett 19(3):2235–2254
Nguyen H et al (2020) Clay-metal oxide hybrid nanocomposites as anti-corrosion coatings for harsh environments. Mater Chem Phys 246:122509
Nguyen TT, Wang C, Kim IS (2022) Chitosan-based nanocomposites in water purification: an eco-friendly approach. Carbohyd Polym 273:118564
Patel A et al (2020) Improved thermal and chemical properties of eco-friendly nanocomposites. Environ Sci Technol 54(18):11323–11332
Patel RN et al (2022) Surface functionalization in nanocomposites for selective adsorption of heavy metals. ACS Appl Mater Interfaces 14(4):4612–4623
PJFA de Sousa, RAM Santos (2019) Green synthesis of nanocomposites for environmental applications: a review. Environ Sci Pollut Res 26:11483–11503
Puar TH, Kiat (2020) Nanocomposite materials for sustainable environmental applications: mechanical properties and durability. J Mater Res Technol 9(6):12956–12968
Rani et al (2023) Ag-Zeolite nanocomposites for combined heavy metal and dye removal in wastewater. ACS Sustain Chem Eng 11(5):1879–1890
Rosenfeld Y (2019a) Life cycle energy impact of traditional automotive materials. J Ind Ecol
Rosenfeld M (2019b) Lightweighting and material efficiency in automotive manufacturing. J Sustain Eng 12(4):355–369
Raza ZA, Abid S, Banat F (2021) Biodegradable polymers in sustainable environmental remediation applications. Environ Sci Pollut Res 28:10521–10538
Shokouhimehr M et al (2019) Electrostatic assembly of nanoparticles: a versatile method for functionalization. ACS Nano 13(8):8648–8664
Singh R, Patel M (2023) Advanced polymeric nanocomposite membranes for water treatment: enhancing permeability, contaminant removal, and resistance properties. J Water Treat Res 15(4):123–140
Singh T et al (2020a) Nanocomposites in environmental remediation: a focus on thermal and mechanical improvements. Compos Sci Technol 200:108479
Singh M et al (2020b) Functionalized nanocomposites with high adsorption capacity for pollutant removal. J Hazard Mater 389:122079
Smith J, Li X (2023) Advanced polymeric nanocomposite membranes for water treatment: a review on enhanced permeability and resistance properties. J Environ Chem 45(3):123–145. https://doi.org/10.1016/j.jenvchem.2023.05.004
Smith R et al (2018) Bioconjugated nanoparticles for targeted applications in biotechnology. ACS Appl Bio Mater 1(3):300–310
Smith LJ, Zhao X, Smith MJ (2021) Cyclodextrin-based nanocomposites for pollution control: advances and future directions. Environ Pollut 284:117430
Sun Y et al (2020) Fenton-like degradation of pollutants using Fe_3O_4 nanocomposites. Environ Sci Technol 54(3):1288–1298
Tokareva O et al (2021) Biomimetic nanocomposites based on spider silk and nanoparticles for enhanced durability. Nat Commun 12(1):1234
Wan X (2020a) Energy efficiency in nanocomposite materials for cars. J Cleaner Prod
Wan L (2020b) Durability and lifecycle benefits of nanocomposites in industrial applications. J Appl Nanomater 8(2):115–130
Wang L et al (2019a) Degradation of organic dyes via redox-active metal nanocomposites. Chemosphere 227:142–150
Wang X et al (2019b) Enhanced thermal stability of silica-CNT hybrid nanocomposites for aerospace applications. Polym Degrad Stab 167:34–42

Wang Y et al (2020a) Polymer-matrix nanocomposites for high-strength applications: a review. Mater Sci Eng R 140:100542

Wang J et al (2020b) TiO_2-functionalized nanocomposites for photocatalytic degradation of organic pollutants. Appl Catal B 260:118191

Wu L et al (2019) Grafting of polymers onto nanoparticles: mechanisms and applications. Macromolecules 52(5):1710–1721

Yao X et al (2018) Bone-mimetic nanocomposites for bone tissue engineering. ACS Biomater Sci Eng 4(4):1169–1181

Yoo H (2018a) Advances in nanocomposite automotive applications and energy savings. Int J Automot Technol

Yoo K (2018b) Nanocomposites in energy-optimized production. Mater Sci Eng, A 650:252–260

Zhang X, Wu Y, Wang T, Li Z, Wang C (2015) Graphene oxide as a versatile nanoplatform for efficient removal of pesticides from aqueous solutions. Chem Eng J 276:187–194

Zhang H et al (2018) PEG-functionalized nanoparticles for targeted drug delivery in cancer therapy. Biomater Sci 6(5):1238–1250

Zhang L et al (2019a) Designing nanocomposites for enhanced reactivity and adsorption in environmental applications. J Mater Chem A 7(20):12455–12470

Zhang H et al (2019b) Photocatalytic degradation of pharmaceuticals by TiO_2-based nanocomposites. J Hazard Mater 365:663–671

Zhang X et al (2019c) Effect of polymer encapsulation on thermal stability of nanocomposites for electronic applications. Thermochim Acta 680:178369

Zhang L et al (2020) Covalent functionalization of nanoparticles for biomedical applications. Adv Mater 32(36):2000193

Zhao et al (2020) Fe_3O_4/activated carbon nanocomposites for tetracycline removal through adsorption and Fenton-like degradation. Water Res 170:115294

Zhao H et al (2022) CNT-graphene hybrids for supercapacitor electrodes: enhanced conductivity and stability. J Power Sources 512:230541

Applications, Challenges and Future Perspectives in Nanocomposites-Based Wastewater Treatment

Shivani Garg, Urvashi Kashyap, Anshul, Nitika Rani, Pooja Arora, and John Paul

Abstract Wastewater treatment by nanocomposites has gained significant attention as an innovative approach to address the growing global water scarcity and pollution challenges. This manuscript discusses nanocomposites' application, challenges and prospects in wastewater treatment. Nanocomposites comprise nanomaterials with various matrices that offer many enhanced properties of pollutant removal efficiency. The nanocomposite's key applications include membrane filtration, disinfection, adsorption, photocatalysis and water quality monitoring. Various types of nanocomposites, including metal/metal oxide, polymer-based, carbon-based, hybrid and magnetic nanocomposites, have been extensively researched for their unique characteristics and effectiveness in the treatment of wastewater. Nanocomposites-based treatments face numerous challenges, viz., environmental and health concerns, cost-effectiveness, scalability, selectivity, disposal issues and stability. To address these challenges, the continuous development of novel nanocomposites and integration with other technologies, along with the application of artificial intelligence (AI) in material design, are explored. The application of AI in material design is highlighted as a potential path for optimizing performance and overcoming challenges. This manuscript also highlights the need for management strategies, conducting comprehensive risk assessments, and addressing ethical concerns to ensure more efficient, cost-effective and sustainable solutions for the global water crisis through nanocomposites-based wastewater treatment.

Keywords Nanocomposites · Wastewater treatment · Membrane filtration · Photocatalysis · Sustainable development · Environmental challenges

---

S. Garg · U. Kashyap · Anshul · N. Rani · P. Arora (✉)
Institute of Environmental Studies, Kurukshetra University, Kurukshetra, Haryana, India
e-mail: pooja.evs09@gmail.com

J. Paul
Faculty of Mechanical and Automotive Engineering Technology, University Malaysia Pahang, Pekan, Pahang, Malaysia

S. Roy et al. (eds.), *Nanocomposites for Sustainable Wastewater Treatment*, Composites Science and Technology, https://doi.org/10.1007/978-981-95-1369-7_11

1 Introduction

The global water crisis, characterized by increasing water scarcity and pollution, presents one of the most pressing challenges of the century. The need for effective wastewater treatment technologies has become more critical than ever due to rapid industrialization, population growth, and the impacts of climate change (Dutta et al. 2021). To address these challenges, nanocomposites have emerged as a promising solution, offering enhanced capabilities in pollution removal and water purification through their unique physicochemical properties (Pandey et al. 2017). Nanocomposites, which combine nanomaterials with various matrices, represent a significant advancement in materials science applied to environmental remediation (Rando et al. 2022). These materials exhibit remarkable characteristics such as superior selectivity in pollutant removal, high surface area-to-volume ratio, improved mechanical strength and enhanced reactivity (Lan et al. 2022). The integration of various nanomaterials, including metal nanoparticles, metal oxides, polymeric substances, and carbon-based materials, has enabled the development of nanocomposite systems with customized properties for specific wastewater treatment applications (Khodakarami and Bagheri 2021). In recent years, significant progress has been made in the development and application of various nanocomposite types for wastewater treatment. Metal and metal oxide nanocomposites, particularly those incorporating silver, titanium dioxide and iron, have shown exceptional antimicrobial and photocatalytic properties (Bashambu et al. 2021). Polymer-based nanocomposites offer advantages in terms of processability and functionality, while carbon-based nanocomposites, including graphene and carbon nanotubes, demonstrate remarkable adsorption capabilities (Idumah and Hassan 2016). Magnetic nanocomposites have gained attention for their ease of separation and recovery from treated water (Mehmood et al. 2021).

The application of nanocomposites in wastewater treatment encompasses various processes, including membrane filtration, adsorption, photocatalytic degradation, and disinfection. These materials have demonstrated effectiveness in removing a wide range of pollutants, including organic compounds, heavy metals, microorganisms and emerging contaminants (Dutta et al. 2019). The integration of artificial intelligence in nanocomposite design and optimization has opened new avenues for enhancing treatment efficiency and developing more sustainable solutions (Alprol et al. 2024). Despite their promising potential, the implementation of nanocomposite-based wastewater treatment faces several challenges. These include concerns about environmental and human health impacts, scalability issues, cost-effectiveness and the need for improved stability and selectivity (Tripathy et al. 2024). The disposal of used nanocomposites and their potential environmental accumulation requires careful consideration and management strategies (Chen et al. 2016). The major challenges in wastewater treatment include increasing micropollutant complexity, high energy consumption, membrane fouling and the need for more sustainable treatment methods (Eniola et al. 2022). Recent perspectives focus on developing energy-efficient hybrid treatment systems, incorporating machine learning for process optimization, and advancing green synthesis methods for nanocomposites (Li et al. 2024;

Chang et al. 2024). Additionally, emerging research emphasizes circular economy approaches in wastewater treatment, including resource recovery and the development of biodegradable nanocomposites (Ribarova et al. 2024). This chapter aims to provide an overview of nanocomposite applications in wastewater treatment, examining their types, applications, challenges and future perspectives. Special attention is given to the recent development in material designs, integration with other technologies, and the role of artificial intelligence in optimizing performance. The chapter also encompassed management strategies, and ethical considerations necessary for sustainable implementation.

2 Types of Nanocomposites Used in Wastewater Treatment

There are many types of nanocomposites used to treat wastewater, such as carbon-based, polymer-based and metal-based. Figure 1 shows the classification framework of different nanocomposites used for wastewater treatment and Fig. 2 shows the overview of the process of treatment of wastewater through nanocomposites. The detail of each method is discussed in the sections and subsections below.

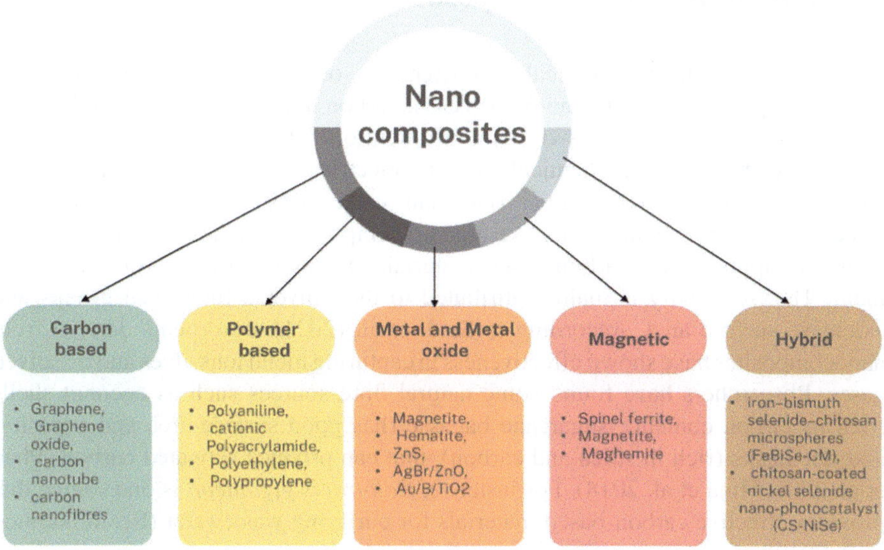

Fig. 1 Classification framework of nanocomposites

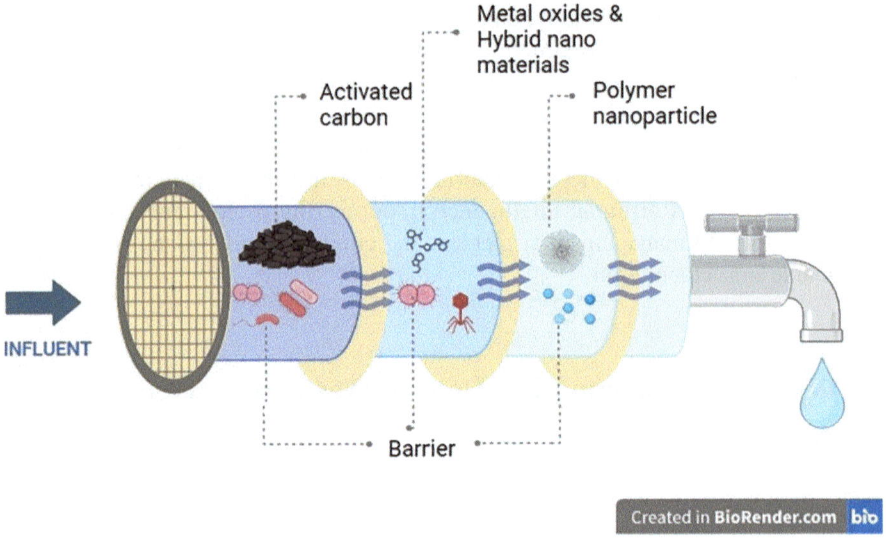

Fig. 2 Sequential action of nanocomposites on foreign materials in wastewater

2.1 *Carbon-Based Nanocomposites*

The distinctive physicochemical characteristics of carbon-based nanomaterials, such as graphene and graphene-oxide (GO), carbon nanotubes (CNTs) and carbon nanofibers (CNFs), have drawn a lot of attention. These characteristics include superior mechanical and chemical stability, electrical conductivity, reinforcement capacity and antifouling qualities (Noamani et al. 2019). Due to their ability to selectively absorb contaminants like organic compounds and heavy metals in water treatment applications, carbon-based materials are favoured for wastewater treatment. This selectivity is mainly attributed to their diverse functional groups and extensive surface area. According to Seidi et al. (2021), adsorbents derived from nanocomposites have shown effectiveness in capturing metal ions, dyes and dissolved gases. Researchers have found some natural fibre sources such as coconut shells (rich in carbon content), sugarcane bagasse (has good surface area and porosity) and rice husks (rich in silica and carbon), that can produce activated carbon (Phan et al. 2006; Hina et al. 2018). For instance, *Calotropis gigantean* is one source that provides effective carbon-based materials for purifying wastewater (Senthilkumaar et al. 2010).

2.2 Metal and Metal Oxide Nanocomposites

Metal oxide nanoparticles are environment-friendly materials used to absorb metals and dyes effectively (Gupta et al. 2016). Noble metal nanoparticles have been used for the removal of pollutants and the detection of biomolecular contaminants, such as pesticides, in water and wastewater (Khodakarami and Bagheri 2021). They come in shapes, like sheets and tubes, and possess a significant surface area and chemical stability that enhances their ability to eliminate pollutants (Ghosh et al. 2022). These nanoparticles are particularly useful in wastewater treatment due to their breakdown and oxidation properties in water. Based on research conducted by Ghosh et al. (2022), it has been found that metal oxides exhibit versatility in their ability to adsorb particles and pollutants from wastewater due to their photocatalytic activity and antibacterial properties.

2.3 Polymer-Based Nanocomposites

Natural fibers composed of cellulose and other components, such as hemicellulose and lignin, are commonly utilized in creating polymer matrices based on their beneficial properties, such as porosity and low density. Additionally, they are valued for being biodegradable and cost-effective (Joshi et al. 2004; Ramamoorthy et al. 2015). In chemical processes, polymers are employed as helping agents, filtration aids, and coagulants and flocculants (Khodakarami and Bagheri 2021). However, to expand their applications further, it is crucial to tackle issues related to their vulnerability to moisture and other limitations (Mohammed et al. 2015). The incorporation of cationic polyacrylamide resulted in a twofold increase in water flux and caused the silica nanoparticles to grow from a few nanometers to approximately 4 μm, improving their settleability and filterability. The use of cationic polymers necessitates the neutralization of the charge on negatively charged colloidal particles within the system. Conductive polymer-based adsorbents, like polypyrrole and polyaniline, are attracting attention for their ability to adsorb metal ions due to the presence of heteroatoms in their structures that facilitate adsorption. These materials are known for being low in toxicity and easy to synthesize (Bhandari et al. 2021).

2.4 Magnetic Nanocomposites

The primary iron-based magnetic nanoparticles used are nanoscale zero-valent iron, Fe3O4 (magnetite), and γ-Fe_2O_3 (maghemite) (Marcelo et al. 2021). In contrast to adsorbents that are not magnetic and less efficient in pollution removal processes, magnetic nanoparticles (MNPs) stand out for their ability to attract and eliminate pollutants with the help of external magnetic fields (Hassanzadeh-Afruzi et al. 2022).

The magnetic properties allow for the easy separation of the material after water treatment, facilitating its recovery and subsequent reuse. The process by which iron-based nanoparticles remove contaminants is influenced by the oxidation states of iron and involves mechanisms such as adsorption, chemical reduction, and reductive precipitation. Research indicates that MNPs such as chitosan-coated ferrosoferric oxide have shown adsorption capabilities and better dispersion in water when used for treating wastewater (Zhang et al. 2010). These tiny particles show promise in cleaning wastewater by eliminating harmful substances such as methylene blue, achieving an effectiveness of over 99% as demonstrated in a recent study (Yadav et al. 2021).

2.5 Hybrid Nanocomposites

The use of hybrid nanocomposites is becoming increasingly crucial in wastewater treatment as they combine organic elements with biological components for better effectiveness. These materials are typically produced using methods such as hydrolysis and condensation by adding polymers to clay minerals. Nanocomposites consisting of TiO_2/Ag/MoS_2/Ag were created for water purification, targeting the degradation of organic pollutants and the reduction of heavy metal ions. These nanocomposites demonstrated enhanced photodegradation efficiency for RhB and effectively photochemically reduced Cr (II) in wastewater at a pH of 3 (Pal et al. 2022). Applications of nanocomposites can be seen in fields such as solar energy systems and battery technology to promote energy conservation (Dmitriev et al. 2020). Hybrid nanocomposites have proven effective in removing pollutants through photocatalytic degradation processes such as the breakdown of methylene blue and ciprofloxacin (Wolski et al. 2021; Abu-Dalo et al. 2021). These advancements showcase the ability of nanocomposites to address wastewater-related challenges efficiently.

3 Applications of Nanocomposites in Wastewater Treatment

As a sophisticated material, a nanocomposite emerges as superior to other materials for wastewater treatment. These are advanced substances that integrate nanoparticles with a matrix substance, exhibiting enhanced characteristics for eliminating pollutants from water (Kanoun et al. 2021; Theron et al. 2008; Cho and Lee 2022). An overview of the applications of nanocomposites in wastewater treatment, specifically highlighting recent advancements and research, is discussed in the following subsections (Fig. 3).

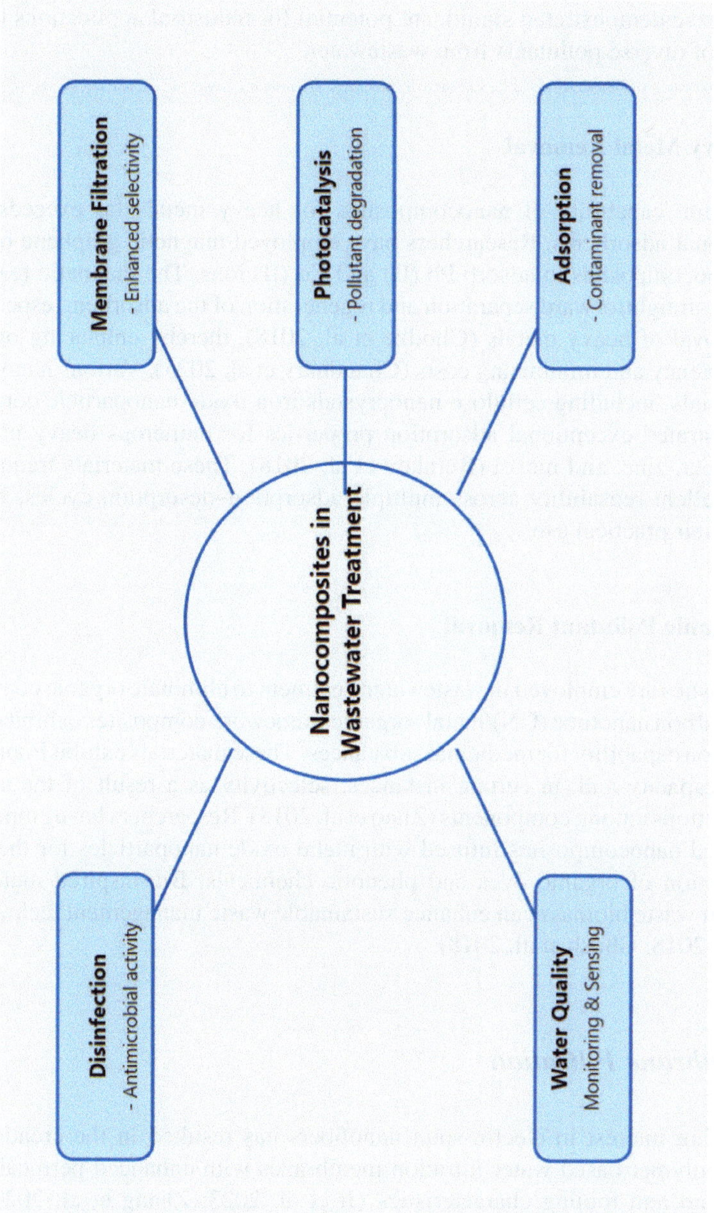

Fig. 3 An overview of major applications of nanocomposites in wastewater treatment

3.1 Adsorption

Due to its extensive surface area-to-volume ratio and mesoporosity, nanocomposite adsorbents have demonstrated significant potential for industrial applications in the elimination of diverse pollutants from wastewater.

3.1.1 Heavy Metal Removal

The adsorption capability of nanocomposites for heavy metals far exceeds that of conventional adsorbents. Researchers have employed magnetic graphene oxide/chitosan nanocomposites to adsorb Pb (II) and Cd (II) ions. The magnetic features facilitate the straightforward separation and regeneration of the adsorbent, especially for the removal of heavy metals (Ghodke et al. 2018), thereby enhancing overall process efficiency and minimising costs (Chaudhary et al. 2025). Various nanocomposite materials, including cellulose nanocrystals/iron oxide nanoparticle composites, demonstrated exceptional adsorption properties for numerous heavy metals, such as copper, zinc, and nickel (Burakov et al. 2018). These materials frequently provide excellent reusability across multiple adsorption–desorption cycles, hence enhancing their practical use.

3.1.2 Organic Pollutant Removal

Nanocomposites are employed in wastewater treatment to eliminate organic contaminants. The carbon nanotube (CN)/metal–organic framework composites exhibit superior adsorption capability for medicinal substances. These materials exhibit improved adsorption capacity and, in certain instances, selectivity as a result of the synergistic interactions among components (Zhao et al. 2018). Researchers have employed biochar-based nanocomposites infused with metal oxide nanoparticles for the efficient adsorption of organic dyes and phenolic chemicals. Bio-inspired materials derived from waste biomass can enhance sustainable waste management techniques (Dutta et al. 2018; Ghosh et al. 2018).

3.2 Membrane Filtration

The increasing interest in electro-spun nanofibers has resulted in the creation of innovative polymer-based water filtration membranes with enhanced permeability, selectivity and anti-fouling characteristics (Ji et al. 2023; Zhang et al. 2024) by integrating several nanoparticles throughout the membrane matrix (Siddique et al.

2021). In recent years, nanomaterials such as graphene oxide (GO), silver nanoparticles, and metal–organic frameworks (MOFs) have enhanced water flux and pollutant rejection in nanocomposite membranes. Functionalized silver nanoparticles and graphene oxide incorporated into polyether sulfone membranes demonstrate exceptional antibacterial properties and elevated flux, making them a viable choice for water purification and wastewater treatment. In forward osmosis procedures, the innovative MOF-based nanocomposite membranes demonstrated improved water flux and reverse solute flux, signifying a novel and promising approach for high-salinity wastewater treatment. These membranes can be tailored to exclude particular impurities and attain elevated water permeability (Ramos-Delgado et al. 2016).

Membrane fouling is a leading challenge in membrane filtration that occurs because of the growing deposition and on the surface of the membranes over time. Recently, nanocomposites have shown great potential to tackle this challenge (Nivedita and Joseph 2021). Anti-biofouling by novel zwitterionic polymer/silica nanocomposite coatings for reverse osmosis membranes with excellent organic fouling resistance and chlorine stability offers the promise of successful membrane-based wastewater reuse properties (Ji et al. 2023; Zhang et al. 2024). Photocatalytic titanium dioxide nanoparticles (TiO_2 NPs) in nanocomposite membranes (also synthesized by alumina nanoparticles (NPs) encapsulation on TiO_2 NPs) exhibited the capability for real time, in-situ fouling mitigation under UV conditions. Such an innovation is expected to decrease maintenance requirements and substantially increase membrane lifetime in treatment systems (Ramos-Delgado et al. 2016).

3.3 Photocatalysis

The nanocomposite photocatalysts are attractive and desirable due to the enhanced degradation of stubborn organic pollutants using light energy. One of the most significant advancements in this field has been the design of nanocomposites capable of harnessing visible light, which constitutes a much larger portion of the solar spectrum compared to UV light. TiO_2/graphene quantum dot nanocomposites have exhibited exceptional photocatalytic performance under visible light, showing significantly improved degradation of dyes and pharmaceutical compounds compared to bare TiO_2 (Choi et al. 2022). Ternary nanocomposites of materials such as g-C_3N_4, $BiVO_4$, and reduced graphene oxide have shown enhanced absorption in the visible region and separation of charges (Oladipo and Mustafa 2023). These materials enable enhanced photodegradation of refractory organic pollutants, further broadening the scope of contaminants to be treated.

The Z-scheme heterojunction nanocomposites are very efficient in improving charge separation and extending light harvesting. Composites such as $BiVO_4$/g-C_3N_4 (where g-C_3N_4 is the graphitic carbon nitride) or CdS quantum dot-loaded WO_3 nanosheets have shown remarkable photocatalytic activity toward the degradation of different kinds of pollutants, including antibiotics, under simulated solar light (Oladipo and Mustafa 2023). The use of advanced nanocomposites has opened the door for improved wastewater treatment methods that could be powered by renewable solar energy.

3.4 Disinfection

Antibacterial nanocomposites have excellent potential applications for water disinfection and microbial removal in treatment systems. Metal nanoparticles, such as silver and copper-based nanocomposites, display high interest in the research community owing to their potential antimicrobial properties against waterborne pathogens. Copper nanoparticle/chitosan nanocomposites have demonstrated a considerable rate of pathogenic inactivation and less harmful effects against non-target organisms (Cobos et al. 2020). Silver nanoparticles and graphene oxide-containing multifunctional nanocomposite coatings showed long-term antibacterial properties and antimicrobial performance in a recent study on water treatment infrastructure over biofilm buildup. The beauty of these coatings is that they help keep treatment systems clean and running smoothly over a long period (Verma et al. 2020). Photocatalytic nanocomposites are appealing because they can kill viruses using light-driven disinfection. For instance, g-C_3N_4 nanosheets modified with Ag_3PO_4 nanoparticles perform remarkably well in the visible-light-driven photocatalytic disinfection of *Escherichia coli* as well as other waterborne pathogens. The incorporation of self-cleaning features, such as photocatalytic properties from titanium dioxide (TiO_2) and hydrophobic coatings, allows surfaces to effectively degrade organic pollutants and repel water, ensuring they remain clean over time (Zhou et al. 2024). These nanostructures showed a lower bandgap and superior UV/Vis-light photocatalytic activity compared to zinc oxide (ZnO) or titanium dioxide (TiO_2) alone and have been shown to effectively inactivate tri-state cross-contaminating bacterial (*Escherichia coli, Pseudomonas aeruginosa*) and viral (Enteric Human Adenovirus serotype 35) pathogens within minutes of exposure (Zhou et al. 2024). The materials appear to hold great potential for solar-driven water disinfection systems in regions with poor access to conventional treatments (Verma et al. 2020).

3.5 Sensing and Monitoring

Nanocomposite-based sensors play a significant role in the in-situ monitoring of water quality parameters and contaminants in treatment systems. Ultra-Sensitive Detection of Heavy Metals in Wastewater Using Nanocomposite Sensors, such as electrochemical sensors (ECSs) based on gold nanoparticles (AuNPs) and reduced graphene oxide, is highly sensitive with good selectivity toward several ions of heavy metals, so that efficacy for point-of-care is quite possible (Zhou et al. 2020). Silver nanoparticle (AgNPs)-decorated cellulose nanocrystals were prepared and demonstrated to be effective for use as colorimetric sensors with a fast time scale ($\leq$ 1 min) and low detection limit (~ 10 ppb) for detecting mercury ions in water. These sensors have the potential to provide low-cost, portable solutions for monitoring drinking water quality in underserved areas (Zhou et al. 2020). Next-generation fluorescent nanocomposite sensors composed of quantum dots (QDs) and molecularly imprinted polymers (MIPs) have shown selective, sensitive detection of an emerging contaminant, perfluoroalkyl substances (PFAS), in water samples. These advanced sensors could serve as early warning systems for water treatment plants by detecting contaminants even at very low concentrations (Liu et al. 2021). Organic pollutants in wastewater can be rapidly sensed and classified by the use of nanocomposites, including electronic nose (e-nose) systems based on metal oxide nanoparticles and conductive polymers. Such systems would offer continuous monitoring of hazardous and complex mixtures of pollutants, leading to more timely responses in treatment steps (Liu et al. 2021).

3.6 Comparison Between Applications of Different Nanocomposites

Various nanocomposite-based methods have emerged as promising solutions for wastewater treatment, each offering distinct advantages and limitations. Table 1 presents a comprehensive comparison of five major application methods, highlighting their removal efficiency, cost considerations and specific nanocomposites employed in each approach, thereby providing insights for selecting optimal treatment strategies.

4 Challenges in Nanocomposite-Based Wastewater Treatment

Nanocomposites-based wastewater treatment faces several challenges that require careful consideration for widespread implementation. Scalability and cost-effectiveness remain significant hurdles, as the production of nanocomposites often involves complex synthesis procedures and expensive raw materials, making large-scale applications economically challenging (Livingston et al. 2020). Environmental and health concerns have emerged as important considerations, particularly regarding the potential release of nanoparticles into treated water and their subsequent impact on aquatic ecosystems and human health through bioaccumulation and biomagnification (Huang et al. 2021). The stability and longevity of nanocomposites present

Table 1 Comparative study of different methods of application in wastewater treatment

Methods	Nanocomposites used	Advantages	Limitations	Removal efficiency	Cost	References
Adsorption	Fe_3O_4/graphene oxide CNT/Metal oxides Clay/bio-based composites	Easy separation High surface area Simple operation Reusability	Saturation pH-dependent Competition between pollutants	85–98% for heavy metals and 90–95% for organics	Moderate	Kolya and Kang (2023), Ajith and Rajamani (2021)
Membrane filtration	GO/polymer membranes CNT/ceramic hybrids MOF composites	Continuous operation High selectivity Low space requirement	Membrane fouling High energy consumption Complex fabrication	95–99% for various pollutants	High	Sun et al. (2024), Esfahani et al. (2019)
Photocatalysis	TiO_2/graphene ZnO/Ag composites $BiVO_4$ hybrids	Solar energy utilization Complete mineralization No secondary pollution	Limited to daylight Slow kinetics Electron–hole recombination	70–90% of organic pollutants	Low-moderate	Jiang et al. (2020), Ren et al. (2021)
Disinfection	Ag/Cu nanocomposites GO/metal oxide hybrids Chitosan-based materials	High antimicrobial activity Low toxicity Long-lasting effect	Bacterial resistance Cost of notable metals Limited scope	99.9% for bacterial removal	High	Omran and Baek (2022), Cobos et al. (2020)
Sensing and monitoring	QD/graphene composites Au/Ag nanostructures Conducting polymer hybrids	High sensitivity Real-time monitoring Selective detection	Complex calibration Limited stability Interference effect	Detection limit-ppb-ppt range	Moderate-High	Zhou et al. (2020), Liu et al. (2021)

another challenge, as these materials may undergo degradation or lose their functional properties during prolonged use, especially during complex treatment conditions or when exposed to complex wastewater matrices (Tripathy et al. 2024). Selectivity and efficiency issues arise from the varied nature of pollutants in wastewater, where nanocomposites might show preferential removal of certain contaminants while performing poorly against others, highlighting the need for versatile and robust materials (Sahu et al. 2023). The regeneration and disposal of used nanocomposites pose significant environmental and economic challenges as the regeneration processes are often energy-intensive treatments, and improper disposal can lead to secondary pollution (Akharame et al. 2019). The recovery of valuable components from the used nanocomposites remains technically challenging and economically unfeasible in many cases (Rial and Ferreira 2022). The interconnected challenges necessitate a holistic approach to developing next-generation nanocomposites, focusing on sustainable synthesis methods, improved stability, enhanced selectivity and environment-friendly disposal strategies.

5 Future Perspectives

The prospects for wastewater treatment utilizing nanocomposite materials suggest significant advancements in various aspects. Newly developed nanocomposite materials signify a progressive area in this domain, particularly with bio-inspired nanocomposites and stimuli-responsive substances exhibiting superior selectivity and efficacy in the extraction of pollutants (Chen et al. 2024). Innovative materials, including metal–organic framework (MOF)-based nanocomposites and systems integrated with quantum dots, are currently being engineered to enhance performance attributes while minimizing environmental repercussions (Poonia et al. 2023). The diverse technologies, especially membrane processes and advanced oxidation systems, bring synergistic effects that improve treatment efficacy and mitigate the constraints inherent to each technology (Dong et al. 2022). Integrating nanocomposites with biological treatment systems and electrochemical processes shows significant potential for the advancement of hybrid treatment solutions (Arif et al. 2024).

Another key development includes smart responsive nanocomposites that are poised with materials that can sense changes in environmental conditions and selectively target specific pollutants (Nazarzadeh Zare et al. 2021). These materials have programmable properties responsive to a wide variety of stimuli, including pH, temperature, and light, to allow for more efficient and controlled treatment processes (Umar et al. 2021). Sustainable and green synthesis approaches are increasingly favoured and focus on bio-based precursors and environmentally friendly production methods. These methodologies utilize renewable resources and minimize energy consumption and the use of harmful chemicals in the synthesis of nanocomposites (Mondal et al. 2020).

Artificial Intelligence (AI) and machine learning (ML) in the nanocomposite design area are indeed some of the remarkable integration developments. They bring out the capability for the prediction of material properties, optimization of synthesis parameters, and development of tailor-made nanocomposites for specific applications (Wu 2023). Machine learning algorithms are applied for the analysis of large material property and performance metric datasets to accelerate the development of new nanocomposites and reduce the time and cost of experiments significantly (Champa-Bujaico et al. 2022).

6 Regulatory and Ethical Considerations

Regulatory and ethical considerations play a crucial role in the implementation of nanocomposite-based wastewater treatment technologies. Current regulations and guidelines vary significantly across different regions and jurisdictions, with many countries still developing comprehensive frameworks for nanomaterial regulation. The European Union's REACH (Registration, Evaluation, Authorization and Restriction of Chemicals) regulation and the US EPA's guidelines provide baseline frameworks for nanomaterial safety assessment and usage in water treatment applications (Allan et al. 2021). However, the rapid development of new nanocomposites often outpaces regulatory updates, creating potential gaps in oversight. Risk assessment and management strategies for nanocomposites require a thorough evaluation of both short-term and long-term impacts. This includes comprehensive toxicological studies, environmental fate analysis, and exposure assessments throughout the material's lifecycle (Schwirn et al. 2020). Standardized protocols for safety evaluation and risk characterization are essential for ensuring consistent assessment across different nanocomposite applications (Huang et al. 2024). The ethical implications of nanotechnology in water treatment encompass various concerns, including equitable access to water treatment technologies, potential environmental justice issues, and the responsibility for long-term environmental impacts. Questions arise regarding the distribution of benefits and risks across different socioeconomic groups and geographical regions (Hussain and da Costa 2022). Further, the transparency in reporting potential risks and limitations of nanocomposite-based treatments, along with public engagement in decision-making processes, becomes crucial for maintaining ethical standards and public trust in these emerging technologies (Srivastava 2023).

7 Conclusion and Outlook

The treatment of wastewater using nanocomposites is an important advancement in water challenges globally. This chapter takes various considerations on nanocomposite-based treatment systems, from basic applications and emerging challenges to future possibilities. Indeed, despite the capabilities that are possessed by these materials in terms of efficient pollutant removal, disinfection, and water quality surveillance, a few significant challenges, such as scalability, cost-effectiveness, and ecotoxicity, remain prevalent. The area certainly offers promising advancement in the future with the combination of artificial intelligence, the development of intelligent responsive materials, and environmentally benign synthesis techniques. The addition of bio-inspired nanocomposites with hybrid treatment systems opens new horizons for better performance with reduced environmental impacts. However, it will rely heavily on solving the regulatory frameworks, standardizing the risk assessment protocols, and satisfying the ethical requirements.

References

Abu-Dalo MA, Al-Rosan SA, Albiss BA (2021) Photocatalytic degradation of methylene blue using polymeric membranes based on cellulose acetate impregnated with ZnO nanostructures. Polymers 13:3451. https://doi.org/10.3390/polym13193451. PubMed: 34641266

Ajith MP, Rajamani P (2021) Nanotechnology for water purification—current trends and challenges. J Nanotechnol Nanomater 2:88–91

Akharame MO, Fatoki OS, Opeolu BO (2019) Regeneration and reuse of polymeric nanocomposites in wastewater remediation: the future of economic water management. Polym Bull 76:647–681. https://doi.org/10.1007/s00289-018-2403-1

Allan J, Belz S, Hoeveler A, Hugas M, Okuda H, Patri A, Rauscher H, Silva P, Slikker W, Sokull-Kluettgen B, Tong W, Anklam E (2021) Regulatory landscape of nanotechnology and nanoplastics from a global perspective. Regul Toxicol Pharmacol 122:104885. https://doi.org/10.1016/j.yrtph.2021.104885. PubMed: 33617940

Alprol AE, Mansour AT, Ibrahim MEED, Ashour M (2024) Artificial intelligence technologies revolutionizing wastewater treatment: current trends and future prospective. Water 16:314. https://doi.org/10.3390/w16020314

Arif M, Raza H, Akhter T (2024) Classification, synthesis, characterization, and applications of metal nanoparticle-containing hybrid microgels: a comprehensive review. RSC Adv 14:24604–24630. https://doi.org/10.1039/d4ra04128c. PubMed: 39108974

Bashambu L, Singh R, Verma J (2021) Metal/metal oxide nanocomposite membranes for water purification. Mater Today: Proc 44:538–545. https://doi.org/10.1016/j.matpr.2020.10.213

Bhandari H, Garg S, Gaba R (2021) Advanced nanocomposites for removal of heavy metals from wastewater. Macromol Symp 397:2000337. https://doi.org/10.1002/masy.202000337

Burakov AE, Galunin EV, Burakova IV, Kucherova AE, Agarwal S, Tkachev AG, Gupta VK (2018) Adsorption of heavy metals on conventional and nanostructured materials for wastewater treatment purposes: a review. Ecotoxicol Environ Saf 148:702–712. https://doi.org/10.1016/j.ecoenv.2017.11.034. PubMed: 29174989

Champa-Bujaico E, García-Díaz P, Díez-Pascual AM (2022) Machine learning for property prediction and optimization of polymeric nanocomposites: a state-of-the-art. Int J Mol Sci 23:10712. https://doi.org/10.3390/ijms231810712. PubMed: 36142623

Chang HL, Liu YL, Keng CJ, Jiang HL, Hu J (2024) Challenges of industrial wastewater treatment: utilizing membrane bioreactors (MBRs) in conjunction with artificial intelligence (AI) technology. J Ind Prod Eng 41:422–427. https://doi.org/10.1080/21681015.2024.2330401

Chaudhary J, Gupta B, Tailor G, Mehta C (2025) Nanomaterials in bioremediation: recent trends and challenges. In: Enzyme immobilization with nanomaterials: applications and challenges, pp 105–126

Choi H, Surendran S, Sim Y, Je M, Janani G, Choi H, Kim JK, Sim U (2022) Enhanced electrocatalytic full water-splitting reaction by interfacial electric field in 2D/2D heterojunction. Chem Eng J 450:137789

Chen D, Zhu H, Yang S, Li N, Xu Q, Li H, He J, Lu J (2016) Micro-nanocomposites in environmental management. Adv Mater 28:10443–10458. https://doi.org/10.1002/adma.201601486. PubMed: 27781315

Chen S, Tong X, Huo Y, Liu S, Yin Y, Tan ML, Cai K, Ji W (2024) Piezoelectric biomaterials inspired by nature for applications in biomedicine and nanotechnology. Adv Mater 36:e2406192. https://doi.org/10.1002/adma.202406192. PubMed: 39003609

Cho S, Lee Y (2022) Recent advances in synthesis and application of polymer nanocomposites for water and wastewater treatment. J Clean Prod 296:126404

Cobos M, De-La-Pinta I, Quindós G, Fernández MJ, Fernández MD (2020) Synthesis, physical, mechanical, and antibacterial properties of nanocomposites based on poly (vinyl alcohol)/graphene oxide–silver nanoparticles. Polymers 12:723. https://doi.org/10.3390/polym12030723. PubMed: 32214025

Dmitriev et al (2020) Hybrid graphene nanocomposites: thermal interface materials and functional energy materials. Intech Open. https://doi.org/10.5772/intechopen.89631

Dong G, Chen B, Liu B, Hounjet LJ, Cao Y, Stoyanov SR, Yang M, Zhang B (2022) Advanced oxidation processes in microreactors for water and wastewater treatment: development, challenges, and opportunities. Water Res 211:118047. https://doi.org/10.1016/j.watres.2022.118047. PubMed: 35033742

Dutta AK, Ghorai UK, Chattopadhyay KK, Banerjee D (2018) Removal of textile dyes by carbon nanotubes: a comparison between adsorption and UV assisted photocatalysis. Physica E 99:6–15. https://doi.org/10.1016/j.physe.2018.01.008

Dutta D, Arya S, Kumar S (2021) Industrial wastewater treatment: current trends, bottlenecks, and best practices. Chemosphere 285:131245. https://doi.org/10.1016/j.chemosphere.2021.131245. PubMed: 34246094

Dutta V, Singh P, Shandilya P, Sharma S, Raizada P, Saini AK, Gupta VK, Hosseini-Bandegharaei A, Agarwal S, Rahmani-Sani A (2019) Review on advances in photocatalytic water disinfection utilizing graphene and graphene derivatives-based nanocomposites. J Environ Chem Eng 7:103132. https://doi.org/10.1016/j.jece.2019.103132

Eniola JO, Kumar R, Barakat MA, Rashid J (2022) A review on conventional and advanced hybrid technologies for pharmaceutical wastewater treatment. J Clean Prod 356:131826. https://doi.org/10.1016/j.jclepro.2022.131826

Esfahani MR, Aktij SA, Dabaghian Z, Firouzjaei MD, Rahimpour A, Eke J, Escobar IC, Abolhassani M, Greenlee LF, Esfahani AR, Sadmani A, Koutahzadeh N (2019) Nanocomposite membranes for water separation and purification: fabrication, modification, and applications. Sep Purif Technol 213:465–499. https://doi.org/10.1016/j.seppur.2018.12.050

Ghodke S, Chaudhary S, Ahmed S, Kaushik M (2018) Advanced engineered nanomaterials for the treatment of wastewater. Handbook of nanomaterials for industrial applications. Elsevier, Amsterdam, pp 959–970

Ghosh G, Basu S, Saha S (2018) Photocatalytic degradation of textile dye using TiO_2-activated carbon nanocomposite. In: AIP conference proceedings. AIP Publishing, 1966. https://doi.org/10.1063/1.5038706

Ghosh N, Das S, Biswas G, Haldar PK (2022) Review on some metal oxide nanoparticles as effective adsorbent in wastewater treatment. Water Sci Technol 85:3370–3395. https://doi.org/10.2166/wst.2022.153. PubMed: 35771052

Gupta VK, Moradi O, Tyagi I, Agarwal S, Sadegh H, Shahryari-Ghoshekandi R, Makhlouf ASH, Goodarzi M, Garshasbi A (2016) Study on the removal of heavy metal ions from industry waste by carbon nanotubes: effect of the surface modification: a review. Crit Rev Environ Sci Technol 46:93–118. https://doi.org/10.1080/10643389.2015.1061874

Hassanzadeh-Afruzi F, Heidari G, Maleki A (2022) Magnetic nanocomposite hydrogel based on Arabic gum for remediation of lead (II). From contaminated water. Mater Chem Horizons 1:107–122

Hina K, Zou H, Qian W, Zuo D, Yi C (2018) Preparation and performance comparison of cellulose-based activated carbon fibres. Cellulose 25:607–617. https://doi.org/10.1007/s10570-017-1560-y

Huang W, Song B, Liang J, Niu Q, Zeng G, Shen M, Deng J, Luo Y, Wen X, Zhang Y (2021) Microplastics and associated contaminants in the aquatic environment: a review on their ecotoxicological effects, trophic transfer, and potential impacts to human health. J Hazard Mater 405:124187. https://doi.org/10.1016/j.jhazmat.2020.124187. PubMed: 33153780

Huang X, Auffan M, Eckelman MJ, Elimelech M, Kim JH, Rose J, Zuo K, Li Q, Alvarez PJJ (2024) Trends, risks and opportunities in environmental nanotechnology. Nat Rev Earth Environ 5:572–587. https://doi.org/10.1038/s43017-024-00567-5

Hussain CM, da Costa GM (2022) Environmental, ethical, and economical issues of nanotechnology. Jenny Stanford Publishing

Idumah CI, Hassan A (2016) Emerging trends in graphene carbon-based polymer nanocomposites and applications. Rev Chem Eng 32:223–264. https://doi.org/10.1515/revce-2015-0038

Ji K, Liu C, He H, Mao X, Wei L, Wang H, Zhang M, Shen Y, Sun R, Zhou F (2023) Research progress of water treatment technology based on nanofiber membranes. Polymers 15:741. https://doi.org/10.3390/polym15030741. PubMed: 36772042

Jiang N, Du Y, Ji P, Liu S, He B, Qu J, Wang J, Sun X, Liu Y, Li H (2020) Enhanced photocatalytic activity of novel TiO_2/Ag/MoS_2/Ag nanocomposites for water-treatment. Ceram Int 46:4889–4896. https://doi.org/10.1016/j.ceramint.2019.10.225

Joshi SV, Drzal LT, Mohanty AK, Arora S (2004) Are natural fiber composites environmentally superior to glass fiber reinforced composites? Compos A 35:371–376. https://doi.org/10.1016/j.compositesa.2003.09.016

Kanoun O, Lazarević-Pašti T, Pašti I, Nasraoui S, Talbi M, Brahem A, Adiraju A, Sheremet E, Rodriguez RD, Ben Ali M, Al-Hamry A (2021) A review of nanocomposite-modified electrochemical sensors for water quality monitoring. Sensors 21:4131. https://doi.org/10.3390/s21124131. PubMed: 34208587

Khodakarami M, Bagheri M (2021) Recent advances in synthesis and application of polymer nanocomposites for water and wastewater treatment. J Clean Prod 296:126404. https://doi.org/10.1016/j.jclepro.2021.126404

Kolya H, Kang CW (2023) Next-generation water treatment: exploring the potential of biopolymer-based nanocomposites in adsorption and membrane filtration. Polymers 15:3421. https://doi.org/10.3390/polym15163421. PubMed: 37631480

Lan D, Zhu H, Zhang J, Li S, Chen Q, Wang C, Wu T, Xu M (2022) Adsorptive removal of organic dyes via porous materials for wastewater treatment in recent decades: a review on species, mechanisms and perspectives. Chemosphere 293:133464. https://doi.org/10.1016/j.chemosphere.2021.133464. PubMed: 34974043

Li X, Su J, Wang H, Boczkaj G, Mahlknecht J, Singh SV, Wang C (2024) Bibliometric analysis of artificial intelligence in wastewater treatment: current status, research progress, and future prospects. J Environ Chem Eng 12:113152. https://doi.org/10.1016/j.jece.2024.113152

Liu X, Zhang Y, Wang S (2021) Silver nanoparticle-based nanocomposites for mercury ion detection in wastewater. Sens Act Part B

Livingston A, Trout BL, Horvath IT, Johnson MD, Vaccaro L, Coronas J, Szekely G (2020) Challenges and directions for green chemical engineering—role of nanoscale materials. Sustain Nanoscale Eng:1–18

Marcelo LR, de Gois JS, da Silva AA, Cesar DV (2021) Synthesis of iron-based magnetic nanocomposites and applications in adsorption processes for water treatment: a review. Environ Chem Lett 19:1229–1274. https://doi.org/10.1007/s10311-020-01134-2

Mehmood A, Khan FSA, Mubarak NM, Tan YH, Karri RR, Khalid M, Walvekar R, Abdullah EC, Nizamuddin S, Mazari SA (2021) Magnetic nanocomposites for sustainable water purification—a comprehensive review. Environ Sci Pollut Res Int 28:19563–19588. https://doi.org/10.1007/s11356-021-12589-3. PubMed: 33651297

Mohammed L, Ansari MNM, Pua G, Jawaid M, Islam MS (2015) A review on natural fiber reinforced polymer composite and its applications. Int J Polym Sci 2015:1–15. https://doi.org/10.1155/2015/243947

Mondal P, Anweshan A, Purkait MK (2020) Green synthesis and environmental application of iron-based nanomaterials and nanocomposite: a review. Chemosphere 259:127509. https://doi.org/10.1016/j.chemosphere.2020.127509. PubMed: 32645598

Nazarzadeh Zare E, Mudhoo A, Ali Khan M, Otero M, Bundhoo ZMA, Patel M, Srivastava A, Navarathna C, Mlsna T, Mohan D, Pittman CU, Makvandi P, Sillanpää M (2021) Smart adsorbents for aquatic environmental remediation. Small 17:e2007840. https://doi.org/10.1002/smll.202007840. PubMed: 33899324

Nivedita S, Joseph S (2021) Performance of polycaprolactone/TiO_2 composite membrane for the effective treatment of dairy effluents. Water Sci Technol 83:2477–2485. https://doi.org/10.2166/wst.2021.143. PubMed: 34032624

Noamani S, Niroomand S, Rastgar M, Sadrzadeh M (2019) Carbon-based polymer nanocomposite membranes for oily wastewater treatment. NPJ Clean Water 2:20. https://doi.org/10.1038/s41545-019-0044-z

Omran B, Baek KH (2022) Graphene-derived antibacterial nanocomposites for water disinfection: current and future perspectives. Environ Pollut 298:118836. https://doi.org/10.1016/j.envpol.2022.118836. PubMed: 35032599

Oladipo AA, Mustafa FS (2023) Bismuth-based nanostructured photocatalysts for the remediation of antibiotics and organic dyes. Beilstein J Nanotechnol 14(1):291–321

Pal K, Chakroborty S, Panda P, Nath N, Soren S (2022) Environmental assessment of wastewater management via hybrid nanocomposite matrix implications—an organized review. Environ Sci Pollut Res Int 29:76626–76643. https://doi.org/10.1007/s11356-022-23122-5. PubMed: 36168009

Pandey N, Shukla SK, Singh NB (2017) Water purification by polymer nanocomposites: an overview. Nanocomposites 3:47–66. https://doi.org/10.1080/20550324.2017.1329983

Phan NH, Rio S, Faur C, Le Coq L, Le Cloirec P, Nguyen TH (2006) Production of fibrous activated carbons from natural cellulose (jute, coconut) fibers for water treatment applications. Carbon 44:2569–2577. https://doi.org/10.1016/j.carbon.2006.05.048

Poonia K, Patial S, Raizada P, Ahamad T, Parwaz Khan AAP, Van Le Q, Nguyen VH, Hussain CM, Singh P (2023) Recent advances in metal organic framework (MOF)-based hierarchical composites for water treatment by adsorptional photocatalysis: a review. Environ Res 222:115349. https://doi.org/10.1016/j.envres.2023.115349. PubMed: 36709022

Ramamoorthy SK, Skrifvars M, Persson A (2015) A review of natural fibers used in biocomposites: plant, animal and regenerated cellulose fibers. Polym Rev 55:107–162. https://doi.org/10.1080/15583724.2014.971124

Ramos-Delgado NA, Gracia-Pinilla MÁ, Mangalaraja RV, O'shea K, Dionysiou DD (2016) Industrial synthesis and characterization of nanophotocatalysts materials: Titania. Nanotechnol Rev 5:467–479. https://doi.org/10.1515/ntrev-2016-0007

Rando G, Sfameni S, Galletta M, Drommi D, Cappello S, Plutino MR (2022) Functional nanohybrids and nanocomposites development for the removal of environmental pollutants and bioremediation. Molecules 27:4856. https://doi.org/10.3390/molecules27154856. PubMed: 35956804

Ren G, Han H, Wang Y, Liu S, Zhao J, Meng X, Li Z (2021) Recent advances of photocatalytic application in water treatment: a review. Nanomaterials 11:1804. https://doi.org/10.3390/nano11071804. PubMed: 34361190

Rial JB, Ferreira ML (2022) Potential applications of spent adsorbents and catalysts: re-valorization of waste. Sci Total Environ 823:153370. https://doi.org/10.1016/j.scitotenv.2022.153370. PubMed: 35093378

Ribarova I, Vasilaki V, Katsou E (2024) Review of linear and circular approaches to on-site domestic wastewater treatment: analysis of research achievements, trends and distance to target. J Environ Manage 367:121951. https://doi.org/10.1016/j.jenvman.2024.121951. PubMed: 39079496

Sahu A, Dosi R, Kwiatkowski C, Schmal S, Poler JC (2023) Advanced polymeric nanocomposite membranes for water and wastewater treatment: a comprehensive review. Polymers 15:540. https://doi.org/10.3390/polym15030540. PubMed: 36771842

Schwirn K, Voelker D, Galert W, Quik J, Tietjen L (2020) Environmental risk assessment of nanomaterials in the light of new obligations under the REACH regulation: which challenges remain and how to approach them? Integr Environ Assess Manag 16:706–717. https://doi.org/10.1002/ieam.4267. PubMed: 32175661

Seidi F, Reza Saeb M, Huang Y, Akbari A, Xiao H (2021) Thiomers of chitosan and cellulose: effective biosorbents for detection, removal and recovery of metal ions from aqueous medium. Chem Rec 21:1876–1896. https://doi.org/10.1002/tcr.202100068. PubMed: 34101343

Senthilkumaar S, Krishna SK, Kalaamani P, Subburamaan CV, Subramaniam NG (2010) Adsorption of organophosphorous pesticide from aqueous solution using waste jute fiber carbon. Mod Appl Sci 4:67

Siddique T, Dutta NK, Choudhury NR (2021) Mixed-matrix membrane fabrication for water treatment. Membranes 11:557. https://doi.org/10.3390/membranes11080557. PubMed: 34436320

Srivastava S (2023) Ethical issues regarding the use of nanobiotechnology-based products. Nanobiotechnology for the livestock industry. Elsevier, Amsterdam, pp 435–473

Sun K, Lyu Q, Zheng X, Liu R, Tang CY, Zhao M, Dong Y (2024) Enhanced water treatment performance of ceramic-based forward osmosis membranes via MOF interlayer. Water Res 254:121395. https://doi.org/10.1016/j.watres.2024.121395. PubMed: 38452527

Theron J, Walker JA, Cloete TE (2008) Nanotechnology and water treatment: applications and emerging opportunities. Crit Rev Microbiol 34:43–69. https://doi.org/10.1080/10408410701710442. PubMed: 18259980

Tripathy J, Mishra A, Pandey M, Thakur RR, Chand S, Rout PR, Shahid MK (2024) Advances in nanoparticles and nanocomposites for water and wastewater treatment: a review. Water 16:1481. https://doi.org/10.3390/w16111481

Umar K, Yaqoob AA, Ibrahim MNM, Parveen T, Safian MTU (2021) Environmental applications of smart polymer composites. In: Smart polymer nanocomposites. Woodhead Publishing, pp 295–312

Verma N, Vaidh S, Vishwakarma GS, Pandya A (2020) Antimicrobial nanomaterials for water disinfection. Nanotoxicity. Elsevier, Amsterdam, pp 365–383

Wolski L, Grzelak K, Muńko M, Frankowski M, Grzyb T, Nowaczyk G (2021) Insight into photocatalytic degradation of ciprofloxacin over CeO_2/ZnO nanocomposites: unravelling the synergy between the metal oxides and analysis of reaction pathways. Appl Surf Sci 563:150338. https://doi.org/10.1016/j.apsusc.2021.150338

Wu XL (2023) Sb(III) removal from aqueous solutions by the mesoporous Fe_3O_4/GO nanocomposites: modeling and optimization using artificial intelligence. J Environ Inform 41

Yadav S, Asthana A, Singh AK, Chakraborty R, Vidya SS, Susan MABH, Carabineiro SAC (2021) Adsorption of cationic dyes, drugs and metal from aqueous solutions using a polymer composite of magnetic/β-cyclodextrin/activated charcoal/Na alginate: isotherm, kinetics and regeneration studies. J Hazard Mater 409:124840. https://doi.org/10.1016/j.jhazmat.2020.124840. PubMed: 33482479

Zhang S, Niu H, Cai Y, Shi Y (2010) Barium alginate caged Fe_3O_4@ C18 magnetic nanoparticles for the pre-concentration of polycyclic aromatic hydrocarbons and phthalate esters from environmental water samples. Anal Chim Acta 665:167–175. https://doi.org/10.1016/j.aca.2010.03.026. PubMed: 20417327

Zhang J, Yu M, Tao S (2024) Advanced electrospinning nanomaterials: from spinning fabrication techniques to electrochemical applications. Nano Res 17(8):7077–7116

Zhao G, Huang X, Tang Z, Huang Q, Niu F, Wang X (2018) Polymer-based nanocomposites for heavy metal ions removal from aqueous solution: a review. Polym Chem 9:3562–3582. https://doi.org/10.1039/C8PY00484F

Zhou T, Chen Z, Wang H (2020) Gold nanoparticle-graphene oxide nanocomposites for lead ion detection in wastewater. Environ Sci Technol

Zhou M, Zhang X, Quan Y, Tian Y, Chen J, Li L (2024) Visible light-induced photocatalytic and antibacterial adhesion properties of superhydrophilic TiO_2 nanoparticles. Sci Rep 14(1):7940

MIX
Papier aus verantwortungsvollen Quellen
Paper from responsible sources
FSC® C105338

If you have any concerns about our products,
you can contact us on
ProductSafety@springernature.com

In case Publisher is established outside the EU,
the EU authorized representative is:
Springer Nature Customer Service Center GmbH
Europaplatz 3, 69115 Heidelberg, Germany

Printed by Libri Plureos GmbH
in Hamburg, Germany